T0256839

Scattering in Quantum Field Theories

PRINCETON SERIES IN PHYSICS

Edited by Phillip W. Anderson, Arthur S. Wightman,
and Sam B. Treiman (published since 1976)

Studies in Mathematical Physics: Essays in Honor of Valentine Bargmann
edited by Elliot H. Leib, B. Simon, and A. S. Wightman

Convexity in the Theory of Lattice Gases *by Robert B. Israel*

Works on the Foundations of Statistical Physics *N. S. Krylov*

Surprises in Theoretical Physics *by Rudolf Peierls*

The Large-Scale Structure of the Universe *by P. J. E. Peebles*

Statistical Physics and the Atomic Theory of Matter, From Boyle and
Newton to Landau and Onsager *by Stephen G. Brush*

Quantum Theory and Measurement *edited by John Archibald Wheeler and
Wojciech Hubert Zurek*

Current Algebra and Anomalies *by Sam B. Treiman, Roman Jackiw, Bruno
Zumino, and Edward Witten*

Quantum Fluctuations *by E. Neslon*

Spin Glasses and Other Frustrated Systems *by Debashish Chowdhury (Spin
Glasses and Other Frustrated Systems* is published in co-operation with
World Scientific Publishing Co. Pte. Ltd., Singapore.)

Weak Interactions in Nuclei *by Barry R. Holstein*

Large-Scale Motions in the Universe: A Vatican Study Week *edited by
Vera C. Rubin and George V. Coyne, S. J.*

Instabilities and Fronts in Extended Systems *by Pierre Collet and
Jean-Pierre Eckmann*

More Surprises in Theoretical Physics *by Rudolf Peierls*

From Perturbative to Constructive Renormalization *by Vincent Rivasseau*

Supersymmetry and Supergravity (2d ed.) *by Julius Wess and Jonathan
Bagger*

Maxwell's Demon: Entropy, Information, Computing *edited by Harvey S.
Leff and Andrew F. Rex*

Introduction to Algebraic and Constructive Quantum Field Theory
by John C. Baez, Irving E. Segal, and Zhengfang Zhou

Principles of Physical Cosmology *by P. J. E. Peebles*

Scattering in Quantum Field Theories: The Axiomatic and Constructive
Approaches *by Daniel Iagolnitzer*

Scattering in Quantum Field Theories

The Axiomatic and Constructive Approaches

Daniel Iagolnitzer

Princeton Series in Physics

PRINCETON UNIVERSITY PRESS
PRINCETON, NEW JERSEY

Copyright ©1993 by Princeton University Press

Published by Princeton University Press, 41 William Street,
Princeton, New Jersey 08540
In the United Kingdom: Princeton University Press,
Chichester, West Sussex

All Rights Reserved

Library of Congress Cataloging-in-Publication Data
Iagolnitzer, Daniel.
　　Scattering in quantum field theories : the axiomatic and constructive
approaches / Daniel Iagolnitzer.
　　　　p.　cm. — (Princeton series in physics)
　　Includes bibliographical references and index.
　　ISBN 0-691-08589-7
　　1. Scattering (Physics) 2. Quantum field theory. I. Title.
II. Series.
QC174.52.S32I24　1992
539.7'58—dc20　　　　　　　　　　　　　　　　　　92-15633

This book has been composed in Computer Modern using TEX

Princeton University Press books are printed on acid-free paper and
meet the guidelines for permanence and durability of the Committee
on Production Guidelines for Book Longevity of the Council on Library
Resources

Printed in the United States of America

10　9　8　7　6　5　4　3　2　1

Contents

PREFACE   ix

INTRODUCTION   xiii
   1. Organization of the book   xiii
   2. Description of contents   xiv
   3. Technical remarks   xx

I. THE MULTIPARTICLE S MATRIX   3
   1. Introduction   3
   2. General S-matrix formalism   6
     2.1 Free-particle states and S matrix   6
     2.2 Cluster property, connected S matrix, unitarity equations 11
   3. Multiple scattering and Landau surfaces   14
     3.1 Definitions and examples   15
     3.2 $+\alpha$-Landau surfaces, causal directions, plus $i\varepsilon$ rules 20
   4. The physical region macrocausal S matrix   24
     4.1 Macrocausality and physical region analyticity 24
     4.2 Macrocausal factorization and local discontinuity formulae 28
     4.3 The $3 \rightarrow 3$ S matrix below the 4-particle threshold 35
   5. The analytic S matrix   36
     5.1 Hermitean analyticity, crossing, and all that 37
     5.2 Analyticity in unphysical sheets   39
     5.3 Basic discontinuity formulae for $3 \rightarrow 3$ processes 42
   6. Analysis of Landau singularities   44
     6.1 Graphs with single and double lines: holonomic cases 44
     6.2 Simplified theory of the m-particle threshold and expansions
        in terms of holonomic contributions   47
     6.3 The nonsimplified theory: outlook and conjectures 51
   Appendix: The multiparticle S matrix in two-dimensional
        space-time   52

II. SCATTERING THEORY IN AXIOMATIC FIELD THEORY   55

Introduction and General Formalism   55
  1. Introduction   55
    1.1 Axiomatic field theory: general preliminaries   55
    1.2 Scattering theory: historical survey   58
    1.3 Description of contents   66
  2. General formalism   67
    2.1 The axiomatic framework   67
    2.2 Fields and particles: preliminary discussion   71
    2.3 Asymptotic states and S matrix (Haag-Ruelle theory)   77

Causality and Analyticity in the Linear Program   81
  3. Causality and local analyticity   81
    3.1 Asymptotic causality properties of chronological N-point
        functions   81
    3.2 Momentum-space analyticity and local decompositions   89
    3.3 Reduction formulae and results on the S matrix   91
  4. The analytic N-point functions   96
    4.1 General primitive results of the linear program   96
    4.2 2-point and 4-point functions   104

Particle Analysis in the Nonlinear Program   108
  5. The nonlinear program-direct methods   108
    5.1 Discontinuity formulae for absorptive parts and a class of
        generalized optical theorems   108
    5.2 4-point function and 2-particle threshold   112
    5.3 The 5-point function in the 3-particle region   115
    5.4 The 6-point function in the 3-particle region: Landau
        singularities and structure equation   119
  6. The nonlinear program based on irreducible kernels   125
    6.1 Some general preliminary theorems   125
    6.2 The 2-particle structure of 4-point and N-point functions   128
    6.3 The 6-point function in the 3-particle region (even theories)   132
  7. Macrocausal properties: further results and conjectures   138
    7.1 Macrocausal factorization: some general results   138
    7.2 General conjectures   141
    7.3 Macrocausal analysis of the 2-particle threshold   145

III. EUCLIDEAN CONSTRUCTIVE FIELD THEORY   149
  1. Introduction   149
    1.1 Historical survey and description of contents   149
    1.2 The Euclidean axioms   154
    1.3 The models   156

2. The perturbative approach 160
   2.1 Perturbative series 160
   2.2 Perturbative renormalization: phase-space analysis and
      effective expansions 164
3. The $P(\varphi)_2$ models 170
   3.1 Preliminaries 170
   3.2 Cluster expansions and related expansions of N-point
      functions 173
   3.3 Infinite-volume limit, Euclidean axioms, Borel summability 180
4. The massive Gross-Neveu model in dimension two 186
   4.1 Preliminary results 186
   4.2 Phase-space analysis and renormalization 192
   4.3 Large-momentum and short-distance properties—Wilson
      short-distance expansion 198
5. Bosonic models: complements 203
   5.1 Ultraviolet limit in (massive) $P(\varphi)_2$ models 203
   5.2 Phase-space analysis in more general models: massive φ_3^4
      and "infrared φ_4^4" (outline) 205

IV. PARTICLE ANALYSIS IN CONSTRUCTIVE FIELD THEORY 208
   1. Introduction 208
      1.1 The perturbative approach 208
      1.2 Constructive and semi-axiomatic approaches 212
   2. Irreducible kernels in super-renormalizable models 217
      2.1 Cluster expansions of order $\nu > 1$ 217
      2.2 The 2-particle irreducible Bethe-Salpeter kernel and the BS
         equation in $P(\varphi)_2$ models 221
   3. Irreducible kernels in nonsuper-renormalizable theories 225
      3.1 Irreducible kernels satisfying regularized equations 225
      3.2 The Bethe-Salpeter and renormalized BS kernels in the
         Gross-Neveu model 229
   4. Two-particle structure in weakly coupled field theories 237
      4.1 Two-particle bound states and asymptotic completeness
         in even theories 237
      4.2 Noneven theories 244
      4.3 Gross-Neveu model and semi-axiomatic approaches 247
   5. Many-particle structure analysis: general results and conjectures 249
      5.1 Structure equations 249
      5.2 Discontinuity formulae of Feynman-type integrals 256
      5.3 Asymptotic completeness relations, S-matrix discontinuity
         formulae, and all that 264

MATHEMATICAL APPENDIX: DISTRIBUTIONS, ANALYTIC
FUNCTIONS, AND MICROLOCAL ANALYSIS   271
   1. Microsupport of distributions   271
   2. Local analyticity properties, general decomposition theorems,
      generalized edge-of-the-wedge theorems   273
   3. Products and integrals of distributions, restrictions to
      submanifolds   274
   4. Holonomicity (introduction)   275
   5. Phase-space decompositions   276

BIBLIOGRAPHY   279

REFERENCES   281

INDEX   287

Preface

Quantum Field Theory has been recognized for a long time as a fundamental theory in High Energy Particle physics. Its main developments have been carried out in the perturbative approach in which quantities of interest are expressed for each given model (corresponding to a specific type of interaction at the microscopic space-time level), as series with respect to a possibly renormalized coupling constant. Although perturbative renormalization provides well-defined quantities at each order, these series are in general divergent, however small the coupling is. A number of works have been devoted to establishing field theory on more rigorous, nonperturbative bases and to getting a better and deeper understanding of its general properties. To that purpose "axiomatic" and "constructive" approaches have been developed, with the first aim of setting a precise formalism to start from: general (model-independent) axioms, possibly completed by further conditions intended to characterize subclasses of theories of interest, and nonperturbative definition of models respectively. From the viewpoint of particle physics, a further ambition is to develop a satisfactory relativistic scattering theory including the analysis of the particle content of the theories and the determination of general properties of multiparticle collision amplitudes. In spite of more general investigations, the best results so far in this domain apply to theories describing systems of massive particles with short-range interactions, to which we shall mainly restrict our attention. This is a strong physical limitation and important aspects of modern particle physics developed in the 1970s and 1980s are absent from these theories. Moreover, the very mathematical existence of such theories, established in space-time dimension 2 or 3, remains doubtful in dimension 4 if we leave aside trivial theories of free particles without interactions. However, many basic features of field theory are present and the deeper analysis that can be carried out for these theories already exhibits a large number of properties of physical and mathematical interest which justify, in our opinion, their study as a first approach to relativistic quantum physics.

Several textbooks in the 1960s and the first part of the 1970s (see bibliography) have already presented important achievements of that period, such as, in the axiomatic approach, the setting of the Wightman axioms or related axioms in the theory of local observables; the introduction of asymptotic states and of the scattering operator S or S matrix; TCP and spin-statistics theorems; crossing and dispersion relations for $2 \rightarrow 2$ collision amplitudes, as well as further results on these amplitudes derived from unitarity; and in the constructive approach, the definition of a first class of nontrivial theories, in particular, the massive weakly-coupled $P(\varphi)_2$ models (polynomial interaction in space-time dimension 2). These models, initially defined in the unphysical Euclidean space-time (imaginary times), are shown to satisfy properties (the "Euclidean axioms") which, by analytic continuation from imaginary to real times, allow one to reconstruct a Wightman field theory in Minkowski space-time. The existence of a basic lowest physical mass $\mu > 0$, close to the bare mass m at small coupling, is also established.

On the other hand, ideas and methods developed by various authors in the 1970s and 1980s have led to appreciable simplifications and improvements of some earlier results and to a number of new developments. Some of them are treated in the standard books of the 1980s by Glimm and Jaffe and by Bogolubov et al. cited in the bibliography. However, from the viewpoint of this book some main ideas and methods are omitted or are only briefly mentioned there. In particular:

- in the axiomatic framework: the derivation, from locality and conditions on the mass spectrum, of general causal and momentum space analyticity properties of N-point Green functions and multiparticle collision amplitudes, and on the basis of the further condition of asymptotic completeness ("nonlinear program"), of a more refined structure "in terms of particles," which is a rigorous version of ideas suggested by perturbation theory, at least in particular situations.

- in the Euclidean constructive framework: developments of "phase-space analysis," in particular in the 1980s, leading to a rigorous "renormalization group" treatment of new models such as, beside $(\varphi^4)_3$, the renormalizable but not super-renormalizable, asymptotically-free massive Gross-Neveu model in dimension 2. (This model is analogous from the viewpoint of perturbative renormalization to $(\varphi^4)_4$ but, in contrast to the latter, is known to exist as a nontrivial theory.) Related results (other models, nonperturbative proofs of short-distance properties such as Wilson short-distance expansion, . . .) have also been achieved.

- in the constructive framework or related semi-axiomatic approaches: developments of scattering theory in the second part of the 1970s, and with more general methods and results in the second part of the 1980s, by analytic continuation (from Euclidean to Minkowski energy-momentum space) of structural equations involving irreducible kernels. These equations are obtained in the constructive framework by refinements of the cluster and phase-space expansions used to define the models. The analysis is then related to one of those considered in the nonlinear axiomatic program and, as the latter, can be considered as a development of the program of "many-particle structure analysis" proposed by K. Symanzik in the 1960s. However, it is more direct, due to the underlying structure of the theories considered, and at the same time yields information on the mass spectrum (including possible bound states), on asymptotic completeness (here part of the results to be established) and on many-particle structure in higher and higher energy regions.

Our aim in this book has been to provide a modern and coherent presentation of these developments, in a self-contained way including needed preliminary bases and relevant earlier results. The recent publication in this series (see bibliography) of the book by V. Rivasseau, which treats the second theme mentioned above (Euclidean definition of models), has led us to make this part in our book somewhat shorter and we refer to his book for more explanations and details on some aspects of the subject. However, our treatment of basic methods and results remains essentially complete and self-contained, and includes a few complementary aspects. In fact, we believe the subject deserves today more than one presentation and it takes all its value in our book, from the viewpoint of particle physics, from subsequent developments.

We always put the emphasis on main ideas, although we have also tried to give a minimal number of technical details that should give the interested reader a basis for further study. As such, the book is addressed to students and researchers in particle or in mathematical physics who wish to understand some basic facts of field theory, but not necessarily to learn the subject thoroughly. The book may also be of interest to mathematicians in several respects since it touches on, with interesting physical applications, subjects such as the theory of analytic functions of several complex variables, microlocal analysis, and functional integrals. No preliminary knowledge is required.

I wish to emphasize the importance of previous collaborations and further discussions with J. Bros and J. Magnen in the elaboration of this book. They have kindly made available to me their knowledge of axiomatic and constructive theories, respectively, to the development of which they so much contributed. I also wish to thank J. Bros, J. Magnen and V. Rivasseau for reading parts of the manuscript and for their useful comments. Finally, I am very grateful to Professor A. Wightman, who played such an important role both at the very origin and in the developments of axiomatic and constructive field theories, for his interest in this work and useful remarks. On the other hand, I am very grateful to J. Ergotte, V. Lambert and S. Zaffanella for their very kind and efficient typing of the book in spite of the presentation of the handwritten manuscript and of the successive modifications.

Paris
December 1991

Introduction

§1. Organization of the book

The aim of this introduction is to present an overall view of the organization and contents of the book, which will complement indications given in the Preface. Contents will be described in Sect. 2 and some brief technical remarks are given in Sect. 3. More details and explanations on the various topics treated, as also an historical survey and references, will be found in the introductory section of each chapter.

The book is composed of four chapters. Chapters II, III and IV are concerned with the first, second and third themes mentioned in the Preface, namely, scattering theory in axiomatic field theory, the Euclidean definition of models in constructive field theory and related results, and scattering theory in constructive and semi-axiomatic approaches respectively. Chapter I first introduces the scattering operator S, or S matrix, independently of field theory, and presents some general properties of multiparticle collision amplitudes that can be expected on various grounds and that we later aim to establish and analyze in the deeper framework of field theory.

Although the situation is more subtle for various reasons, the general idea is roughly to introduce more and more structure from I to IV and to derive corresponding results. I presents some properties and results independently of field theory, which is introduced in II and III. II starts from model independent axioms and conditions, first in the framework of the "linear program" (locality, conditions on the mass spectrum, invariance properties) and then in the "nonlinear program" in which asymptotic completeness is added as a supplementary condition. The more refined underlying structure that can be extricated from the definition of models (III), or which is assumed in semi-axiomatic approaches, then allows a further analysis in IV.

Although they have close links, as will appear, each chapter has its own autonomous interest and can be read to a large extent independently of the others. Another way of reading the book, which has its own value, is to start with the perturbative approach, namely, Sect. 1.3 and Sect. 2

of Chapter III for the Euclidean definition of models, and Sect. 1.1 of Chapter IV for various aspects of scattering theory in that approach.

An appendix will briefly present some simple mathematical notions and results of "microlocal analysis" (following the terminology proposed by M. Sato, which is attractive but has no link with microlocality in field theory), which are relevant in various parts of the book.

§2. Description of contents

The S matrix. Chapter I introduces the scattering operator S, or S matrix, independent of field theory, for a given set of stable (massive) physical particles. The general formalism is described in Sect. 2. Preliminary definitions and results on Landau surfaces, useful also in later chapters, are given in Sect. 3. Sect. 4 then presents asymptotic causal properties in terms of real intermediate particles (macrocausality and macrocausal factorization), which apply to collision amplitudes between initial and final displaced wave functions and are expressed mathematically as "essential support" or "microsupport" properties of momentum-space scattering functions. Corresponding results on the physical-region analytic structure of the multiparticle S matrix follow: analyticity outside $+\alpha$-Landau surfaces, existence (for each given process) of a function analytic in a domain of the complex mass-shell to which the physical scattering function is equal when away from these surfaces and from which it is a "plus $i\varepsilon$" boundary value at $+\alpha$-Landau points (apart from some "exceptional" points where a sum of several boundary values is needed), and local discontinuity formulae. Standard analyticity properties on the complex mass-shell (hermitean analyticity, crossing, dispersion relations, analyticity in unphysical sheets) are then mentioned in Sect. 5.1, and 5.2 mainly in the case of $2 \rightarrow 2$ processes. Basic discontinuity formulae (or "generalized optical theorems") needed in multiparticle dispersion relations are described for $3 \rightarrow 3$ processes in Sect. 5.3. Holonomicity properties, or more generally expansions of the (nonholonomic) S matrix in terms of elementary holonomic contributions (where holonomicity is understood in the sense of M. Sato, see Appendix) are finally discussed in Sect. 6 in relation to the analysis of Landau singularities, in particular at the 2-particle threshold (in the two different cases of even or odd dimension of space-time) or at the m-particle threshold in a simplified theory. More general expansions in terms of (holonomic) Feynman-type integrals will appear in field theory in Chapters II and IV.

An appendix will mention specific features in 2-dimensional space-time. As will be explained, macrocausal factorization can then lead, under some restrictive conditions (satisfied in a particular class of models of field theory), to a factorization of the momentum-space multiparticle S matrix itself into a product of two-body scattering functions and to related "factorization equations" of interest in theories with several types of particles.

Our purpose in Chapter I is mainly descriptive and we do not try to start there from a minimal number of principles or axioms in a pure S-matrix approach such as the heuristic program of S-matrix theory based on "maximal analyticity" ideas or the more recent (and more precise) analysis of the physical region structure based on unitarity and macrocausality.

Axiomatic field theory. Scattering theory in axiomatic field theory is treated in Chapter II. After a more general presentation of the axiomatic approach in Sect. 1.1, our analysis is mainly restricted, as announced in the preface, to theories intended to describe systems of massive particles with short-range interactions, and characterized in the axiomatic framework by a special condition on the mass spectrum. An historical survey of the subject is given in Sect. 1.2. We then describe results of the "linear program" obtained from the Wightman axioms, or related ones, completed most of the time by the further specific condition on the mass spectrum. The analysis is based in particular on the interplay of support properties in space-time and in energy-momentum space provided by locality and the spectral condition respectively. The introduction of "asymptotic states," that is, physical states (with "all their evolution") that can be interpreted as free-particle states before or after interactions, and hence of the S matrix (Haag-Ruelle theory) is recalled in Sect. 2, which concludes our presentation of the general formalism. Results of the linear program to be described will then include:

(i) the derivation, in Sect. 3, of general asymptotic causality properties of chronological N-point functions, and in turn of the S matrix. Connected scattering functions of $m \rightarrow n$ processes (with $m+n = N$) are shown in the course of the analysis to be, in their respective physical regions, restrictions to the mass-shell (in the sense of distributions) of connected, amputated chronological N-point functions $T_N(p_1, \ldots, p_N)$ in energy-momentum space. The above causality properties entail the existence, for each N, of an analytic function from which T_N is the boundary value at all real points, from directions depending on the real region considered, but the analyticity domain obtained in complex space does not in general intersect the complex mass-shell. They, moreover, yield local decompositions of T_N, near any physical point, as a (finite) sum of boundary values of functions analytic in specified domains which do intersect locally the complex mass-shell. Corresponding decompositions of the S matrix follow. The sum reduces to one term if $m = n = 2$ ($N = 4$) and more generally in some (limited) parts of the $m \to n$ physical regions.

(ii) the definition, in Sect. 4, of "analytic N-point functions," in a global "primitive domain" in complex energy-momentum space which admits real $m \to n$ physical regions on its boundary. Various boundary values of interest at real points are introduced, including "cell" boundary values, as also the chronological function mentioned in (i). The intersection of the primitive domain with the complex mass-shell is again empty. The situation is improved by techniques of holomorphy envelopes, with corresponding local and global results on the S matrix at $N = 2$ and $N = 4$ (including crossing) recalled in Sect. 4.2. Similar results cannot be expected at $N > 4$, in which case the best results in the linear program are so far the local results described in (i).

A unified presentation of the above results will be given in the second part of Chapter II. As will be explained, they are still remote (already for the 4-point function, for which analytic continuations into unphysical sheets is not established, and in a more crucial way for N-point functions, $N > 4$) from the structure "in terms of particles" one aims to establish. For example, results of (i) do express a basic and general idea of causality, but not yet in terms of real intermediate particles. Correspondingly, Landau singularities are not extricated and the sums involved in local decompositions do not reduce in general to one term (whereas this is expected apart from some exceptional $+\alpha$-Landau points). At the global level, crossing (i.e. analytic continuation between crossed physical regions *on the complex mass-shell)* is not established at $N > 4$. We then describe in the last part of Chapter II improved, but so far partial, results obtained from the further axiom of "asymptotic completeness" (completeness of asymptotic states in the Hilbert space of states) and regularity conditions: the nonlinear program. Useful discontinuity formulae for "absorptive parts," as also a class of generalized optical theorems (in a weaker off-shell version), which directly follow from asymptotic completeness in the general framework previously established, are first presented in Sect. 5.1. Results to be next described, derived either by "direct methods" in Sect. 5, or through the introduction of "irreducible" kernels in Sect. 6, include structure equations in terms of particles in low-energy 2- and 3-particle regions, which apply

both to N-point functions and to the S matrix, and exhibit them locally
as a sum of Feynman-type integrals with corresponding Landau singularities. Partial crossing properties have also been established for the 5-
and 6-point functions, as will be mentioned in Sect. 1.2, but this will
not be treated in this book. (In contrast to the 4-point function, it is
believed that recourse to the nonlinear program is needed in general for
$N > 4$; see further comments in Sect. 1.2.)

The last section (Sect. 7) describes further results and general conjectures on macrocausal properties in terms of intermediate particles.

The program of Sect. 6 based on irreducible kernels seems potentially
best adapted for many particle structure analysis in higher and higher
energy regions. However, problems occur in the axiomatic framework
in which these kernels have to be defined from N-point Green functions via (possibly regularized) integral Bethe-Salpeter type equations,
and in which their irreducibility properties, in the sense of analyticity in complex energy-momentum space in a region around Euclidean
space (imaginary energies) that increases with degrees of irreducibility,
have to be derived from asymptotic completeness equations in successive
steps. This program will be reconsidered in Chapter IV in constructive
theory, where general irreducible kernels (satisfying required analyticity
properties) will be introduced in a more direct way, independently of
asymptotic completeness which will then be part of the results to be
established. A semi-axiomatic framework, in which the existence and
suitable properties of these kernels are assumed, will also be considered
there.

Euclidean constructive field theory. Chapter III is devoted to the definition of models in Euclidean space-time, which turns out to be the natural region where they can first be defined. Various preliminaries such as
the Euclidean axioms (Sect. 1.2) and a heuristic introduction to models
(Sect. 1.3) will be included in the introductory section (Sect. 1). The
perturbative approach is then outlined in Sect. 2. Perturbative renormalization is, in particular, outlined in Sect. 2.2 in a way which is an
introduction to methods used later in constructive theory. We start constructive field theory with the massive (weakly-coupled) $P(\varphi)_2$ models.
They are treated in Sect. 3, where simple cluster expansions allowing one
to define the infinite-volume limit are introduced. (The analysis of the
ultraviolet limit in these models is left to Sect. 5.1.) We then describe in
Sect. 4 the more recent definition, through phase-space analysis, of the
(nonsuper-renormalizable) massive weakly-coupled Gross-Neveu model
in dimension 2 (GN_2). The definition of massive $(\varphi^4)_3$ will be outlined in
Sect. 5.2. Although it is still super-renormalizable, in contrast to GN_2,

and is thus simpler from the viewpoint of renormalization, simplifications occur for different reasons in fermionic models like GN_2 which are therefore treated first. Sect. 5.2 will also briefly mention the treatment of other Euclidean bosonic models (which are not of direct interest in this book) such as "infrared φ_4^4" with fixed ultraviolet cut-off.

The emphasis will be on the description of the main methods that have been developed in constructive field theory, namely the cluster and phase-space expansions, in a way that will allow one, through refinements of the analysis, to go beyond the mere definition of the models in various directions, such as on the one hand the derivation of large momentum and short-distance properties in the Gross-Neveu model, including Wilson-Zimmerman short-distance expansion, and on the other hand results on scattering theory to be presented in Chapter IV. Results of interest to be outlined also include Borel summability properties in the coupling explained in Sect. 3.3 for φ_2^4, the divergent perturbative series then appearing as asymptotic series, and the derivation of Euclidean axioms.

Particle analysis in constructive theory and related semi-axiomatic approaches. As already mentioned in the Preface, the Euclidean axioms allow one to reconstruct Wightman theories in Minkowski space-time. However, the procedure used to that purpose, based on analytic continuation from imaginary to real times, gives no information on the particle content of the theory (mass spectrum). Results on the latter, as also on asymptotic completeness and particle structure, will be derived in Chapter IV from an alternative procedure, namely analytic continuation, from Euclidean to Minkowski energy-momentum space, of structure equations involving irreducible kernels. These equations will be either integral Bethe-Salpeter type equations, or expansions, with convergence properties at small coupling, of N-point functions in terms of Feynman-type integrals with irreducible kernels at each vertex. We first give the Euclidean definition of these kernels in constructive theory, using methods developed in the second part of the 1980s (based on cluster expansions "of order > 1"), which are more direct and general in various respects than those used previously in the second part of the 1970s. We start in Sect. 2 with $P(\varphi)_2$ models in which these kernels, including the "2-particle irreducible" Bethe-Salpeter kernel and generalizations, have a simple perturbative content. We then consider non-super-renormalizable theories in Sect. 3 and define (in Sect. 3.1) irreducible kernels satisfying regularized structure equations involving, as in the axiomatic framework, fixed ultraviolet cut-off factors in relevant integrals (although these equations do apply to the actual theory without cut-off). These kernels are not intrinsic and do not have a simple perturbative content but are best suited for purposes of Chapter IV, as in the axiomatic framework: ultraviolet problems do not occur in view of regularization. The actual Bethe-Salpeter and renormalized BS kernel will also be introduced in the Gross-Neveu model in Sect. 3.2. Exponential decay in Euclidean space-time, depending on degrees of irreducibility, follows from the definition of these various kernels and yields, by Fourier-Laplace transformation, corresponding analyticity in complex energy-momentum space around the Euclidean region.

Besides the determination of the basic physical mass μ of the theory, which appears as a pole of the 2-point function, the best results on particle analysis, presented in Sect. 4, then apply to the 2-particle structure. Methods first developed in the axiomatic framework, namely, Fredholm theory in complex space, show in general analyticity of the 4-point Green function in a 2-sheeted (d even) or multisheeted (d odd) domain around the 2-particle threshold (where d is the dimension of space-time), up to possible poles to be interpreted as 2-particle bound states or resonances (or anti-bound states). A more refined analysis then allows one to control possible poles in even (Sect. 4.1) or noneven (Sect. 4.2) weakly-coupled theories in dimensions 2 and 3 (and in hypothetical theories in dimension 4) and to complete the proof of asymptotic completeness in the low-energy 2-particle region. Poles corresponding to bound states may occur near the 2-particle threshold, however small the coupling is, due to a kinematical threshold effect in dimensions 2 and 3 (but not in dimension 4).

In the case of non-super-renormalizable theories, the analysis makes recourse to irreducible kernels satisfying regularized equations. In the Gross-Neveu model, the Bethe-Salpeter or renormalized BS kernel can alternatively be used as explained in Sect. 4.3; ultraviolet problems are then treated by using large momentum properties of these kernels in Euclidean directions and "subtracted" Bethe-Salpeter equations or related expansions in terms of the renormalized BS kernel. This approach is of interest more particularly in a semi-axiomatic framework in which such kernels are considered as the basic quantity.

In Sect. 5, we finally describe the situation in more general energy regions (in Minkowski space). Results obtained so far, in general then, rely on the expansions in terms of Feynman-type integrals already mentioned. As they stand, they apply to theories without bound states and for smaller and smaller values of the coupling as the energy increases. They remain of a more heuristic nature, because in particular of possible divergences of these expansions, even at small coupling, away from Euclidean space. (Such divergences already occur in the analysis of 2-particle structure and are responsible for the occurrence of bound or anti-bound states mentioned above in dimensions 2 and 3.) We first present the expansions, established in constructive theory in Euclidean space, that are relevant in the study of each energy region. Conjectures, only partly established so far, on the analytic structure of individual integrals away from Euclidean space are then described. They include various discontinuity formulae that allow one in turn to derive, at least formally, unitarity and asymptotic completeness relations for Green functions, as also various S-matrix discontinuity formulae or macrocausal properties in terms of particles.

§3. Technical remarks

(i) Bibliography and references

For conciseness, there will be no attempt to quote all works which have contributed to the developments of field theory or more specifically of the topics treated in this book. We have cited (a) works upon which our presentation is based, (b) works on specific points that we have mentioned but not treated in detail and (c) some, but not systematically all, important works that have contributed to the development of the topics treated. In particular, references to most of the fundamental works in field theory that have preceded research in axiomatic and constructive field theory, as also to many important works on general aspects of the latter, are absent. For more complete references, the reader may consult the books cited in the bibliography and the works that will be explicitly quoted.

Books in the short bibliography are listed in historical order with four separate sections (axiomatic field theory, constructive field theory, S matrix, general books on field theory), and a last section on proceedings of Conferences in axiomatic and constructive field theory. Other references are listed in alphabetical order. References to books cited in the bibliography are underlined, for example, [SW] refers to the book of Streater and Wightman, to be found in one of the sections of the bibliography (in this case, axiomatic field theory).

(ii) Notations

We have tried to use a coherent system of notations throughout the book. However, this is not always the case because we have also tried to adopt usual and convenient notations in the context of each chapter. For instance, initial energy-momenta are noted p_i in Chapter 1 (apart from Sect. 5), but $-P_i$ in most other parts of the book. $S(p_1, \ldots, p_N)$ will denote, in Chapters I and II, the S-matrix kernel in Minkowski energy-momentum space, but $S(x_1, \ldots, x_N)$ in Chapters III and IV will denote the "Schwinger" N-point function in Euclidean space-time. Moreover, Euclidean space-time or energy-momentum variables are considered in Chapter III as real, whereas time and energy variables should be purely imaginary in accordance with previous notations in Minkowski space, that is, the respective definitions of these variables differ by a factor i. Correspondingly, x^2 and p^2 will then be equal to $x_{(0)}^2 + \vec{x}^2$ and $p_{(0)}^2 + \vec{p}^2$ respectively, instead of $x_{(0)}^2 - \vec{x}^2$ and $p_{(0)}^2 - \vec{p}^2$ in Minkowski space (with a different convention on the global sign). In Chapter IV, space-time variables are always Euclidean, but energy-momentum variables are either Euclidean or Minkowskian. Finally, diagrams denoting convolution integrals contain, by definition, a global energy-momentum conservation (emc) δ-function in Chapters I and II; however, these functions will not be included in Chapter II-6 and Chapter IV in diagrams denoting $\mathcal{G}$-convolution integrals (or related quantities) defined in the space of real *or complex* energy-momenta satisfying emc. We use, as much as possible, the sign $\sim$ for the Fourier transform of a function (e.g. $\tilde{F}(p_1, \ldots, p_N)$ is the Fourier transform of $F(x_1, \ldots, x_N)$) but this sign is sometimes omitted.

(iii) Numberings

Sections in each chapter are numbered from 1 to the last. Subsections are numbered in each section from 1 to the last. The number of the section is included: for example, Sect. 4.3 is the third subsection of Sect. 4 in the chapter under consideration. If, in some chapter, reference is made to a section or a subsection of another chapter, the latter is indicated explicitly.

Equations, figures, theorems, etc. are numbered in each chapter from 1 to the last, independently of the section. The number of the chapter is included: for example, (3.17) is the 17th equation of Chapter III and similarly, Figure 2.10 is the 10th figure of Chapter II.

(iv) Errors

It is too much to hope that the author's efforts to get all signs and constants right have been successful. Communications regarding errors would be gratefully received.

Scattering in Quantum Field Theories

CHAPTER I

The Multiparticle S Matrix

§1. Introduction

The aim of this chapter is to introduce the scattering operator S, or S matrix, independently of field theory and to present general properties that can be expected on various rigorous or heuristic grounds, with emphasis on various aspects of interest developed in the 1970s and 1980s, which will not be found in standard books of field theory. As mentioned in the Preface, we restrict our attention to systems of massive particles with short-range interactions. Our purpose is mainly descriptive. Links between various properties will be mentioned but we do not try to start from a minimal number of principles or axioms in a pure S-matrix approach. For simplicity and conciseness, we mainly consider a theory with only one type of stable particle, of mass $\mu > 0$, and without spin, unless otherwise stated. Modifications in more general cases are unessential with regard to most of the topics to be discussed. Momentum-space collision amplitudes or "scattering functions" are *a priori* defined for on-shell values of the initial and final energy-momenta ("on-shell context"). However, they will also occasionally be considered, as will be the case in field theory, as on-shell restrictions (in the sense of distributions) of quantities defined in the ambient space obtained by removing the mass-shell constraints ("off-shell context").

The S-matrix formalism is outlined in Sect. 2 where a first set of usual properties is presented: unitarity, energy-momentum conservation and cluster property, either in space-time or in energy-momentum space. Sect. 3 provides preliminary definitions and results on classical multiple scattering (Landau surfaces, causal directions, etc.) that will be useful both below and in later chapters. The analysis is mainly based on the original work of [CS], but takes also into account considerations issued from the more refined analysis given on some points in the 1970s, in particular in [I,KS1,KKS]. Macrocausal properties and related momentum-space analyticity properties of scattering functions are then presented in Sect. 4 for general processes with m initial and n final particles following original works of [CS,IS] and the somewhat more refined analysis given in [I]. The macrocausal properties under consideration (macrocausality and macrocausal factorization), which are important refinements of the space-time cluster property of Sect. 2, are (in the on-shell context) asymptotic properties of transition amplitudes between suitable initial and final wave functions, in the limit when initial and final particles are infinitely displaced from each other. Macrocausality (Sect. 4.1) asserts exponential fall-off (in a specified sense) if initial and final particles cannot be linked causally via real intermediate (stable) particles in accordance with classical ideas, whereas macrocausal factorization gives information in the causal case. Both are conveniently expressed mathematically as "essential support" or "microsupport" properties of scattering functions. (The definition of this notion and relevant results are given in the Mathematical Appendix.) Macrocausality entails in particular the existence, for each $m \to n$ process, of a unique analytic function defined in a domain of the complex mass-shell, to which the physical region (connected) scattering function $T_{m,n}$ is equal outside $+\alpha$-Landau surfaces of (connected) multiple scattering graphs, and from which it is a "plus $i\epsilon$" boundary value at almost all $+\alpha$-Landau points. This is no longer true, in the on-shell context, at points that lie at the intersection of several $+\alpha$-Landau surfaces with conflicting plus $i\epsilon$ rules, which are already encountered in simple cases such as $3 \to 3$ processes below the 4-particle threshold. They are exceptional in space-time dimensions $d > 2$, but will play a more important role at $d = 2$. Macrocausal factorization yields on the other hand, and is equivalent in usual situations, to local discontinuity formulae around the physical region Landau singularities (Sect. 4.2). If the graph that gives rise to the $+\alpha$-Landau surface considered has no more than one line between any pair of vertices, the discontinuity is expressed as an on-mass-shell convolution integral of a product of scattering functions associated to each vertex, as an extension of a formula due to Cutkosky on discontinuities of Feynman integrals. A more complicated conjecture, following [Sta] and references therein, applies for graphs with sets of multiple lines. Structure equations also follow in simple cases in given parts of the physical region which may include $+\alpha$-Landau points (as mentioned above) where the plus $i\epsilon$ rule fails: they express the scattering function as a sum of (independent) contributions associated with $+\alpha$-Landau surfaces encountered in that region. Results will be stated in particular (Sect. 4.3) for $3 \to 3$ processes below the 4-particle threshold (and away from 2- and 3-particle thresholds), in which case relevant graphs are graphs with only one internal line and "triangle" graphs.

A more detailed analysis when graphs with sets of multiple lines are involved will be given later through the introduction of "irreducible kernels," either (in simplest cases) in Sect. 6 below in a pure S-matrix approach, or in the framework of field theory in the next chapters.

Sect. 5, which has a more qualitative character, presents general analyticity properties believed to hold on the complex mass-shell, away from the physical region, following various investigations of the 1960s and 1970s described in several of the books cited in the bibliography and in [Sta]. Statements in Sects. 5.1 and 5.2 apply mainly to 2-body processes. Hermitean analyticity and crossing are presented in Sect. 5.1. The crossing property links, via analytic continuation in the complex mass-shell, scattering functions of various crossed processes. Analyticity in "unphysical sheets" is introduced in Sect. 5.2. It is in particular explained how "local maximal analyticity" around the 2-particle threshold, with possible poles in unphysical sheets (interpreted in, for example, the second sheet as associated to unstable particles), is generated by the unitarity equation. (As mentioned later, there are only two sheets locally if the dimension d of space-time is even, but an infinite number of sheets otherwise.)

As is well known and recalled in Sect. 5.1, Cauchy's theorem and analyticity properties in the physical sheet entail so called "dispersion relations" for 2-body processes (2 initial and 2 final particles). Generalizations of the fixed momentum transfer relations of the 2-body case have also been proposed. Global discontinuity formulae that generalize those arising from the unitarity equation in the 2-body case are then involved. They express differences of analytic continuations of scattering functions, "above" or "below" various normal thresholds, in terms of on-mass-shell convolution integrals of products of (physical) scattering functions. Such formulae, also called generalized optical theorems, have been proposed in particular for 6-particle functions ($3 \rightarrow 3$ and $2 \rightarrow 4$ processes) as described in Sect. 5.3 in the $3 \rightarrow 3$ case, following [Sta].

Results on the nature of Landau singularities are given in Sect. 6 in particular in connection with their possible "holonomic" structure. (Holonomy is here understood in the mathematical sense of M. Sato, recalled in Sect. 4 of the Appendix.) We start in Sect. 6.1 with simple cases corresponding to graphs without multiple lines and such that individual scattering functions involved at each vertex in the local discontinuity formula are under control from the outset: for example, they remain analytic in integration domains. Singularities generated are then poles, logarithms, powers, etc., and are indeed holonomic. Refined situations [KS1] in which singularities are still holonomic are also mentioned. However, the situation is different in general, and in fact already in the simplest case of the 2-particle threshold in a $2 \rightarrow 2$ process. The discontinuity, provided by the unitarity equation, then involves at each

vertex the 2-body scattering function (or its hermitean conjugate) under study. This case, as also more generally the m-particle threshold,
$m \geq 2$, of a $m \rightarrow m$ process in a simplified theory "with no subchannel interaction," is treated in Sect. 6.2 following [BI2]. The traditional
2-sheeted square-root type singularity of the case $m = 2$, d even is still
obtained if $(m - 1)d - m$ is even. However, if $(m - 1)d - m$ is odd,
for example, $m = 2$, d odd or $m = 3$, d arbitrary, the singularity is
more complicated: an infinite number of sheets and moreover nonholonomicity. Local expansions of interest in terms of specified holonomic
contributions involving an irreducible kernel U will then be described.
Their link with analogous expansions in terms of Feynman-type integrals with irreducible Bethe-Salpeter type kernels at each vertex, in the
context of field theory, will be explained in Chapter II. In the nonsimplified theory, the subject will be discussed again in the framework of
field theory in Chapter IV where more general expansions, involving a
more general class of Bethe-Salpeter type kernels, will be presented.

An appendix describes specific phenomena in 2-dimensional space-
time. It is in particular explained by following [I3] how, under some
supplementary conditions, macrocausality and macrocausal factorization yield, on-shell, a factorization property of the *momentum space*
multiparticle S matrix itself into a product of 2-body scattering functions, a property introduced on various grounds in various works of the
second half of the 1970s for a class of models of field theory. As also
explained, related "factorization equations" on products of 2-body scattering functions are moreover obtained in a theory with several types of
particles.

§2. General S-matrix formalism

2.1. Free-particle states and S matrix

The basic assumption of the S-matrix formalism (for the class of theories describing systems of massive particles with short-range interactions) is that each physical system "sub specie aeternitatis" (i.e. "with
all its evolution") is asymptotically tangent, before and after interactions, to (generally different) free-particle states (without interactions)
called respectively its incoming and outgoing states. This situation is
schematically represented in Figure 1.1.

Free-particle states are assumed to be rays (i.e. vectors defined up to
scalar multiples) in an Hilbert space $\mathcal{F}$, called Fock space, which is a
direct sum of n-particle spaces $\mathcal{F}_n$, $n = 1, 2, \ldots$; each n-particle space is

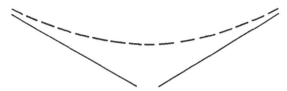

Figure 1.1 The dotted curved line represents a physical system (with all its evolution). Straight lines represent the systems of free particles (with all their evolution, without interactions) with which it coincides asymptotically, before and after interactions.

itself a (suitably symmetrized) tensorial product of one-particle spaces introduced below. Given two states $\hat{\varphi}$, $\hat{\psi}$, the probability of "detecting the state $\hat{\varphi}$ in the state $\hat{\psi}$" is assumed to be equal to the ray product $\hat{\psi}.\hat{\varphi}$, namely the scalar product squared $|\langle\psi/\varphi\rangle|^2$ of unit-norm representative vectors. On the other hand, given any element g of the Poincaré (in homogeneous Lorentz) group (semi-direct product of subgroups of space-time translations and Lorentz transformations), it is assumed that there is a well-defined correspondence which, to each state $\hat{\varphi}$, associates the transformed state $\hat{\varphi}^{(g)}$. The further assumption of Poincaré invariance of detection probabilities ($\hat{\psi}^{(g)}.\hat{\varphi}^{(g)} = \hat{\psi}.\hat{\varphi}$, $\forall\, g$), together with some technical conditions, then yields, in view of a well-known theorem by Wigner, the existence in $\mathcal{F}$ of a unitary representation of the (restricted) Poincaré group, or more precisely of its covering group. One-particle states are assumed to belong to spaces of irreducible unitary representations. The latter are characterized by two numbers: the "mass," strictly positive in the cases we consider and the spin s, which is integer or half-integer. If we restrict our attention to a theory with only one type of particle of spin zero and mass μ, the one-particle space $\mathcal{F}_1$ is the space of square integrable wave functions φ (with complex values) of a d-dimensional "energy-momentum" variable p restricted to the positive energy mass-shell. If $(p)_0$ and $\vec{p}$ denote the energy and $(d-1)$-dimensional momentum components of p, the latter is defined by the equation:

$$p^2 = \mu^2, \quad (p)_0 > 0 \qquad (1.1)$$

where $p^2 \equiv (p)_0^2 - \vec{p}^2$, $\vec{p}^2 = \sum_{\nu=1}^{d-1}(p)_\nu^2$.

The scalar product in $\mathcal{F}_1$ is defined as:

$$\langle\psi/\varphi\rangle = \int \bar{\psi}(p)\varphi(p)\,\mathrm{d}\mu(p), \qquad (1.2)$$

where $d\mu$ is the usual invariant measure on the positive energy mass-shell, namely,

$$d\mu(p) = \delta_+ \left(p^2 - \mu^2\right) dp \equiv \delta(p^2 - \mu^2)\theta((p)_0)dp, \qquad (1.3)$$

where $\theta(x) = 1$ if $x > 0$, $\theta(x) = 0$ if $x < 0$. If ψ and φ are expressed in terms of the momentum $\vec{p}$, $d\mu(p)$ reduces to $d\vec{p}/\left(2\omega\left(\vec{p}\right)\right)$, $\omega\left(\vec{p}\right) = (\vec{p}^2 + \mu^2)^{1/2}$.

The action in $\mathcal{F}_1$ of space-time translations is given, for any space-time vector $a = (a)_0, \vec{a}$, where $(a)_0$ and $\vec{a}$ are the time and space components of a, by the formula

$$\varphi^{(a)}(p) = e^{i\,p.a}\varphi(p), \qquad (1.4)$$

where $p.a = (p)_0(a)_0 - \vec{p} \cdot \vec{a}$ (space and time variables are dual to momentum and energy respectively).

For reasons omitted here, n-particle spaces are constructed in usual theories as tensorial products of one-particle spaces with symmetrization or antisymmetrization with respect to identical particles, called respectively bosons and fermions. For further reasons, particles with integer or half-integer spin are bosons and fermions respectively ("spin-statistics theorem"). For a theory with only one type of particle of spin zero, $\mathcal{F}_n$ is thus the space $(\mathcal{F}_1)_s^{\otimes n}$ of symmetric square integrable functions φ_n of n on-shell energy-momentum variables $p_1, \dots, p_n$; the scalar product of functions φ, ψ in $\mathcal{F}_n$ is given by the formulae

$$\langle \psi/\varphi \rangle = \int \bar{\psi}\left(p_1, \dots, p_n\right) \varphi\left(p_1, \dots, p_n\right) d\mu\left(p_1, \dots, p_n\right), \qquad (1.5)$$

where $d\mu\left(p_1, \dots, p_n\right)$ denotes on-shell integration over $p_1 \dots, p_n$.

Finally, a general vector φ in $\mathcal{F}$ is a collection $(\varphi_1, \varphi_2, \dots, \varphi_n, \dots)$ of vectors φ_n in each $\mathcal{F}_n$ (with the condition $\|\varphi\| < \infty$, where $\|\varphi\|^2 = \Sigma \|\varphi_n\|^2$), that is,

$$\mathcal{F} = \bigoplus_{n=0}^{\infty} (\mathcal{F}_1)_s^{\otimes n}. \qquad (1.6)$$

By the assumption mentioned at the beginning, each physical system yields two rays $\hat{\varphi}_{\text{in}}$, $\hat{\varphi}_{\text{out}}$ in $\mathcal{F}$ which are respectively its incoming and outgoing states. Conversely, at least in a theory without superselection rules (as we consider here for simplicity), each ray in $\mathcal{F}$ is assumed to represent either the incoming state of a physical system, or alternatively the outgoing state of a (generally different) physical system. This set of assumptions entails the existence of a well-defined one-to-one correspondence $\hat{S} : \hat{\varphi} \to \hat{S}\hat{\varphi}$, between rays of $\mathcal{F}$, which to each $\hat{\varphi}$ representing the incoming state of a physical system associates the ray $\hat{S}\hat{\varphi}$ that represents its outgoing state.

This correspondence can be represented in many ways by an operator S between vectors of $\mathcal{F}$. The property of *conservation of probabilities*:

$$\left(\hat{S}\hat{\psi}\right) \cdot \left(\hat{S}\hat{\varphi}\right) = \hat{\psi} \cdot \hat{\varphi} \qquad (1.7)$$

(together with some further considerations omitted here) ensures, in view of Wigner's theorem, that $\hat{S}$ can be represented by a *unitary* operator S which is unique up to multiplication by a phase (i.e. a complex scalar of modulus one). Unitarity includes in particular *linearity*.

The transition amplitude $\langle\psi|S|\varphi\rangle$ from $|\varphi\rangle$ to $|\psi\rangle$ is a well-defined number whose modulus squared $|\langle\psi|S|\varphi\rangle|^2$ is (if the norms of $|\varphi\rangle$ and $|\psi\rangle$ are unity) the *transition probability* from the initial state $\hat{\varphi}$ (represented by $|\varphi\rangle$) to the final state $\hat{\psi}$ (represented by $|\psi\rangle$), i.e. the probability of detecting the outgoing state $\hat{S}\hat{\varphi}$ (of the system whose incoming state is $\hat{\varphi}$) in the state $\hat{\psi}$.

For given numbers m, n of initial and final particles, $S_{m,n}(p_1, \ldots, p_m; p_{m+1}, \ldots, p_{m+n})$ is the momentum-space kernel of the restriction $S_{m,n}$ of S from $\mathcal{F}_m$ to $\mathcal{F}_n$: the *transition amplitude* $\langle\psi_n|S|\varphi_m\rangle$ from a vector φ_m in $\mathcal{F}_m$ to a vector ψ_n in $\mathcal{F}_n$ is written (formally) as

$$\langle\psi|S|\varphi\rangle = \int S_{m,n}(p_1, \ldots, p_{m+n})\, \bar{\psi}_n(p_{m+1}, \ldots, p_{m+n})\, \varphi_m(p_1, \ldots, p_m)$$
$$d\mu(p_1, \ldots, p_m)\, d\mu(p_{m+1}, \ldots, p_{m+n}). \qquad (1.8)$$

As the kernel of a bounded operator, $S_{m,n}(p_1, \ldots, p_{m+n})$ is also a well-defined *tempered distribution* in the space $\mathbb{R}^{(m+n)(d-1)}$ of initial and final energy-momenta $p_k, k = 1, \ldots m + n$, satisfying the (positive-energy) mass-shell constraints for each k. The right-hand side of (1.8) is still a well-defined number if the product of square integrable functions is replaced by any test function $f_{m+n}(p_1, \ldots, p_{m+n})$ which no longer necessarily has a product form with respect to initial and final variables but belongs to the Schwartz space $\mathcal{S}_{m+n}$ of C^∞ (i.e. infinitely differentiable) functions with rapid decrease at infinity (i.e. faster than any inverse power) as well as their derivatives.

Translation invariance and energy-momentum conservation. Energy-momentum conservation can be viewed as the assumption that the transition probability (hence the transition amplitude) vanishes if the energy-momentum conservation (emc) constraint

$$\sum_{i=1}^{m} p_i = \sum_{j=m+1}^{m+n} p_j \qquad (1.9)$$

cannot be satisfied with $(p_1, \ldots, p_m)$ and $(p_{m+1}, \ldots, p_{m+n})$ in the supports of φ_m and ψ_n respectively. As a consequence, $S_{m,n}$ can be written (in a well-defined sense) in the form

$$S_{m,n}(p_1, \ldots, p_{m+n}) = \delta^d \left(\sum_{i=1}^{m} p_i - \sum_{j=m+1}^{m+n} p_j \right) s_{m,n}(p_1, \ldots, p_{m+n}),$$

$$(1.10)$$

where $s_{m,n}$ is a distribution defined in the *physical region* $M_{m,n}$ of the $m \rightarrow n$ process, namely the real submanifold of $\mathbb{R}^{d(m+n)}$ defined by the positive energy mass-shell conditions (1.1) on each p_k and the emc equation (1.9). More precisely, the right-hand side of (1.10) makes sense only, in the on-shell context, away from points $p_1 = p_2 = \ldots = p_{m+n}$ (all p_k colinear, hence equal for a theory with only one mass μ) where the manifold $M_{m,n}$ is no longer regular but has a conical singularity. *A priori* possible derivatives of the emc δ-function are excluded in (1.10) because S is a bounded operator.

Alternatively, emc and Eq. (1.10) can be directly derived from (and are equivalent to) the property of translation invariance of transition probabilities, i.e. invariance under replacement of initial and final states by those obtained after space-time translation with a common space-time vector $a = ((a)_0, \vec{a})$. The proof uses the fact that the translation of a state φ_m gives the translated wave function:

$$\varphi_m^{(a)}(p_1, \ldots, p_m) = \varphi_m(p_1, \ldots, p_m) e^{ia \cdot (p_1 + \cdots + p_m)}. \qquad (1.11)$$

An intermediate step is the derivation of translation invariance of transition amplitudes (without *a priori* possible phases).

Invariance of transition probabilities under more general Poincaré transformations, in particular Lorentz transformations, also yields invariance of transition amplitudes, with consequences on the kernels $s_{m,n}$. We only note that $S_{1,1}(p_1; p_2)$ is of the form

$$S_{1,1}(p_1; p_2) = 2\gamma \left(\vec{p}_1^2 + \mu^2 \right)^{1/2} \delta(\vec{p}_1 - \vec{p}_2), \qquad (1.12)$$

where γ is a phase that does not depend on $p_1 = p_2$ but may *a priori* depend on the particle considered in a theory with several types of particles. We shall assume here that $\gamma = 1$.

Tangent and cotangent space at physical region points. For the purposes of Sect. 3, we give some geometrical remarks on the physical-region manifold $M_{m,n}$. First, given any (regular) point $P = (P_1, \ldots, P_{m+n})$ of $M_{m,n}$, the *tangent space* $T_P M$ to M at P (leaving the indices m, n implicit) is (by differentiation of the mass-shell and emc equations) the space of vectors $(dq_1, \ldots, dq_{m+n})$ in $\mathbb{R}^{(m+n)d}$ subject to the conditions:

$$\sum_{i=1}^{m} dq_i = \sum_{j=m+1}^{m+n} dq_j \qquad (1.13a)$$

$$P_k \cdot dq_k = 0, \qquad k = 1, \ldots, m+n, \qquad (1.13b)$$

where $P \cdot dq = (P)_0 (dq)_0 - \vec{P}.d\vec{q}$ for each index k. If the scalar product of a point $p = (p_1, \ldots, p_{m+n})$ and a point $u = (u_1, \ldots, u_{m+n})$ in the dual space is defined as

$$u.p = \sum_{k=1}^{m+n} \epsilon_k p_k.u_k, \qquad (1.14)$$

where $\epsilon_k = -1$ if k is initial and $\epsilon_k = +1$ if k is final, the *cotangent space* $T_P^* M$ to M at P, dual by definition to $T_P M$, is thus the set of points $u = (u_1, \ldots, u_{m+n})$ defined modulo addition of vectors of the form $t = (t_1, \ldots, t_{m+n})$, $t_k = \lambda_k P_k + a$ where λ_k is an arbitrary real scalar and a is independent of k (so that $t.dq = 0$ for any dq in $T_P M$ in view of (1.13)).

As easily seen, a point in $T_P^* M$ is equivalently characterized by a relative configuration (i.e. a configuration defined modulo global space-time translation) of *trajectories* (= straight lines in space-time) associated to each initial and final particle and parallel to the respective energy-momenta P_k: to each point u is associated the set of trajectories (P_k, u_k) passing through u_k and parallel to P_k.

2.2. Cluster property, connected S matrix, unitarity equations

It is natural physically to introduce the notion of a free-particle state $\hat{\varphi}$ composed of free-particle substates $\hat{\varphi}_{(1)}, \ldots, \hat{\varphi}_{(N)}$. If the latter belong to subspaces of $m_1, \ldots, m_N$ particles respectively, $\hat{\varphi}$ will be an m-particle state, $m = m_1 + \cdots + m_N$, represented by the suitably symmetrized product of individual wave functions. For more general substates, a more detailed analysis still leads us to assume that $\hat{\varphi}$ is represented by a suitably symmetrized tensorial product of individual representative vectors.

The *space-time cluster property* is the assertion that the transition probability W from initial to final states composed of displaced sub-states $\hat{\varphi}_1^{(a_1)}, \ldots, \hat{\varphi}_N^{(a_N)}$ and $\hat{\psi}_1^{(a_1)}, \ldots, \psi_N^{(a_N)}$, respectively, factorizes into the product of partial transition probabilities W_k from $\hat{\varphi}_k$ to $\hat{\psi}_k$ (independent of a_k in view of translation invariance) in the limit when all differences $|a_k - a_{k'}|$, $k \neq k'$, $|a|^2 = (a)_0^2 + \vec{a}^2$, tend to infinity:

$$R = W - \prod_k W_k \to 0. \tag{1.15}$$

This corresponds physically to the idea that the overall process should factorize in a first approximation into independent subprocesses remote from each other in space-time (see further discussion in Sect. 4). It can be shown that this property yields a similar factorization property of *transition amplitudes* without phases (excluded by a refined use of linearity), and in turn that *connected* transition amplitudes *vanish* in the limit when one $|a_k - a_{k'}|$ or more tend to infinity.

Connected kernels $S_{m,n}^c(p_1, \ldots, p_{m+n})$ are defined inductively by the formula

$$S_{m,n}(p_1, \ldots, p_{m+n}) = S_{m,n}^c(p_1, \ldots, p_{m+n}) + \sum_\pi \prod_k S_{m_k,n_k}^c(p_{\pi_k}), \tag{1.16}$$

where the sum in the right-hand side runs over nontrivial partitions π of initial and final indices $1, \ldots, m+n$ into subsets π_k with m_k and n_k initial and final indices and where p_{π_k} is the set of initial and final energy-momenta corresponding to indices in π_k. The trivial partition gives the first term in the right-hand side of (1.16). The definition is completed by $S_{1,0} = S_{0,1} = 0$. On the other hand,

$$S_{1,1}^c(p, q) = S_{1,1}(p, q) = \mathbb{1}_{1,1}(p, q)$$
$$= 2\left(\vec{p}^2 + \mu^2\right)^{1/2} \delta\left(\vec{p} - \vec{q}\right) \tag{1.17}$$

and the relations $S_{m,0} = S_{0,n} = 0$, as also $S_{1,n} = S_{m,1} = 0$ if $m > 1$ or $n > 1$, which correspond to the stability of the particles and are consequences of energy-momentum conservation, yield the same relations for connected kernels.

It can be checked that they are also kernels of bounded operators $S_{m,n}^c$ and still contain a global emc δ-function:

$$S_{m,n}^c(p_1, \ldots, p_{m+n}) = \delta^d\left(\sum_{i=1}^m p_i - \sum_{j=m+1}^{m+n} p_j\right) T_{m,n}(p_1, \ldots, p_{m+n}),$$
$$\tag{1.18}$$

where $T_{m,n}$ is again a distribution defined in the physical region $M_{m,n}$. The above decrease property of connected amplitudes between initial and final wave functions entails that connected kernels contain no δ-function of partial energy-momentum conservation between subgroups of initial and final particles, in contrast to nonconnected kernels which, in view of (1.16), are sums of terms which do contain products of partial emc δ-functions.

The decomposition (1.16) of the nonconnected kernels as a sum of products of connected kernels with no partial emc δ-function is called *momentum-space cluster property*. The distributions $T_{m,n}$ will be called "scattering functions" of the corresponding $m \to n$ processes. They are expected to satisfy the simplest properties and to be (in general) the basic quantities of interest.

Unitarity ($SS^{-1} = SS^\dagger = \mathbb{1}$) yields an infinite system of nonlinear equations between nonconnected kernels derived, for example, from the relations (valid for any given m, n)

$$\sum_{r' \geq 2} S_{m,r'} \left(S^{-1} \right)_{r',n} = \mathbb{1}_{m,n} \qquad (1.19)$$

between operators, with $S^{-1} = S^\dagger$. Kernels of terms in the left-hand side are on-mass-shell convolution integrals, for example:

$$\left(S_{2,3} \left(S^\dagger \right)_{3,2} \right) (p_1, \ldots, p_4) \qquad (1.20)$$

$$= \int S_{2,3}(p_1, p_2; k_1, k_2, k_3)(S^\dagger)_{3,2}(k_1, k_2, k_3; p_3, p_4) \mathrm{d}\mu(k_1, k_2, k_3),$$

where $\mathrm{d}\mu$ is the on-mass-shell integration measure defined previously. The sum in the left-hand side of (1.19) runs over terms with $r' \geq 2$. If $s = (p_1 + \cdots + p_m)^2 = (p_{m+1} + \cdots + p_{m+n})^2$ denotes the squared center-of-mass energy of the process (s reduces to the squared energy $((p_1)_0 + \cdots + (p_m)_0)^2$ in a reference frame in which $\vec{p}_1 + \cdots + \vec{p}_m = 0$, and is Lorentz invariant), the sum is, on the other hand, limited to $r' < r$ in the region $s < (r\mu)^2$. (The mass-shell and emc constraints on internal energy-momenta, i.e. $k_\ell^2 = \mu^2$, $(k_\ell)_0 > 0$, $\ell = 1, \ldots, r'$, $\sum k_\ell = p_1 + \cdots + p_m$, yield $s \geq (r'\mu)^2$ so that terms with $r' \geq r$ vanish in that region.) Finally, the kernel of the identity $\mathbb{1}_{m,n}$ from $\mathcal{F}_m$ to $\mathcal{F}_n$ vanishes if $m \neq n$ and is, at $m = n$, a sum of products of kernels $\mathbb{1}_{1,1}$ of the form (1.17) relating one initial and one final energy-momenta.

All terms are well-defined as kernels of (products of) bounded operators and are thus also tempered distributions. They all contain an overall emc δ-function and, in view of (1.16), can be decomposed as sums of terms containing various products of partial emc δ-functions. Unitarity equations in terms of connected kernels, in which all terms contain an overall emc δ-function but *a priori* no partial emc δ-function, can then be derived. We indicate below simple examples. Plus and minus bubbles refer to connected kernels of S and of $-S^{-1} = -S^\dagger$ respectively. All terms in the right-hand side are again on-mass-shell convolution integrals, well-defined kernels of bounded operators and tempered distributions.

(a) $m = n = 2$

$$\underline{\textcircled{+}} - \underline{\textcircled{-}} = \underline{\textcircled{+}\,\textcircled{-}} + \underline{\textcircled{+}\textcircled{-}} + \cdots, \quad (1.21)$$

where the sum in the right-hand side runs over terms with less than r internal lines in the region $s < (r\mu)^2$.

(b) $m = n = 3, s < 9\mu^2$

$$\text{≡}\textcircled{+}\text{≡} - \text{≡}\textcircled{-}\text{≡} = \text{≡}\textcircled{+}\,\textcircled{-}\text{≡} + \text{≡}\textcircled{+}\textcircled{-}\text{≡}$$
$$+ \sum_{i=1,2,3} {}_i\,\overline{\textcircled{+}\,\textcircled{-}} + \sum_{j=4,5,6} \overline{\textcircled{+}\,\textcircled{-}}\,{}_j$$
$$+ \sum_{\substack{i=1,2,3 \\ j=4,5,6}} {}_i\,\overline{\textcircled{+}\textcircled{-}}\,{}^j \quad . \quad (1.22)$$

We note that terms in the last sum vanish, in view of emc constraints, outside a codimension one surface, for example, the surface $k^2 = \mu^2$, $k = p_1 + p_2 - p_4$ if $i = 3, j = 4$:

$$\begin{smallmatrix}1\\2\\3\end{smallmatrix}\overline{\textcircled{+}\,\textcircled{-}}\begin{smallmatrix}4\\5\\6\end{smallmatrix} = \left[T_{2,2}(p_1, p_2; p_4, k) T_{2,2}^{(-)}(k, p_3; p_5, p_6) \delta_+(k^2 - \mu^2) \right]$$
$$\times \delta\left[(p_1 + p_2 + p_3) - (p_4 + p_5 + p_6) \right]. \quad (1.23)$$

§3. Multiple scattering and Landau surfaces

In Sect. 3.1, we present definitions and examples of Landau surfaces and their $+\alpha$ parts, as also of causal directions at $+\alpha$-Landau points. A further analysis of $+\alpha$-Landau surfaces and causal directions is given in Sect. 3.2, at the end of which plus $i\epsilon$ directions at $+\alpha$-Landau points, dual by definition to causal directions, are introduced.

More precisely, Landau and $+\alpha$-Landau surfaces will be defined either in the physical region of a given process, or in the ambient space obtained by removing the mass-shell, but not the emc constraints ("on-shell" and "off-shell" contexts). Laudau surfaces were initially introduced as possible singularities of Feynman integrals of perturbation theory. The $+\alpha$-Landau surfaces will also appear later as singularities of physical region scattering functions, while "mixed-α" parts of Landau surfaces will be possible singularities of unitarity-type integrals and of analytic continuations of scattering functions in unphysical sheets. However, it turns out that, according to the original definitions, some mixed-α surfaces may cover the full physical region or open subsets of it, in connection with the possibility of "$u = 0$" diagrams (see Sect. 3.1), so that information on singularities stated in terms of these surfaces is of little interest. On the other hand, singularities are not restricted to the surfaces defined simply by removing "$u = 0$" diagrams. A more detailed analysis [KKS] leads to introduce more appropriate "modified Landau surfaces" as will be briefly mentioned. The problem is much less crucial for $+\alpha$-Landau surfaces; it then appears essentially in the on-shell context and at M_0 points (some initial or some final energy-momenta colinear). It is believed that it can be taken into account by a simple modification, given in Sect. 3.1, of the definition of causal directions at these points.

3.1. *Definitions and examples*

A *multiple scattering graph* G is composed of interaction vertices, oriented internal lines joining pairs of vertices, and initial and final external lines incoming at, or outgoing from, some vertex. There are at least two incoming and two outgoing lines at each vertex (apart from $1 \to 1$ vertices in the case of connected parts of G with only one incoming and one outgoing line). Each line is associated with a physical particle.

A *space-time diagram* $\mathcal{D}$ associated to G is a representation of G in which each vertex is a point in space-time and lines have well-defined on-shell energy-momenta (with positive energies) to which they must be parallel. They are issued from and/or end at respective space-time vertices. Finally, energy-momentum conservation must be satisfied at each vertex. The mass-shell constraints on external (but not on internal) lines are removed in the off-shell context.

A *classical, or causal, multiple scattering diagram* $\mathcal{D}_+$ is such that the further causality requirement $(b_\ell)_0 \geq (a_\ell)_0$ is satisfied for each internal line ℓ issued from a vertex a_ℓ and ending at a vertex b_ℓ.

The *trajectory* of a line ℓ in $\mathcal{D}$ or $\mathcal{D}_+$ will always mean, by definition, the full straight line in space-time parallel to its energy-momentum and passing through the vertex or vertices where the line is incoming or outgoing.

A *nontrivial* diagram is such that there are at least two noncoincident vertices (i.e. vertices that do not occupy the same space-time position).

A $u = 0$ *diagram* in the off-shell context is such that all vertices involving external lines lie at the same position in space-time. In the on-shell context, it is simply a diagram such that initial and final trajectories all pass through a common point.

Given a graph G with m initial and n final lines, the Landau surface $L(G)$, and its part $L_+(G)$, also called $+\alpha$-Landau surface of G for reasons that appear below, are the sets of points $p = (p_1, \ldots, p_{m+n})$ (satisfying the emc and, in the on-shell context, the mass-shell constraints) such that there exists at least one nontrivial diagram, $\mathcal{D}$ or $\mathcal{D}_+$ respectively, associated to G and with external energy-momenta p_k. Equivalently, $L(G)$ and $L_+(G)$ are the sets of points p such that there exist real scalars α_ℓ, or positive scalars α_ℓ respectively, not all zero, and energy-momenta k_ℓ for each internal line ℓ of G satisfying the *Landau equations*: mass-shell conditions on each k_ℓ (and $(k_\ell)_0 > 0$), energy-momentum conservation at each vertex, and loop equations

$$\sum_{\ell \in z} \epsilon_\ell \alpha_\ell k_\ell = 0 \qquad (1.24)$$

for each closed loop z of G. If one moves in a given sense along z, $\epsilon_\ell = +1$ if this sense coincides with the orientation of ℓ, and $\epsilon_\ell = -1$ if they are opposite. Each set $\{\alpha_\ell, k_\ell\}$ satisfying the Landau equations can in fact be put in a 1-1 correspondence with a diagram defined modulo space-time translation via the rule $b_\ell - a_\ell = \alpha_\ell k_\ell$ for each internal line ℓ.

The "mixed-α part" of $L(G)$ is the complement of $L_+(G)$: it corresponds to α'_ℓs that cannot all have the same sign.

An example of a Landau "surface" whose mixed-α part occupies the full physical region is that associated with a graph G of the form:

In fact, given any physical point $p = (p_1, \ldots, p_6)$, there always exist on-shell energy-momenta k_1, k_2, k_3 such that $k_1 = k_2$, $p_1 + p_2 + p_3$ $(= p_4 + p_5 + p_6) = k_1 + k_2 + k_3$. A nontrivial diagram $\mathcal{D}$ is then obtained by putting the two space-time vertices involving external lines at the origin and the two others elsewhere in the direction of $k_1 = k_2$, the three sets of two internal lines having energy-momenta equal to $k_1 = k_2$ and the last internal line ℓ having an energy-momentum equal to k_3. (It is issued from, and ends at the origin: $\alpha_\ell = 0$.) The diagram $\mathcal{D}$ thus exhibited is a $u = 0$ diagram, either in the on-shell or off-shell context. An alternative definition of Landau surfaces in which one would simply require the existence of non $u = 0$, nontrivial diagrams would not be useful (singularities of relevant quantities are *not* restricted to such surfaces). The analysis [KKS] from various viewpoints then leads one to define modified Landau surfaces through limiting procedures involving sequences of modified diagrams in which some of the constraints are removed, and are recovered only in the limit: singularities of relevant quantities are believed to be restricted to these surfaces.

The diagram $\mathcal{D}$ exhibited above is a "mixed-α" diagram. If we restrict our attention to $+\alpha$-Landau surfaces, $u = 0$ nontrivial diagrams appear essentially in the on-shell context, and only at M_0 points. An example is obtained for the graph , when $p_4 = p_5$: the space-time vertex involving 1, 2, 3 is put, for example, at the origin and the second one at a point λp_4, $\lambda > 0$; internal energy-momenta are taken equal to $p_4 = p_5$. This is a $u = 0$ diagram in the on-shell context (though not in the off-shell context) since external trajectories all pass through the origin. It is believed that there is no need to modify the definition of $+\alpha$-Landau surfaces, but this $u = 0$ problem, at M_0 points, will lead to a modified definition (at these points) of causal directions as will be explained below.

Given a physical $+\alpha$-Landau point P of a surface $L_+(G)$, a *causal direction* at P relative to G is, in the off-shell context, a direction in the space of points $u = (u_1, \ldots, u_{m+n})$ defined modulo a global space-time translation $(u_k \rightarrow u_k + a, \ k = 1, \ldots, m + n)$, such that there exists (for any representative point u) a non $u = 0$ diagram $\mathcal{D}_+$ associated to G, with external energy-momenta P_k, and such that each u_k lies at the interaction vertex of $\mathcal{D}_+$ involving the external line k. In the on-shell context, and if P is not a M_0 point, it is by definition a direction in the cotangent space $T_P^* M$ (see end of Sect. 2.2) corresponding to the set of external trajectories of a non $u = 0$ diagram $\mathcal{D}_+$ associated to G (modulo global space-time translations and dilations by positive coefficients): each point u_k of the off-shell context is replaced by the trajectory (P_k, u_k).

The introduction of limiting procedures does not seem to modify this definition at non M_0 points. It does modify it at M_0 points. A (modified) causal direction at a M_0 point P will be defined in the on-shell context in a way similar to above except that one admits new diagrams $\mathcal{D}_+$ with vertices "at infinity" in some direction in space-time. Incoming and outgoing trajectories at such a vertex must be parallel but need not coincide, though the analysis [KKS] leads one to impose some constraints, as shown in the example of Figure 1.2, where $d_2 = -d_1$ in the equal mass case.

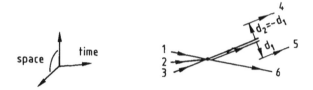

Figure 1.2 A modified causal diagram $\mathcal{D}_+$. The vertex involving external lines 4, 5 is at infinity in the direction of $p_4 = p_5$.

However, the notion of causality (distinction between past and future) is largely lost in such situations in the on-shell context.

Examples. We give below some simple examples of $(+\alpha)$-Landau surfaces and causal directions for *connected* graphs G. (The analysis for nonconnected graphs follows easily.)

(i) Graphs ⬚⬚⬚ with n sets of r internal lines

The loop equations give $\alpha_1 k_1 = \ldots = \alpha_r k_r$ for any set of r internal lines joining two vertices. The mass-shell constraints on internal lines and the emc conditions $\left(k_\ell^2 = \mu^2, (k_\ell)_0 > 0, \Sigma k_\ell = p_1 + p_2\right)$ then yield $\alpha_1 = \ldots = \alpha_r$ and $k_1 = \ldots = k_r = (p_1 + p_2)/r$, so that

$$(p_1 + p_2)^2 = (r\mu)^2. \tag{1.25}$$

Conversely, one checks that (1.25) is the equation of $L(G) = L_+(G)$. $L_+(G)$ is a smooth (analytic) submanifold of codimension 1 in the off-shell context, but also in the on-shell context if $r \geq 3$. There is correspondingly one (and only one) causal direction at any point P of $L_+(G)$: $u_1 = u_2 = a$, $u_3 = u_4 = b$, $b - a = \lambda(P_1 + P_2)$, $\lambda > 0$, in the off-shell context. In the on-shell context, initial and final trajectories meet respectively at a, b. Internal trajectories in diagrams $\mathcal{D}_+$ are all parallel to $P_1 + P_2$ and successive vertices $a_0 = a, a_1, \ldots, a_n = b$ satisfy the conditions $(a_{k+1})_0 \geq (a_k)_0$.

For $r = 2$, $L(G)$ is, in the on-shell context, the set of M_0 points p such that $p_1 = p_2 = p_3 = p_4$. This set has a codimension > 1 in $M_{2,2}$ if the space-time dimension is > 2. The set of causal directions (corresponding to parallel, but not coincident trajectories) has a corresponding dimension > 1.

(ii) Graph 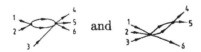 with one internal line

In the absence of loop equation, $L(G) = L_+(G)$ is given by the equation:

$$k^2 = \mu^2, (k)_0 > 0 \text{ where } k = p_1 + p_2 - p_4. \tag{1.26}$$

(iii) Triangle graph

There is one loop equation:

$$-\alpha_1 k_1 + \alpha_2 k_2 + \alpha_3 k_3 = 0. \tag{1.27}$$

By considering scalar products with k_1, k_2 and k_3 respectively and using the relations $k_1^2 = k_2^2 = k_3^2 = \mu^2$, (1.27) yields a system of three equations involving $z_{1,2} = (k_1 k_2)/\mu^2$, $z_{1,3} = (k_1 k_3)/\mu^2$, $z_{2,3} = (k_2 k_3)/\mu^2$, where these three variables can equally be expressed in terms of the *external* energy-momenta: $2\mu^2 z_{1,2} (= 2k_1 k_2 = (k_1 + k_2)^2 - 2\mu^2) = (p_1 + p_2)^2 - 2\mu^2$, and similarly $2\mu^2 z_{1,3} = (p_4 + p_5)^2 - 2\mu^2$, $2\mu^2 z_{2,3} = -(p_3 - p_6)^2 + 2\mu^2$. The existence of (non all zero) $\alpha_1, \alpha_2, \alpha_3$ implies that the determinant of the system vanishes. This gives the equation of $L(G)$:

$$z_{1,2}^2 + z_{1,3}^2 + z_{2,3}^2 + 2z_{1,2}z_{1,3}z_{2,3} - 1 = 0. \tag{1.28}$$

A section of $L(G)$ is shown in Figure 1.3, at $z_{2,3}$ fixed, in the space of the variables $z_{2,3}-1$, $z_{1,3}+1$, $z_{1,2}+1$. This section is a hyperbola tangent to the axes OA, OB at points A, B corresponding to solutions such that $\alpha_3 = 0$ and $\alpha_2 = 0$ respectively. The axes OA, OB represent the sections of the Landau surfaces $(p_1+p_2)^2 = 4\mu^2$ and $(p_4 + p_5)^2 = 4\mu^2$ of the graphs

obtained by contraction of one of the lines of G.

The arc AB corresponds to $L_+(G)$ while remaining parts of the curve correspond to "mixed-α" parts of $L(G)$.

Given any physical region point P of $L_+(G)$ there is only one solution $\{\alpha_\ell, K_\ell\}$, $\alpha_\ell \geq 0$, up to multiplication of all α_ℓ by $\lambda > 0$: the $\alpha_\ell's$ are solutions of the system of three equations derived from

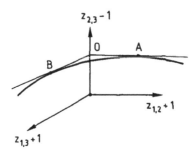

Figure 1.3 A section of the Landau surface $L(G)$ of the triangle graph.

(1.27) as mentioned above and the $K'_\ell s$ are easily determined in turn. There is correspondingly only one diagram $\mathcal{D}_+$ (up to translations and dilations) and one causal direction at P relative to G (if we leave here aside, in the on-shell context, the M_0 points corresponding to A or to B in Figure 1.3).

3.2. $+\alpha$-Landau surfaces, causal directions, plus $i\varepsilon$ rules

We start with a preliminary remark: an infinite number of different graphs may give rise to the same $+\alpha$-Landau surface and produce moreover the same causal directions at all points of that surface. First, if $P \in L_+(G)$, P also belongs to the surfaces $L_+(G')$ of any graph G' obtained from G by replacing any vertex v by a suitable subgraph G_v: to see that $P \in L_+(G')$, it is sufficient to choose the coefficients α_ℓ of the new lines equal to zero. Moreover, these α_ℓ are not always necessarily zero. Let, for example, G be a graph including a set of $r \geq 2$ internal lines between two vertices and such that there is a diagram $\mathcal{D}_+$ in which these two vertices are represented by two different points a, b $((b)_0 > (a)_0)$. Then P also belongs to $L_+(G')$ for any graph G' obtained by replacing this set by any more complicated subgraph with successive subinteractions as in the example (i) of Sect. 3.1.

For each class of graphs obtained by procedures such as above and giving rise to the same surface, we consider the simplest "elementary" one, for example, the graph ⨂ in the case of the surface (1.25) with $r = 3$.

The $+\alpha$-Landau surfaces for a $2 \to 2$ process are the surfaces $s = (r\mu)^2$, $r = 2, 3, \ldots$ already mentioned. For more general $m \to n$ processes, physical region $+\alpha$-Landau points are contained in the union L_+ of surfaces $L_+(G)$, which are almost everywhere smooth (analytic) codimension 1 submanifolds of the physical region $M_{m,n}$, and are not dense in it but divide it into sectors.

Given a point P of a smooth codimension one surface $L_+(G)$, there is a unique set $\{\alpha_\ell, k_\ell\}$ (up to multiplication of all α_ℓ by $\lambda > 0$), hence a unique $\mathcal{D}_+$ associated to G (up to translations and dilations), and a unique causal direction at P associated to G, either in the on-shell or off-shell context, conormal at P to $L_+(G)$ as shown schematically in Figure 1.4.

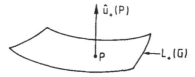

Figure 1.4 Causal direction $\hat{u}_+(P)$ at a point P of a smooth codimension one $+\alpha$-Landau surface $L_+(G)$.

In the neighborhood of P, $L_+(G)$ divides the physical region into two sides. One of them, well characterized by the causal direction, and also [Ph1] by convexity properties of the $+\alpha$-Landau surface in the off-shell context, is called the *physical side* of $L_+(G)$. It is also in usual cases (graphs G with at least one loop) the set of physical region points p such that there exist internal k_ℓ satisfing the mass-shell and emc constraints, but not necessarily the loop equations. In the $2 \to 2$ case, the physical side of the surface $s = (r\mu)^2$ is $s \geq (r\mu)^2$.

One encounters situations where some surfaces are tangent along a common submanifold with common or also possibly, in the on-shell context, opposite physical sides as schematically represented in Figure 1.5.

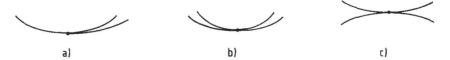

a)   b)   c)

Figure 1.5 Examples of tangent $+\alpha$-Landau surfaces (on-shell context).

The situation of Figure 1.5a occurs in particular for surfaces $L_+(G)$, $L_+(G')$ of graphs G, G' such that G' is obtained from G by contraction: two vertices joined by some internal lines are identified (i.e. replaced by a unique vertex) and these lines are removed. Points P considered are obtained as the limit of points p of $L_+(G)$ when coefficients α_ℓ of these lines tend to zero. This is the type of situation shown in Figure 1.3 at points A, B. (More precisely, this example is somewhat different in the on-shell context because these points are M_0 points: the tangency takes place only in the space of variables $z_{i,j}$. However, similar examples involving non M_0 points are obtained easily.)

A simple example of the situation shown in Figure 1.5c in a $3 \to 3$ process is obtained at physical-region points P such that $P_1 = P_6$, $P_2 = P_4$, $P_3 = P_5$. These points lie on the two surfaces $L_+(G_1)$, $L_+(G_2)$ of the graphs

In either case, the internal energy-momentum K is equal to $P_1 = P_6$. In the off-shell context, the causal directions at P associated with G_1 and G_2 correspond to points $(u_1, \ldots, u_6)$ such that $u_1 = u_2 = u_4$, $u_3 = u_5 = u_6$, $u_3 - u_1 = \lambda P_1$, $\lambda > 0$ and $u_1 = u_3 = u_5$, $u_2 = u_4 = u_6$, $u_2 - u_1 = \lambda P_1$, $\lambda > 0$ respectively. They are different and not opposite. However, in the on-shell context, the external trajectories $1, 6$ of respective diagrams $\mathcal{D}_+$ and the internal trajectory coincide, and there is also coincidence of the external trajectories $2, 4$ and $3, 5$ respectively. But the respective positions are such that the two directions determined by relative configurations of external trajectories are opposite.

If, as above, $P_1 = P_6$, $P_2 = P_4$, $P_3 = P_5$ and if, moreover, P_1, P_2, P_3 lie in a common plane, P always lies on eight surfaces $L_+(G_\beta)$ of two triangle graphs such as (in a given part of the physical region)

and of the six graphs obtained by removing one of the interaction vertices (two of which are the graphs G_1, G_2 above). In the off-shell context, there are in general eight different causal directions (e.g. $u_1 = u_2$, $u_3 = u_6$, $u_4 = u_5$, $u_4 - u_1 = \lambda P_2$, $u_4 - u_3 = \lambda' P_3$, $u_3 - u_1 = \lambda'' P_1$, $\lambda, \lambda', \lambda'' > 0$ for the first triangle graph above, and those already mentioned for G_1 and G_2). In the on-shell context, there is, as easily seen, a common causal direction for one of the triangles and the three corresponding graphs with one internal line, and a common opposite causal direction for the four other graphs.

We now give some examples of $+\alpha$-Landau surfaces which are no longer smooth manifolds at some points or have codimension > 1. An example of the latter in a $4 \to 4$ process is the surface $L_+(G)$ of the tree graph G shown in Figure 1.6a, which is the intersection of the (codimension one) surfaces of the graphs G_1, G_2 obtained from G by contraction of one of the internal lines and shown in Figure 1.6b and c.

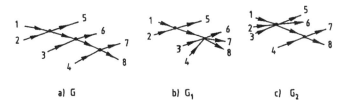

Figure 1.6 A tree graph G and two contracted graphs G_1, G_2.

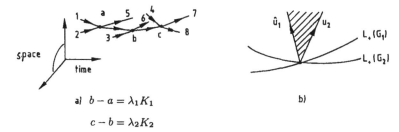

al $b - a = \lambda_1 K_1$

$c - b = \lambda_2 K_2$

Figure 1.7 Causal diagrams (Fig. a) and causal directions (Fig. b) associated to the graph G; λ_1, λ_2 are arbitrary positive scalars.

$L_+(G_1)$, $L_+(G_2)$ and $L_+(G)$ are defined respectively by the equations $k_1^2 = \mu^2$, $k_2^2 = \mu^2$ and $k_1^2 = k_2^2 = \mu^2$ where $k_1 = p_1 + p_2 - p_5$, $k_2 = p_7 + p_8 - p_4$. Either in the off-shell context, or at a non M_0 point P of $L_+(G)$ in the on-shell context, there are, as before, unique causal directions $\hat{u}_1, \hat{u}_2$ at P associated to G_1 and G_2 respectively. On the other hand, possible diagrams $\mathcal{D}_+$ associated to G are shown in Figure 1.7a. Causal directions at P associated to G are all those of the closed convex salient cone, with apex at the origin obtained by linear combinations with ≥ 0 coefficients of extremal points of the half-lines along $\hat{u}_1$ and $\hat{u}_2$ respectively and corresponding to various ratios of λ_1 and λ_2 (see Figure 1.7b).

Points P of a codimension one surface $L_+(G)$ where this surface is no longer smooth are similarly those that lie on several surfaces $L_+(G)$, $L_+(G')$, $L_+(G'' \ldots)$, where G', $G'', \ldots$ are various contractions of G, P being the limit of points p of $L_+(G)$ when various sets of coefficients α_ℓ tend to zero. Surfaces $L_+(G), L_+(G'), \ldots$ will be said to be "related" at P. The causal set of directions associated at P to such a family of related surfaces is still shown to be a closed convex salient cone (with apex at the origin) constructed from the causal directions associated to the smooth codimension one surfaces of the family.

Plus iε directions. Plus $i\epsilon$ directions at a $+\alpha$-Landau point P are those dual to the set of causal directions at P. They are considered in the space of imaginary energy-momentum variables satisfying emc, in the off-shell context, or in the space $iT_P M$, where $T_P M$ is the tangent space at P to the physical region M, in the on-shell context (Eqs. (1.13a) and (1.13b)). The plus $i\epsilon$ directions at P associated with a given $+\alpha$-Landau surface or a given family of related surfaces are those of the open half-space, or open cone with apex at the origin, dual to the causal direction, or to the cone of causal directions, at P. If P belongs to several nonrelated surfaces, plus $i\epsilon$ directions at P are those that belong to the intersection of the cones of plus $i\epsilon$ directions associated to each family of related surfaces. The intersection is empty in the on-shell context (no plus $i\epsilon$ direction) if there are opposite causal directions at P.

Let P lie, for example, on a unique (analytic, codimension one) $+\alpha$-Landau surface and let us consider, in the on-shell context, a system of real analytic local coordinates, denoted here $q = (q_1, \ldots, q_\rho)$, $q_i \in \mathbb{R}$, $i = 1, \ldots, \rho$, $\rho = (d-1)(m+n) - d$, of the physical region M in the neighborhood of P, such that P is represented by $q = 0$ and $L_+(G)$ is tangent to the real axis $q_1 = 0$ with physical side in the region $q_1 \geq 0$. The causal direction $\hat{u}_+$ at P is that of the positive real axis $x_1 > 0$ in the dual system of variables $x_1, \ldots, x_\rho$, $x_i \in \mathbb{R}$, $i = 1, \ldots, \rho$ and plus $i\epsilon$ directions are those of the (open) half-space Im $q_1 > 0$ (see Figure 1.8).

Plus $i\epsilon$ directions thus depend in general on the $+\alpha$-Landau surface and on the point considered on that surface. In the $2 \to 2$ case, they will always become the well-known directions Im $s > 0$, where $s = (p_1 + p_2)^2 = (p_3 + p_4)^2$ is the squared center-of-mass energy of the channel.

§4. The physical region macrocausal S matrix

4.1. Macrocausality and physical region analyticity

We consider for simplicity initial and final states composed of one-particle states with displaced wave functions

$$\varphi_k^{(a_k)}(p_k) = \varphi_k(p_k)e^{ip_k \cdot a_k},$$

where $k = 1, \ldots, m$ and $k = m+1, \ldots, m+n$ respectively. The space-time cluster property of Sect. 2.2 entails, as we have seen, that the connected transition amplitude $S^c\left(\left\{\varphi_k^{(a_k)}\right\}\right)$ vanishes in the limit when one or more $|a_k - a_{k'}| \to \infty$ and in turn that the scattering function

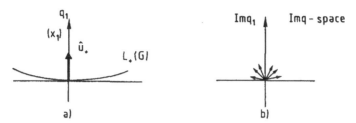

Figure 1.8 A smooth surface $L_+(G)$ in a system of real analytic local coordinates (Fig. a). The causal direction $\hat{u}_+$ at $q = 0$ and the plus $i\epsilon$ directions are shown in Fig. a and b respectively.

$T_{m,n}(p_1, \ldots, p_{m+n})$ contains no partial emc δ-function. No further information on regularity or analyticity properties is obtained at that stage, this being linked to the absence of information on the rate of fall-off in the space-time cluster property. A strong rate of fall-off cannot be expected for several reasons: particles are not necessarily localized in space-time in a sufficiently sharp way, even in an asymptotic sense, and particles can still be linked causally even when they are displaced from each other. This may occur even if relative displacements are all space-like because particles cannot be localized, even approximately, in bounded regions in space-time but will "occupy" in the best case velocity cones or trajectories. More precisely, we admit that space-time localization properties of a free particle with wave function φ are given, at least in an asymptotic sense, by fall-off properties of the space-time wave function

$$\tilde{\varphi}(x) = \int \varphi(p)\, e^{-i\, p.x} \delta_+(p^2 - \mu^2) \mathrm{d}p$$

$$= \int \varphi(p)\, e^{-i\, p.x} \mathrm{d}\vec{p}/2(\vec{p}^2 + \mu^2)^{1/2}, \qquad (1.29)$$

where $(p)_0 = \omega(\vec{p}) = (\vec{p}^2 + \mu^2)^{1/2}$ in the last expression.

If φ is C^∞ with compact support around a point P, one sees easily that $\tilde{\varphi}$ has a rapid fall-off in all directions outside the *velocity cone* $V(\varphi)$, which is the set of lines in space-time issued from the origin and parallel to points p in the support of φ. To see this, one may, for example, replace the variables $\vec{p}$ by the new variable $p.\hat{x}$, where $\hat{x}$ is the unit vector in the direction of x, and $d - 2$ other variables conveniently chosen among the $d - 1$ components of $\vec{p}$. Such a change of variables is always possible outside $V(\varphi)$, and $\tilde{\varphi}(\tau\hat{x})$ then is put in the form of the Fourier transform of a C^∞ function in the variable $p.\hat{x}$: rapid fall-off in τ follows. In $V(\varphi)$, the above change of variables is no longer legitimate (the jacobian being always singular). However, $\tilde{\varphi}$ can be written as the Fourier transform of a function with a singularity of the type $(p.\hat{x} - \mu)^{1/2} \theta(p.\hat{x} - \mu)$. A fall-off like $1/\tau^{(d-1)/2}$ follows when $\tau \to \infty$.

If we next consider a displaced wave function $\varphi^{(a)}$ with, for example, $a = \tau u$, where u is a given space-time translation and τ will tend to infinity, $\tilde{\varphi}^{(\tau u)}(x) = \tilde{\varphi}(x - \tau u)$ so that the asymptotic localization will become the velocity cone $V_{\tau u}(\varphi)$ with apex at the point τu instead of the origin. More precisely, $\tilde{\varphi}^{(\tau u)}(\tau \hat{x})$ has a rapid decrease with τ if $\hat{x}$ does not belong to $V_u(\varphi)$ (hence $\tau \hat{x} \notin V_{\tau u}(\varphi)$).

Wave functions of the form mentioned above are not fully satisfactory for our purposes. In order to get analyticity properties of scattering functions, statements of macrocausality in terms of exponential fall-off (rather than rapid fall-off) properties will be needed. Such statements can be expected only if initial and final particles have asymptotic localization properties in space-time modulo exponential fall-off. Since one wishes to preserve also strict localization, or at least asymptotic localization (modulo exponential fall-off) in energy-momentum, it is convenient (following [O,IS]: see related discussion in field theory in Sect. 2.2 of Chapter II) to consider, for each index k, a family of gaussian-type displaced wave functions, *with widths shrinking with τ*, of the form (up to possible minor changes)

$$\varphi_{k,\tau}^{(P_k, u_k)}(p_k) = \chi_k(p_k)\, e^{-\gamma \tau (\vec{p}_k - \vec{P}_k)^2} e^{i\tau p_k . u_k}, \qquad (1.30)$$

where P_k is a given on-mass-shell energy-momentum, χ_k is locally analytic around P_k, and $\gamma > 0$. When $\tau \to +\infty$, φ is more and more localized around P_k and it can be shown that $\tilde{\varphi}$ is asymptotically localized along the classical trajectory $(P_k, \tau u_k)$ parallel to P_k and passing through τu_k modulo well-specified exponential fall-off, with a rate of exponential fall-off of $\tilde{\varphi}_{k,\tau}^{(P_k, u_k)}(\tau \hat{x})$ at least proportional to γ, for small $\gamma > 0$, if $\hat{x} \notin (P_k, u_k)$.

Macrocausality, in its simplest form, is the assertion that the transition probability between initial and final wave functions of the form (1.30) should fall off exponentially in the $\tau \to \infty$ limit (with a rate of fall-off at least proportional to γ for small γ) if the initial and final particles cannot be linked causally via a network of intermediate real stable particles, i.e. more precisely if there exists no classical multiple scattering diagram $\mathcal{D}_+$ with initial and final trajectories (P_k, u_k). This assertion corresponds physically to the previous asymptotic localization properties of the particles and to the idea of short-range of interactions. (All effects of transfers of energy-momentum that cannot be associated with stable real particles should fall off exponentially with distance, hence with τ.) It excludes à la Martin pathologies (infinite number of unstable particles with arbitrarily small widths, which would produce arbitrarily small rates of exponential fall-off). The physical analysis is more delicate at M_0 points. We shall here simply admit that the statements of macrocausality still hold with the definition of classical diagrams given at these points in Sect. 3 in the on-shell context (and including possibly vertices "at infinity").

A somewhat stronger statement of macrocausality is the assertion that the transition probability should factorize (modulo a remainder with suitable exponential fall-off) into the product of partial transition probabilities between subgroups of initial and final wave functions, if there exist one or several diagrams $\mathcal{D}_+$ with external trajectories (P_k, u_k), but if these diagrams cannot link together trajectories corresponding to different subgroups, they may connect only trajectories inside each subgroup. We note that the u_k need not coincide within each subgroup.

This property might be considered as a special case of the macrocausal factorization property introduced in Sect. 4.2. It corresponds to the situation in which subgroups *cannot* be linked causally via real intermediate stable particles. (The analysis of Sect. 4.2 will apply more specifically to the case when such links are possible.) It is shown to be in turn equivalent (apart possibly from exceptional cases) to the assertion that the *connected* amplitude between the wave functions (1.30) falls off exponentially in the $\tau \to \infty$ limit if there exists no *connected* diagram $\mathcal{D}_+$ with external trajectories (P_k, u_k). It follows in turn that

$$\mathrm{ES}_P\,(T_{m,n}) \subset C_+(P) \tag{1.31}$$

at any physical region point $P = (P_1, \ldots, P_{m+n})$, where ES is the essential support or microsupport defined in Sect. 1 of the Appendix and $C_+(P)$ is the set of causal directions at P associated to connected graphs, as defined in Sect. 3.

In view of results described in the Appendix, the above property entails the analyticity of $T_{m,n}$ outside $+\alpha$-Landau surfaces of connected graphs (since $C_+(P)$ is then empty) and shows that $T_{m,n}$ is at almost all $+\alpha$-Landau points P the boundary value of an analytic function from the plus $i\epsilon$ directions dual to $C_+(P)$. Points left out are either M_0 points or points that lie at the intersection of several (nonrelated) $+\alpha$-Landau surfaces with conflicting plus $i\epsilon$ rules (occurrence of opposite causal directions). These points lie in submanifolds of the physical region of codimension > 1, at least if the dimension of space-time is > 2, so that elementary arguments of analytic continuation provide the existence of a unique analytic function $\underline{T}_{m,n}$ to which $T_{m,n}$ is equal outside $+\alpha$-Landau points and from which it is a plus $i\epsilon$ boundary value at

$+\alpha$-Landau points, apart from those mentioned above. $\underline{T}_{m,n}$ is *a priori* analytic in a domain of the complexified mass-shell $\underline{M}_{m,n}$(defined by the mass-shell and emc constraints $p_k^2 = \mu^2$, $\Sigma p_i = \Sigma p_j$ in the space of complex variables $p_k \in \mathbf{C}^d$).

In the simplest case of a $2 \to 2$ process, the situation is close to that shown in Figure 1.9, if $T_{2,2}$ is expressed in terms of Lorentz invariants, in the complex plane of the variable s $(= (p_1 + p_2)^2)$.

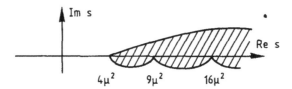

Figure 1.9 The form of the analyticity domain associated to macrocausality in a $2 \to 2$ process. Points $s = 4\mu^2$, $9\mu^2, \dots$ represent the $+\alpha$-Landau surfaces.

Decompositions of $T_{m,n}$ into sums of boundary values of analytic functions are also obtained in parts of the physical region containing previous $+\alpha$-Landau points where the plus $i\epsilon$ rule fails. We shall later discuss more precise decompositions of interest that can be obtained from stronger assumptions either in the S-matrix framework or, in following chapters, in field theory.

4.2. Macrocausal factorization and local discontinuity formulae

Macrocausal factorization. Macrocausal factorization will apply to transition amplitudes between wave functions of the form (1.30) in causal cases, i.e. when there exist one or more diagrams $\mathcal{D}_+$ with external trajectories (P_k, u_k). We consider cases such that there is only one diagram $\mathcal{D}_+$, associated to a connected multiple scattering graph G, or in which all possible diagrams can be obtained from a unique $\mathcal{D}_+$ by replacing one of its vertices by (connected or nonconnected) subdiagrams. We first define the following distribution $D_{m,n}$ associated to G in the space of external energy-momenta $p_1, \dots, p_{m+n}$ satisfying the positive-energy, mass-shell constraints (indices m, n will sometimes be left implicit).

   (i) If there is no set of multiple lines in G, that is, of two or more lines between some pairs of vertices, each vertex of G is replaced by a (nonconnected) S-matrix kernel and D is the corresponding on-shell convolution integral.

For example, if G is the triangle graph of the example (iii) in Sect. 3.1,

$$D(p_1, \ldots, p_6) = \quad {}^{1}_{2}{}_{3} \; \boxed{\cdots} \; {}^{4}_{5}{}_{6}$$

$$\equiv \int S_{2,2}(p_1, p_2; k_1, k_2) S_{2,2}(k_2, p_3; k_3, p_6)$$

$$\times S_{2,2}(k_1, k_3; p_4, p_5) \prod_{\ell=1}^{3} \delta_+ \left(k_\ell^2 - \mu^2\right) dk_\ell, \quad (1.32)$$

where plus boxes represent nonconnected S-matrix kernels.

(ii) If there are sets α of multiple lines between some vertices, a box S_α^{-1} (where α denotes the number of lines in a theory with only one type of particle) is included on each set: S_α^{-1} is the inverse, in the subspace $\mathcal{F}^{(\alpha)}$ of the Fock space $\mathcal{F}$ (direct sum of n-particle spaces with $n \geq \alpha$), of the restriction of S to $\mathcal{F}^{(\alpha)}$. If $\alpha = 2$, S_2^{-1} coincides with $(S^{-1})_{2,2}$ but this is not true at $\alpha > 2$. For example, in the case of the graph $\rangle\!\!=\!\!\langle$, D is the double on-shell convolution integral

$$D(p_1, \ldots, p_4) = \quad {}^{1}_{2} \; \boxed{+} \; \boxed{S_i} \; \boxed{+} \; {}^{3}_{4}. \quad (1.33)$$

In view of the emc δ-functions contained in each vertex, D can also be written as the product of an overall emc δ-function with a distribution d defined in the physical region.

Macrocausal factorization is the assertion that, apart from cases mentioned below, the transition amplitude between initial and final wave functions (1.30) is equal to $D\left(\left\{\varphi_{k,\tau}^{(P_k, u_k)}\right\}\right)$ modulo a remainder that falls off exponentially in the $\tau \to \infty$ limit in the same sense as in Sect. 4.1. In terms of microsupports, one then has

$$s_{m,n} \simeq d_{m,n} \quad \text{at} \quad (P, \hat{u}), \quad (1.34)$$

where $\hat{u}$ is the causal direction associated in $T_P^* M$ to the set of trajectories (P_k, u_k) and (1.34) means that $\hat{u} \notin \mathrm{ES}_P(s_{m,n} - d_{m,n})$. In the case of graphs G without sets of multiple lines, this property corresponds to the idea that the overall process should factorize, in the $\tau \to \infty$ limit, into successive subprocesses with free (on-shell) intermediate particles. For graphs G with sets of multiple lines, the interpretation is less transparent because particles of these sets, which "travel together," can *a priori* have arbitrary subinteractions that are not necessarily well separated from each other in space-time. The inclusion of the factors S_α^{-1} can be justified from (partial) heuristic S-matrix arguments and in a deeper way in field theory (see Chapter IV). It can be interpreted as corresponding to the fact that "interactions between lines of the set would otherwise be counted twice" (once at each of the two vertices linked by these sets).

Exceptional cases are such that the external trajectories can also be obtained (possibly in the sense of limiting procedures) from "mixed-α" diagrams associated to G (or to graphs in which some vertices of G are replaced by subgraphs G_v), with "minus-α lines" corresponding to some of the original lines of G. In such cases, the term $D\left(\left\{\varphi_{k,\tau}^{(P_k,u_k)}\right\}\right)$ would correspond not only to the possibility of causal diagrams but also to diagrams in which some intermediate particles would satisfy anticausal properties.

The connected S matrix. In usual cases, macrocausal factorization yields a corresponding property for the connected S matrix in which only connected diagrams $\mathcal{D}_+$ are relevant. If there is only one (elementary) connected diagram $\mathcal{D}_+$, the following result is obtained:

$$T_{m,n} \simeq (d_c)_{m,n} \text{ at } (P,\hat{u}), \tag{1.35}$$

where d_c is defined in the same way as d, except that kernels associated to each vertex of G in the definition of D_c are here *connected* S-matrix kernels (represented by bubbles). Operators S_α^{-1} on sets of multiple lines are unchanged.

A similar result is also obtained in many cases if there are several connected diagrams $\mathcal{D}_+$ associated to G or to graphs obtained from G with connected *or* nonconnected subgraphs G_v; d_c is then defined as the sum of corresponding distributions $d_c^{(1)}, d_c^{(2)}, \ldots$ For example, if the external trajectories are such that diagrams involved can be associated with the graphs $\overset{1}{\underset{3}{\overset{2}{\rightthreetimes}}}\overset{4}{\underset{6}{}}$ and $\overset{1}{\underset{3}{\overset{2}{\rightthreetimes}}}\overset{4}{\underset{6}{}}$, with trajectories $1,2$ that meet each other in space-time in either case, but either interact or not at their meeting point, then

$$T_{3,3} \underset{(P,\hat{u})}{\simeq} \left[+ \right] \Big/ \delta(\text{emc}), \tag{1.36}$$

where the notation in the right-hand side means that the overall δ-function $\delta\left(p_1 + p_2 + p_3 - p_4 - p_5 - p_6\right)$ has been factored out.

We shall assume that this property holds more generally if P lies on several $+\alpha$-Landau surfaces of connected graphs that are not related at P but are tangent at P with the same causal direction $\hat{u}$.

If P lies on several unrelated surfaces, with different causal directions $\hat{u}_1, \hat{u}_2, \ldots$, (1.35) applies independently for each direction. If P lies on related surfaces, the situation is different. For example, in the case of Figures 1.6 and 1.7, the following results are obtained:

$$\hat{u}_1 \notin \mathrm{ES}_P(T - d_c^{(1)}), \tag{1.37}$$

$$\hat{u}_2 \notin \mathrm{ES}_P(T - d_c^{(2)}), \tag{1.38}$$

$$\hat{u} \notin \mathrm{ES}_P(T - d_c), \tag{1.39}$$

where $d_c^{(1)}$, $d_c^{(2)}$ and d_c are associated to the graphs G_1, G_2 and G of Figure 1.6 respectively and $\hat{u}$ in (1.39) is any direction inside the cone of causal directions shown in Figure 1.7b.

Properties of the distribution d_c. In view of the derivation of local discontinuity formulae of scattering functions, we present some properties that can be proved or conjectured in many simple cases on the distribution d_c associated to a graph G, in the neighborhood of a point P that lies on the unique surface $L_+(G)$. Indications on cases in which these properties apply and on their origin will be given below.

(i) $d_c = 0$ on the nonphysical side of $L_+(G)$

(ii) d_c is analytic (locally) on the physical side of $L_+(G)$. Directions in the microsupport of d_c at a point of $L_+(G)$ are the causal direction $\hat{u}_+$ and the opposite direction $\hat{u}_-$. Correspondingly, d_c can be decomposed locally as a sum (or difference) of two terms:

$$d_c = (d_c)_+ - (d_c)_- \tag{1.40}$$

each of which is analytic outside $L_+(G)$, and along $L_+(G)$ is the boundary value of an analytic function from plus $i\epsilon$ and (opposite) minus $i\epsilon$ directions respectively. The terms $(d_c)_+$ and $(d_c)_-$ are unique modulo addition of a common analytic background.

Since $d_c = 0$ on the nonphysical side of $L_+(G)$, $(d_c)_+ = (d_c)_-$ there also, and these terms are thus obtained on the physical side via plus $i\epsilon$ and minus $i\epsilon$ analytic continuations of this common function around $L_+(G)$.

(iii) In some simple cases of graphs without sets of multiple lines, $(d_c)_+$ can be written as a Feynman-type integral with Feynman propagators $[k_\ell^2 - \mu^2 + i\epsilon]^{-1}$ on internal lines of G, off-shell extensions of scattering functions at each vertex and possibly suitable cut-off factors that modify this integral only by locally analytic backgrounds. (In the framework of field theory, natural extensions of scattering functions will be well-defined N-point Green functions of the theory.)

As announced, we now give some indications on these properties:

(i) Property (i) is always valid as a consequence of the mass-shell constraints on internal lines and of the emc δ–functions of the kernels involved at each vertex.

(ii) We first consider graphs G without sets of multiple lines. Starting from macrocausality, which yields information on the microsupports of individual "bubbles" at each vertex, theorems on products and integrals stated in Sect. 3 of the Appendix entail that, if P is not a $u = 0$ point (see below), the only possible directions in the microsupport of d_c at P are those determined by external trajectories of diagrams associated to G, or to graphs in which some vertices of G are replaced by connected subgraphs G_v; lines of subgraphs G_v must be represented in these diagrams by "$+\alpha$-lines" (satisfying the causality requirement $(b_\ell)_0 \geq (a_\ell)_0$), but there is no such constraint on original internal lines of G. In the cases we consider, such diagrams are in particular the causal diagram $\mathcal{D}_+$ at P, with only $+\alpha$ lines, associated to G itself, as also the opposite diagram $\mathcal{D}_-$, with only "minus α" lines. They yield respectively the directions $\hat{u}_+$ and $\hat{u}_-$. Property (ii) is thus proved if P is not a $u = 0$ point of d_c and if there is no other direction generated by other diagrams.

P is here a $u = 0$ point of d_c (in the on-shell context) if there exist diagrams as above whose external trajectories all pass through a common point. Results in this case, which are still analogous, can be conjectured on the basis of further regularity assumptions.

Eq. (1.40) and the information given below it are direct consequences of the decomposition theorems of Sect. 2 of the Appendix, if $\hat{u}_+$ and $\hat{u}_-$ are the only singular directions.

If the graph G has sets of multiple lines, the situation is more complicated both because of the $u = 0$ problem, which is crucial in this case, and because of the presence of the factors S_α^{-1}. However, it still seems reasonable to conjecture property (ii) in usual cases.

(iii) Property (iii) is obtained in simple cases by direct inspection of the Feynman-type integral associated to G. From results of the Appendix, this integral is shown to satisfy all properties of the terms $(d_c)_+$ that have been mentioned in (ii). In particular, its discontinuity around $L_+(G)$ is equal to d_c. The discontinuity uniquely determines $(d_c)_+$ modulo a locally analytic background, so that property (iii) follows.

However, the result is certainly not valid in general cases, and in particular cannot be expected for graphs with sets of multiple lines. Instead, expansions in terms of an infinite number of Feynman-type integrals involving irreducible kernels at each vertex will be presented in such cases in field theory.

Local discontinuity formulae. We study below the local structure of the scattering function T in the neighborhood of points P that lie on a unique surface $L_+(G)$. Another simple case will also be indicated later. From Sect. 4.1, T is known to be locally the boundary value of an analytic function $\underline{T}$ from plus $i\epsilon$ directions. T can correspondingly be analytically continued from the nonphysical side of $L_+(G)$ to the physical side via "plus $i\epsilon$" paths of analyticity. The following properties are simple consequences of macrocausality *and* macrocausal factorization in cases when property (ii) applies.

(i)′ There also exists an analytic continuation of T around $L_+(G)$ via (opposite) minus $i\epsilon$ paths of analyticity. The corresponding minus $i\epsilon$ boundary value of this minus $i\epsilon$ continuation will be denoted $T^{(L)}$.

(ii)′ $$T - T^{(L)} = d_c \tag{1.41}$$

(iii)′ $$T \simeq (d_c)_+ \tag{1.42}$$

in the neighborhood of P, where $\simeq$ means modulo a locally analytic background.

Proof. From macrocausality and property (ii) of d_c, the only possible singular directions at P of T or d_c, hence $T - d_c$, are $\hat{u}_+$ and $\hat{u}_-$. Macrocausal factorization ensures that $\hat{u}_+$ is not a singular direction at P, so that the only singular direction is $\hat{u}_-$. $T - d_c$ is thus locally the minus $i\epsilon$ boundary value T' of an analytic function. Since $d_c = 0$ on the nonphysical side, $T = T'$ there also and properties (i)′ and (ii)′ follow (with $T' = T^{(L)}$). Property (iii)′ is then straightforward (write, for example, $T - (d_c)_+ = T^{(L)} - (d_c)_-$ and use the edge-of-the-wedge theorem).

Other forms of discontinuity formulae. Other forms of the discontinuity around a given surface $L_+(G)$ associated to a graph including sets of multiple lines can also be conjectured. In the latter, the factor S_α^{-1} is removed and scattering functions at some vertices are replaced by suitable analytic continuations (which go beyond physical region properties studied so far).

Physical region poles. In simple cases, such as those of Sect. 4.3 below, in which scattering functions involved at each vertex remain analytic in integration domains, the discontinuity formula (1.41) gives direct information on the nature of the singularity. For a graph with one internal line, for example, $\underset{3}{\overset{1}{\underset{2}{\times}}}\overset{4}{\underset{6}{}}{}^5$, there is no integration, in view of emc δ−functions at each bubble, and one has:

$$d_c(p_1,\ldots,p_6) = T_{2,2}(p_1,p_2;p_4,k)T_{2,2}(k,p_3;p_5,p_6)\delta_+(k^2 - \mu^2), \quad (1.43)$$

where $k = p_1 + p_2 - p_4$. In this case, the discontinuity is concentrated on the surface $L_+(G)$ and the singularity of $T_{3,3}$ is easily shown to be a pole with a factorized residue:

$$T(p_1,\ldots,p_6) = \frac{1}{2i\pi}\frac{a(p_1,\ldots,p_6)}{k^2 - \mu^2 + i\epsilon}, \quad (1.44)$$

where a is locally analytic and

$$a(p_1,\ldots,p_6)|_{k^2=\mu^2} = T_{2,2}(p_1,p_2;p_4,k)T_{2,2}(k,p_3;p_5,p_6). \quad (1.45)$$

The nature of the singularity associated to a triangle graph will be studied in Sect. 6.1.

If P lies on the related sufaces $L_+(G)$, $L_+(G_1)$, $L_+(G_2)$ of the graphs of Figure 1.6 and if scattering functions involved at each vertex of the graph G in d_c are analytic, macrocausality and macrocausal factorization in the form of Eqs. (1.37), (1.38) and (1.39) entail a simple extension of (1.44), namely, with $k_1 = p_5 - p_1 - p_2$, $k_2 = p_7 + p_8 - p_4$,

$$(2i\pi)^2 T = \frac{a(p_1,\ldots,p_8)}{(k_1^2 - \mu^2 + i\epsilon)(k_2^2 - \mu^2 + i\epsilon)}, \quad (1.46)$$

where a is locally analytic, factorizes at $k_1^2 = k_2^2 = \mu^2$ as the product of three $2 \to 2$ scattering functions (associated to the vertices of G) and satisfies further factorization properties at $k_1^2 = \mu^2$ (but possibly $k_2^2 \neq \mu^2$) or $k_2^2 = \mu^2$ (but possibly $k_1^2 \neq \mu^2$). For example, at $k_2^2 = \mu^2$, $a(p_1,\ldots,p_8) = [T_{3,3}(p_1,\ p_2,p_3;p_5,p_6,\ k_2)(k_1^2 - \mu^2)] \times T_{2,2}(k_2,\ p_4;p_7,$

p_8). (The first bracket does not vanish at $k_1^2 = \mu^2$; the factor $k_1^2 - \mu^2$ cancels the pole of $T_{3,3}$.)

4.3. The $3 \to 3$ S matrix below the 4-particle threshold

Properties discussed in Sect. 4.2 give local information in the neighborhood of $+\alpha$-Landau points. We are now interested in global decompositions of the scattering function T, in bounded parts of the physical region, as a sum of contributions associated with the various $+\alpha$-Landau singularities encountered. The obtention of such decompositions of interest is complicated in general by the possible occurrence of $+\alpha$-Landau surfaces that can be related in various ways at different points. We leave aside such cases below. On the other hand, we do *not* exclude here cases when some $+\alpha$-Landau surfaces can be tangent along some submanifolds with either common or opposite plus $i\epsilon$ directions. These various situations do occur (see Sect. 3.2) in the case, to which we shall restrict our attention, of a $3 \to 3$ process below the 4-particle threshold, and more precisely in the part

$$\mathcal{R} = \left\{ p = (p_1, \ldots, p_6); \ p \in M_{3,3}, \ (3\mu)^2 < s < (4\mu)^2, p \notin M_0 \right\} \quad (1.47)$$

of the $3 \to 3$ physical region, where $s = (p_1 + p_2 + p_3)^2$ is, as usual, the squared center-of-mass energy of the $3 \to 3$ channel. (The condition $p \notin M_0$ excludes 2-particle thresholds $s_{ij} = 4\mu^2$ where $s_{ij} = (p_i + p_j)^2$ and i, j are both initial or both final.) There are 18 (unrelated) $+\alpha$-Landau surfaces $L_+(G_\beta)$ in that region, nine of which are associated with graphs with one internal line and nine with triangle graphs.

To each surface is associated the distribution $d_{c,\beta}$ defined in Sect. 4.2. For any one of these, it is easily seen that all $2 \to 2$ scattering functions involved at each vertex do remain analytic in integration domains since the squared energies of the $2 \to 2$ channels remain between $(2\mu)^2$ and $(3\mu)^2$, and in turn all properties (i), (ii) and (iii) of Sect. 4.2 are satisfied. Moreover, the decomposition (1.40) can be proved for each term $d_{c,\beta}$ in the whole region $\mathcal{R}$: terms $(d_{c,\beta})_+$ and $(d_{c,\beta})_-$ are analytic in $\mathcal{R}$ outside $L_+(G_\beta)$ and are along $L_+(G_\beta)$ plus $i\epsilon$ and minus $i\epsilon$ boundary values respectively.

The following decomposition of the scattering function $T_{3,3}$ can then be stated in $\mathcal{R}$:

$$T \simeq \sum_{\beta=1}^{18} (d_{c,\beta})_+, \quad (1.48)$$

where $\simeq$ means modulo an analytic background in $\mathcal{R}$. Eq. (1.48) can be established from macrocausality and macrocausal factorization which ensure, as easily checked, that $\mathrm{ES}_P(T - \Sigma(d_{c,\beta})_+)$ is empty, $\forall P \in \mathcal{R}$. If we admit (as can probably be established) that the terms $(d_{c,\beta})_+$ can also be expressed in $\mathcal{R}$ as well-defined Feynman-type integrals, (1.48) also yields:

$$\begin{array}{c} \hspace{1cm} \simeq \sum_{\substack{i=1,2,3 \\ j=4,5,6}} {}_i \underline{\hspace{1cm}}^j + \sum_{\substack{i=1,2,3 \\ j=4,5,6}} {}_i \underline{\hspace{1cm}}_j \, , \end{array} \qquad (1.49)$$

where terms in the right-hand side denote Feynman-type integrals (and $\simeq$ means again modulo an analytic background, after factorizing out the overall emc δ-function from all terms).

Decompositions of the form (1.48) can be obtained to some extent or conjectured in more general cases. Decompositions of the form (1.49) can be expected only when graphs giving rise to the $+\alpha$-Landau surfaces do not include sets of multiple lines and under supplementary conditions. Expansions in terms of Feynman-type integrals with irreducible kernels at each vertex will be more generally presented later in field theory.

§5. The analytic S matrix

This section is a brief presentation, of a more heuristic character, of analyticity properties on the complex mass-shell. For rigorous results in field theory, see Chapter II. It is believed that scattering functions can be analytically continued far away from the physical region, on the complex mass-shell, the only singularities encountered being those required by internal consistency arguments. These provide also (at least from a heuristic viewpoint in a pure S-matrix approach) several important properties that we briefly describe below.

For several purposes, it is convenient to modify some sign conventions: energy-momenta variables of a given process will be denoted $\epsilon_k p_k$ instead of p_k; $\epsilon_k = -1$ if k is initial; and $\epsilon_k = +1$ if k is final. With this convention, $(p_k)_0$ is negative if k is initial and emc reads:

$$\sum_{k=1}^{m+n} p_k = 0. \qquad (1.50)$$

Given a total number N of particles, the complex mass-shell $\mathbf{M}_N$ is the set of *complex* $p = (p_1, \ldots, p_N)$, $p_k \in \mathbf{C}^d$, $k = 1, \ldots, N$ satisfying mass-shell and emc constraints ($p_k^2 = \mu^2$, $\forall\ k$, $\Sigma\ p_k = 0$). It has several real parts which are disconnected from each other and correspond to the choice of m indices, $2 \leq m \leq N - 2$, among $1, \ldots, N$ such that $(p_k)_0 < 0$ and n indices such that $(p_k)_0 > 0$, with $m + n = N$. Each real region thus obtained is the physical region of a corresponding process with m initial and n final particles.

On the other hand, Lorentz invariance will be used to re-express the scattering function in term of variables

$$s_I = \left(\sum_{i \in I} p_i \right)^2. \tag{1.51}$$

where I is a subset of indices. These variables are not independent, for example, $s_{CI} \equiv s_I$ where CI is the complement of I in $(1, \ldots, N)$. On the other hand, $s_i = p_i^2$ is restricted on-shell to μ^2 if I is a subset of $(1, \ldots, N)$ with only one element i. There are also other constraints. If $N = 4$, the only relevant subsets have two indices, previous variables s_I are

$$s = (p_1 + p_2)^2 = (p_3 + p_4)^2, \tag{1.52a}$$
$$t = (p_1 + p_3)^2 = (p_2 + p_4)^2, \tag{1.52b}$$
$$u = (p_1 + p_4)^2 = (p_2 + p_3)^2 \tag{1.52c}$$

and satisfy the constraint

$$s + t + u = 4\mu^2 \tag{1.53}$$

in the equal mass case (otherwise $s + t + u = \sum_{k=1}^{4} \mu_k^2$). The scattering function $T_{2,2}$ can then be expressed in terms of two of these, for example, s, t. The physical region of a $2 \rightarrow 2$ process with initial indices 1,2 (and final indices 3,4) is defined by the relation $s \geq 4\mu^2$ and further constraints omitted here.

5.1. Hermitean analyticity, crossing, and all that

Hermitean analyticity. We have seen in Sect. 4 that $T_{2,2}$ is in the physical region $(s > 4\mu^2)$ the boundary value of an analytic function $\underline{T}_{2,2}$ from plus $i\epsilon$ directions which, in terms of Lorentz invariants, are the directions Im $s > 0$. By unitarity $(S^{-1} = S^\dagger)$, $T_{2,2}^{(-)}$ is similarly the boundary value of an analytic function $\underline{T}_{2,2}^{(-)}$ from the directions Im $s < 0$. As a matter of fact, it is believed that $\underline{T}_{2,2}$ and $-\underline{T}_{2,2}^{(-)}$ are analytic continuations of each other around $s = 4\mu^2$. More precisely, there is a common function, analytic locally in a cut-plane (in the complex variable s) with the cut along $s > 4\mu^2$, whose boundary values at $s > 4\mu^2$ from the directions Im $s > 0$ and Im $s < 0$ are $T_{2,2}$ and $-T_{2,2}^{(-)}$ respectively. This function is analytic (and purely imaginary) in the unphysical region s real, $s < 4\mu^2$. (Hermitean analyticity follows for the function obtained by multiplication by the factor i.) In view of this property, the unitarity equation (1.21) then appears as a formula for the discontinuity of $\underline{T}_{2,2}$ around $s = 4\mu^2$.

Crossing. In view of the analysis of Sect. 4, there exist, for each $N \geq 4$, several *a priori* independent analytic functions defined in domains of the complex mass-shell, depending on initial and final indices, from which scattering functions are boundary values in their respective physical regions (apart from exceptional $+\alpha$-Landau points at $N > 4$).

The crossing property asserts the existence, for each N, of a unique function $\underline{T}_N$ which is a common analytic continuation of all of them. More precisely, it asserts that $\underline{T}_N$ is analytic in a one-sheeted domain of the complex mass-shell called "physical sheet," which admits real physical regions on its boundary and from which scattering functions are obtained, in their physical regions, via the relevant plus $i\epsilon$ rules at $+\alpha$-Landau points. (The analyticity of scattering functions outside $+\alpha$-Landau surfaces entails on the other hand the existence of analytic continuations in unphysical sheets.)

In the case of $2 \to 2$ processes ($N = 4$), there are only two independent variables, for example, among the variables s, t, u of (1.52). So, keeping any one fixed leaves a single complex variable. If the fixed invariant is $t = t_0$, the physical sheet is shown, for $-4\mu^2 < t_0 < 4\mu^2$, in the complex s plane in Figure 1.10. There are two cuts, s real $\geq 4\mu^2$, corresponding to the physical region of the process $(1, 2) \to (3, 4)$, and u real $\geq 4\mu^2$, that is, $s \leq -t_0$ (since $s = 4\mu^2 - u - t_0$ in view of (1.53)), corresponding to the physical region of the process $(1, 4) \to (2, 3)$.

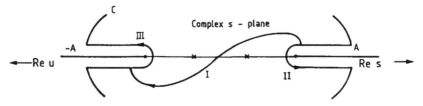

Figure 1.10 The physical sheet at $N = 4$ and t fixed, with the cuts $s \geq 4\mu^2$ and $u \geq 4\mu^2$. Crosses indicate poles at $s = \mu^2$ and $u = \mu^2$. I denotes a path of "crossing" between the two physical regions $s \geq 4\mu^2$ and $u \geq 4\mu^2$, while II and III denote respective paths of "hermitean analyticity." C denotes a contour used to obtain "dispersion relations" ($A \to \infty$).

Scattering functions are obtained from $\underline{T}_4$ in their respective physical regions from the directions Im $s > 0$ and Im $s < 0$ (Im $u > 0$) respectively. Various arguments indicate on the other hand the presence of poles at $s = \mu^2$ and $u = \mu^2$, also shown in Figure 1.10.

Dispersion relations (2 → 2 processes). "Dispersion relations" at fixed t are obtained by integrating $\underline{T}_4$ along a contour of the type C shown in Figure 1.10. If A were taken to infinity and if $\underline{T}_4$ vanished with sufficient rapidity at infinity, Cauchy's theorem would allow one to express $\underline{T}_4$ at any point of the physical sheet in terms of its discontinuities along the two cuts and its residues at the poles. Discontinuities along the cuts are expressed in terms of physical scattering functions via the unitarity equation (1.21). The reader is referred to standard textbooks (see bibliography) for a more complete analysis and various extensions of such relations and for "double dispersion relations."

Some indications relative to multiparticle dispersion relations ($N > 4$) will be given in Sect. 5.3.

Antiparticles, TCP, spin-statistics. In more general theories, the (heuristic) arguments of internal consistency which have yielded previous properties, such as crossing, in S-matrix theory also provide the existence of antiparticles (which coincide with the original particles if there are no quantum numbers, as in the theories considered previously), as also the TCP theorem and the usual spin-statistics connection. The latter results are on the other hand precise consequences of basic axioms in field theory (see [SW]).

5.2. Analyticity in unphysical sheets

Unphysical sheet analyticity and the occurrence of poles in unphysical sheets, to be interpreted as associated to unstable particles, can be analyzed in a precise way in the simple situations considered below. The general multisheeted structure of the functions $\underline{T}_N$, which must take into account singularities required by internal consistency arguments (including branch points arising from unstable particles), is not known so far in detail, even from a heuristic viewpoint.

We consider a $2 \to 2$ process $(1, 2) \to (3, 4)$, or more generally (in view of the purposes of Sect. 6) a $m \to m$ process, $m \geq 2$, in a simplified theory in which the unitarity equation in the low energy region $((m\mu)^2 < s < (M\mu)^2$, for some $M > m)$ reads

$$T_+ - T_- = T_+ *_+ T_-, \qquad (1.54)$$

where $*_+$ in the right-hand side is an on-mass-shell convolution integral over m internal energy-momenta with an emc δ-function between initial, or final, and internal energy-momenta. Equivalently, for example, at $m = 2$, $T_+ *_+ T_-$ is the first term of the right-hand side of (1.21) after factorization of its overall emc δ-function. Eq. (1.54) is the actual unitarity equation of the nonsimplified theory if $m = 2$, at $4\mu^2 < s < 9\mu^2$, whereas all terms in (1.22) apart from the second one have been removed from the right-hand side at, for example, $m = 3$.

The scattering function T will be expressed below in terms of the variable $s = (p_1 + p_2)^2$ or $s = (p_1 + \cdots + p_m)^2$ and of angular variables Ω', Ω'' associated with the initial and final particles. At $m = 2$, Ω' and Ω'' reduce to angles θ', θ'' that define the directions of $\vec{p}_2 = -\vec{p}_1$ and $\vec{p}_3 = -\vec{p}_4$ in a reference frame in which $\vec{p}_1 + \vec{p}_2 = \vec{p}_3 + \vec{p}_4 = 0$. As a matter of fact, $T_{2,2}$ depends only, in view of Lorentz invariance, on the angle θ between $\vec{p}_1$, $\vec{p}_3$ but this fact will not be used. The right-hand side of (1.54) then denotes convolution, for $s > (m\mu)^2$, with an integration measure $\alpha(s, \Omega)d\Omega$ (over intermediate angular variables Ω) arising from the elimination of mass-shell and emc δ-functions:

$$(T_+ *_+ T_-)(s; \Omega', \Omega'') \equiv \int T_+(s; \Omega', \Omega)T_-(s; \Omega, \Omega'')\alpha(s, \Omega)d\Omega. \quad (1.55)$$

The explicit calculation shows that α is of the form

$$\alpha(s, \Omega) = \sigma^\beta \hat{\alpha}(s, \Omega), \quad (1.56)$$

where $\sigma = s - (m\mu)^2$, $\beta = [(m-1)d - m - 1]/2$ and $\hat{\alpha}$ is locally analytic in s; β is integer or half-integer according to the values of m and of the dimension d of space-time. At $m = 2$, β is integer if d is odd and is half-integer if d is even. As in the case $m = 2$, $T = T_+$ and T_- are assumed, at $m > 2$, to be analytic continuations of each other around $s = (m\mu)^2$, and the physical sheet analytic function, denoted below $\underline{T}_+$, from which T_+ and T_- are plus $i\epsilon$ and minus $i\epsilon$ boundary values respectively, is more precisely assumed for our purposes to be analytic in a cut neighborhood of $s = (m\mu)^2$, with the cut along $s \geq (m\mu)^2$.

Analyticity at real physical region points reached from above or below the cut (at $4\mu^2 < s < 9\mu^2$) is known if macrocausality (and unitarity) are assumed. However, only a weaker regularity property, for example, the continuity of the boundary values T_+, T_-, will be needed and used in the following argument, which shows that (1.54) implies "local maximal analyticity" in a multisheeted domain around $s = (m\mu)^2$, up to possible poles in s.

Let $*$ denote the analytic continuation of the operation $*_+$ in s, defined as the convolution with the same measure $\alpha(s, \Omega)d\Omega$ but now for complex values of s; $\underline{*}_+$ will denote its values, for s in the cut domain, obtained by turning in the anti-clockwise sense, and $\underline{*}_-$ its values in the next sheet; $\underline{*}_+ = \underline{*}_-$ if β is integer and $\underline{*}_- = -\underline{*}_+$ if β is half-integer, in view of the behaviour of σ^β. The equation

$$\underline{T}_+ - \underline{T}_- = \underline{T}_+ \, \underline{*}_+ \, \underline{T}_- \qquad (1.57)$$

is considered as a Fredholm resolvent equation with respect to angle variables, s being considered as a complex parameter, and defines $\underline{T}_-$ locally in the cut-plane in terms of $\underline{T}_+$ as an analytic function in s, up to possible poles which are the zeroes of the denominator $D(s)$ in Fredholm solutions. (The numerator $N(s, \Omega', \Omega'')$ and $D(s)$ can be checked to depend analytically on s.)

The boundary value of $\underline{T}_-$ at $s > (m\mu)^2$ from *above* the cut is on the other hand equal to T_- in view of (1.54) and the uniqueness of Fredholm solutions. As a consequence, T_- is both, at $s > (m\mu)^2$, the plus $i\epsilon$ boundary value of $\underline{T}_-$ and the minus $i\epsilon$ boundary value of $\underline{T}_+$. By the edge-of-the-wedge theorem, T_- is therefore locally analytic at s real $> (m\mu)^2$ and $\underline{T}_-$ is an analytic continuation of $\underline{T}_+$ in a different sheet.

This continuation is obtained here by turning around $s = (m\mu)^2$ in the anti-clockwise sense, but a similar argument, with $\underline{*}_+$ replaced by $-\underline{*}_-$ in (1.57), allows one to define an analytic continuation in the usual "second sheet" (obtained from T_+ by turning in the clockwise sense), again up to possible poles in s.

Analytic continuation (with possible poles in s) in *a priori* further and further sheets around $s = (m\mu)^2$ in either sense is obtained similarly by an inductive procedure. We shall see in Sect. 6.2 that there are only two sheets if β is half-integer (e.g. $m = 2$, d even), but an infinite number of sheets if β is integer (e.g. $m = 2, d$ odd).

Remark. A regularity condition (e.g. continuity) on the boundary values T_+, T_- is indeed needed, as first emphasized (at $m = 2$) by A. Martin (see [Ma] and references therein), to use Fredholm theory as indicated. It allows one to exclude a la Martin pathologies that might otherwise prevent analytic continuation, for example, in a second sheet, and would correspond physically to the accumulation of an infinite number of unstable particle poles below the cut (in the second sheet) arbitrarily close to the real s-axis. (Such phenomena have been implicitly excluded in the macrocausality condition; see Sect. 4.1.)

5.3. Basic discontinuity formulae for $3 \to 3$ processes

Generalizations of the fixed momentum transfer (fixed t) dispersion relations have been proposed for $N > 4$; see [Sta] and references therein. (We here consider the actual, nonsimplified theory.) They make use of the Bergman-Weil formula which is a generalization of Cauchy's theorem. Discontinuity formulae that generalize the unitarity equation (1.21) are then needed. For each given $m \to n$ process, they now involve various boundary values of analytic continuations of $\underline{T}_N$, including those corresponding in the physical region to the scattering function $T_{m,n} = T_{m,n}^{(+)}$ and to $T_{m,n}^{(-)}$, but also many others. Discontinuities (i.e. differences between various boundary values) will still be expressed in terms of on-mass-shell convolution integrals involving only physical S matrices (or their hermitean conjugates). We present them below at $N = 6$, which is the case that has been examined most thoroughly, and more precisely for $3 \to 3$ processes (formulae for $2 \to 4$ or $4 \to 2$ processes are similar). Properties below have been given in [Sta] (and references therein) on the basis of heuristic S-matrix arguments. They will be discussed again in field theory in Chapter IV. The $3 \to 3$ channel considered, for example, $(1,2,3) \to (4,5,6)$, will be denoted t and is the analogue of the channel $(1,2) \to (3,4)$ of the case $N = 4$. However, there are now 15 other basic related channels to consider, namely the three channels denoted i, $i = 1,2,3$, corresponding to subsets of two initial particles: for example, $i = 2$ refers to the channel $(1,3;2,4,5,6)$, three analogous channels f, $(f = 4,5$ or $6)$ and nine crossed channels denoted (if); for example, $i = 3$, $f = 4$ refers to the channel $(1,2,4;3,5,6)$. To each set S of channels chosen among the 16 channels introduced above will be associated a function $T^{(S)}$. For a subset of sets S, called "good sets," which include the sets corresponding to cells of field theory (see Chapter II) but also others, $T^{(S)}$ is expected to be the boundary value of an analytic continuation of $\underline{T}_N$ (starting from a common region at small unphysical values of s) beneath all cuts associated with channels $g \in S$ and above all those associated with channels $g \notin S$, that is, with minus $i\epsilon$ and plus $i\epsilon$ distortions, respectively, in the space of imaginary parts Im s_g of the variable s_g, around the normal threshold singularities $s_g = (\ell\mu)^2$ associated with g; $T^{(S)}$ is equal to $T^{(+)}$, or $T^{(-)}$, if S is the empty set, or the set of all channels $g_1, \ldots, g_{16}$, respectively. The way analytic continuation has to be made around other Landau singularities is also well determined. In fact, the way in which the function has to be continued around a normal threshold specifies the way in which it has to be continued around any other singularity that emerges from it or into which it merges. For example, in the $3 \to 3$ case, if we consider the singularity of a triangle graph and the 2-particle normal threshold associated with a contracted graph (see Figure 1.3 in Sect. 3.1), a consistent rule must hold in the neighbourhood of the points where the two surfaces are tangent.

The good sets S are, in the $3 \to 3$ case, those for which there is no pair i, f such that either $(if) \in S$, $t \in S$, $i \notin S$, $f \notin S$ or $(if) \notin S$, $t \notin S$, $i \in S$, $f \in S$. They include 26,018 sets S (out of the $2^{16} = 56,536$ possible sets), among which there are 2,282 sets corresponding to all cells of field theory. Examples of good sets which do not correspond to cells are all sets S composed of only one channel (i) or one channel (f).

It is believed that in the case of sets S corresponding to cells, the boundary value F^S occurring in field theory (and obtained *a priori* from the complex off-shell cell domain) coincides, when restricted to the mass-shell, with that introduced above.

Functions $T^{(S)}$ are also introduced for "bad" sets S and will satisfy weaker analyticity properties mentioned later. Explicit expressions of all terms (and hence of all relevant discontinuities) will be provided by the formula

$$T^{(S)} = \sum_{S' \subset S} (-1)^{|S'|} T_{S'} \tag{1.58}$$

with explicit formulae for all functions T_S. In particular $T_S \equiv 0$ whenever two channels g_i, g_j of S are overlapping, i.e. each one of the two sets of indices that define g_i has a nonempty intersection with each one of the sets that define g_j. This last fact (which ensures that generalized "Steinmann relations" are satisfied; see Steinmann relations in field theory in Sect. 4.1 of Chapter II) limits nonzero functions T_S to those associated with the empty set, the 16 sets S with a single channel g_i, the sets S with two channels of the form $\{(i), (f)\}$, $\{(i), t\}$, $\{t, (f)\}$, $\{(i), (if)\}$, $\{(if), (f)\}$ and finally the sets S with three channels of the form $\{(i), t, (f)\}$, $\{(i), (if), (f)\}$. We give below examples of explicit formulae of these terms T_S (leaving implicit that global emc δ-functions should be factored out from right-hand sides):

$$T_t \left(= T^{(+)} - T^{(t)} \right) = \tag{1.59}$$

$$T_{(if)} = \tag{1.60}$$

$$T_{(i),(f)} = \tag{1.61}$$

where terms in the right-hand sides are sums of on-mass-shell convolution

integrals corresponding, for each set , to all possible sets of intermediate particles in the energy region considered, and the plus box in
the middle in (1.61) denotes the contribution to the full nonconnected
S matrix in which lines i and f cannot go straight through (i.e. are
connected to nontrivial bubbles). Remaining nonzero T_S are defined
similarly. The following property is satisfied for all S:

$$T^{(S)} = \bar{T}^{(\bar{S})}, \tag{1.62}$$

where $\bar{S}$ is the set of all channels that do not belong to S and, for any
S, $\bar{T}^{(S)}$ is defined in the same way as $T^{(S)}$ but with exchange of plus
and minus bubbles $\left(\bar{T}^{(S)} = -(T^{(S)})^\dagger\right)$.

For bad sets S, $T^{(S)}$ cannot be expected to be obtained by analytic
continuation beneath cuts associated to S and above cuts associated to
$\bar{S}$. However, $T^{(S)} + D^{(S)}$ and $\bar{T}^{(\bar{S})} + \bar{D}^{(\bar{S})}$ are always expected to satisfy,
respectively, minus $i\epsilon$ rules of analytic continuation around thresholds
associated to S, and plus $i\epsilon$ rules around thresholds associated to $\bar{S}$.
Here,

$$D^{(S)} = 0 \qquad\qquad \text{if } t \in \bar{S}, \tag{1.63}$$

$$D^{(S)} = \sum_{\substack{(i,f);(if)\in S \\ (i)\in\bar{S},(f)\in\bar{S}}} \text{} \qquad \text{if } t \in S, \tag{1.64}$$

and $\bar{D}^{(S)} = -(D^{(S)})^\dagger$. If S is a good set, $D^{(S)}$ and $D^{(\bar{S})}$ always vanish,
so that $T^{(S)} + D^{(S)}$ ad $\bar{T}^{(\bar{S})} + \bar{D}^{(\bar{S})}$ reduce to $T^{(S)}$ and $\bar{T}^{(\bar{S})}$, which are
equal in view of (1.62). Analyticity properties of $T^{(S)} = \bar{T}^{(\bar{S})}$ follow.

§6. Analysis of Landau singularities

6.1. Graphs with single and double lines: holonomic cases

The triangle graph. We first consider below a physical region point P of
a unique (smooth, codimension one) surface $L_+(G)$ associated to a graph
G *without* sets of multiple lines. We, moreover, assume that scattering
functions at each vertex remain analytic in the integration domain. This
is the case in Sect. 4.3 for a triangle graph G. In the neighborhood of a
point P of $L_+(G)$ the discontinuity d_c can be written in the form:

$$d_c(p) = \int a(p,k) \prod_{j=1}^{2d+3} \delta\left(f_j(p,k)\right) \mathrm{d}k, \tag{1.65}$$

where $p = (p_1, \ldots, p_6)$, $k = (k_1, k_2, k_3)$, $dk = dk_1 dk_2 dk_3$, the $f'_j s$ are the arguments of the three mass-shell δ-functions and of the $2d$ emc δ-functions associated with two of the bubbles (where d is the dimension of space-time), and the function a, which is the product of the three scattering functions at each vertex, remains analytic in the integration domain when p varies in the neighborhood of P.

Given a system of real analytic local coordinates $q = (q_1, \ldots, q_\rho)$ of $M_{3,3}$ near P, with $q_1 = 0$ and $q_1 > 0$ representing locally $L_+(G)$ and its physical side, it can be checked [Ph2] that one can always choose $2d + 2$ from among the $2d + 3$ functions f_j, relabeled below $f_1, \ldots, f_{2d+2}$, and $d - 2$ other analytic functions $x_1(q, k), \ldots, x_{d-2}(q, k)$ such that:

(i)
$$f_{2d+3}(q, k) = q_1 - \sum_{i=1}^{d-2} [x_i(q, k)]^2, \tag{1.66}$$

(ii) the change of variables $q, k \to q, f_1, \ldots, f_{2d+2}, x_1, \ldots, x_{d-2}$ produces a regular jacobian.

Hence, by elimination of the first $2d + 2$ δ-functions

$$d_c(q) = \int a'(q, x)\delta\left(q_1 - \sum x_i^2\right)dx, \tag{1.67}$$

where $x = (x_1, \ldots, x_{d-2})$, and a' (which is the product of a with the jacobian, taken at $f_1 = \ldots = f_{2d+2} = 0$) is analytic in the integration domain.

The presence of the δ-function $\delta\left(q_1 - \sum x_i^2\right)$ in the right-hand side of (1.67) is not surprising and is in fact fully consistent with the fact that d_c is identically zero (see Sect. 4.2) on the nonphysical side ($q_1 < 0$) of $L^+(G)$. The fact that the integral in the right-hand side is a well-defined distribution (locally) is already known from various more general considerations. It follows here from the fact that $\delta\left(q_1 - \sum x_i^2\right)$ is itself a well-defined distribution of q_1, x. By definition,

$$\langle d_c, \phi \rangle = \int \delta\left(q_1 - \sum x_i^2\right) a'(q, x)\phi(q)dq\,dx$$
$$\equiv \int \phi\left(\sum x_i^2, q_2, \ldots, q_\rho\right) a'\left(\sum x_i^2, q_2, \ldots, q_\rho, x\right) dq_2 \ldots dq_\rho\,dx \tag{1.68}$$

for any C^∞ test function ϕ with support in a sufficiently small neighborhood of $q = 0$.

In the case $d = 2$, that is, $d - 2 = 0$, there are no variables x, and d_c is explicitly of the form:

$$d_c(q) = a'(q)\delta(q_1). \tag{1.69}$$

When $d \geq 3$, one checks as explained below that

$$d_c(q) = \theta(q_1)q_1^\beta a''(q), \qquad \beta = \frac{d-4}{2}, \tag{1.70}$$

where $a''(q)$ is locally analytic at $q = 0$. The result is first easily obtained in the region $q_1 > 0$ by replacing $x_1, \ldots, x_{d-2}$ by the new variables $t = \sum x_i^2$ and angular variables. This change of variables introduces the factor $t^\beta \equiv q_1^\beta$. At $q_1 < 0$, one has obviously $d_c = 0$; one finally checks easily that $\langle d_c, \phi \rangle$ is indeed equal to $\int \theta(q_1)q_1^\beta a''(q)\phi(q)\, dq$ even when the support of ϕ contains $q = 0$. QED

The edge-of-the-wedge theorem shows that two distributions that admit the same discontinuity differ at most by a locally analytic function. Hence one obtains the following results for the scattering function T near $q = 0$:

$$\beta = -1 \qquad\qquad T(q) = a_1(q)\frac{1}{q_1 + i\epsilon} + a_2(q), \qquad (1.71)$$

$$\beta = 0, 1, 2, \ldots \qquad T(q) = a_1(q)q_1^\beta \ell n\, q_1 + a_2(q), \qquad (1.72)$$

$$\beta = -\frac{1}{2}, \frac{1}{2}, \frac{3}{2}, \ldots \quad T(q) = a_1(q)q_1^\beta + a_2(q), \qquad (1.73)$$

where a_1, a_2 are locally analytic functions. The difference of the cases (1.72) and (1.73) is simply that q_1^β is analytic in the first case, and hence does not change after turning around $q_1 = 0$, whereas it changes its sign if β is half-integer.

More general results. The same analysis applies equally to more general graphs G with only single lines, at generic points P of real analytic codimension-one surfaces $L_+(G)$, such that all scattering functions associated with each bubble remain analytic in integration domains. The results are still obtained in the same way with

$$\beta = \frac{d\ell - m - 1}{2}, \tag{1.74}$$

where ℓ is the number of independent closed loops and m is the number of internal lines ($\beta \geq -1$ in the cases under consideration). Once the square-root nature of 2-particle thresholds, in even space-time dimension, has been established (see Sect. 6.2) results of the same type can also be obtained for graphs G with sets of doubles lines [KS1]. It is shown in fact that the boxes $S_\alpha^{-1} \equiv S^{-1}$ on each set of double lines can be removed, the scattering functions at each vertex being then replaced by functions that remain analytic in the integration domains.

More refined results of [KS1] include, for example, the following local form in the neighborhood of a point P where the Landau surface of the triangle meets that of a ("self-energy") graph G_1 obtained by contraction of one line (see example (iii) in Sect. 3.1). In a coordinate system where $L_+(G_1)$ is given by $q_1 = 0$ and $L_+(G)$ by $q_1 = q_2^2$, one has

$$T = \left[a_1(q)\sqrt{q_1 + i\epsilon} + a_2(q) \right] \log \left(\sqrt{q_1 + i\epsilon} + q_2 \right)$$
$$+ a_3(q)\sqrt{q_1 + i\epsilon} + a_4(q), \qquad (1.75)$$

where $a_1, \ldots, a_4$ are analytic at $q = 0$.

6.2. Simplified theory of the m-particle threshold and expansions in terms of holonomic contributions

We consider here either a $2 \to 2$ process, or the simplified theory of $m \to m$ processes of Sect. 5.2, and wish to determine the nature of the singularity at the 2-particle threshold $s = 4\mu^2$, or the m-particle threshold $s = (m\mu)^2$. According to the results of Sect. 5.2, the scattering function T can be analytically continued (as a meromorphic function) in an *a priori* multisheeted domain around $s = (m\mu)^2$. We below denote by $T_0, T_1, T_2, \ldots$ the successive determinations obtained at s real $> (m\mu)^2$ after r turns, for example, in the anticlockwise sense, around $s = (m\mu)^2$, $r \geq 0$, with $T_0 = T_+$, $T_1 = T_-$.

The following purely algebraic lemma will be useful.

Lemma 1.1. Let $A_0, A_1, \ldots, A_n$ be suitable kernels and let $o_{(0)}, o_{(1)}, \ldots, o_{(n-1)}$ denote linear operations with associativity properties. Then, the successive relations

$$A_k = A_{k+1} + A_k o_{(k)} A_{k+1}, \quad k = 0, 1, \ldots, n-1 \qquad (1.76)$$

yield:

$$A_0 = A_n + A_0 \left(o_{(0)} + o_{(1)} + \cdots + o_{(n-1)} \right) A_n. \qquad (1.77)$$

Proof. For $n = 2$, write $A_0 = A_1 + A_0 o_{(0)} A_1$, replace A_1 by $A_2 + A_1 o_{(1)} A_2$ and use

$$A_0 o_{(0)} \left(A_1 o_{(1)} A_2 \right) = \left(A_0 o_{(0)} A_1 \right) o_{(1)} A_2$$
$$= A_0 o_{(1)} A_2 - A_1 o_{(1)} A_2. \qquad (1.78)$$

The induction on n is trivial. QED

In the present application, we start from the relation (1.54)

$$T_0 = T_1 + T_0 *_{(0)} T_1, \qquad (1.79)$$

where $*_{(0)} \equiv *_+$. By analytic continuation around $s = (m\mu)^2$, it gives:

$$T_1 = T_2 + T_1 *_{(1)} T_2. \qquad (1.80)$$

If β is half-integer, $*_{(1)} = -*_{(0)}$ in view of the change of sign of the factor σ^β in (1.56) so that Lemma 1.1 gives $T_0 = T_2$. Hence, one obtains a (holonomic) two-sheeted square-root type singularity.

If β is integer, one has instead $*_{(1)} = *_{(0)}$ and more generally $*_{(r)} = *_{(0)}$, $\forall\, r$, so that analytic continuation provides the successive relations

$$T_r = T_{r+1} + T_r * T_{r+1} \qquad (1.81)$$

with here $* \equiv *_{(0)}$, and by Lemma 1.1:

$$T_0 = T_r + r\, T_0 * T_r, \quad \forall\, r. \qquad (1.82)$$

A purely algebraic argument [II] shows that $\underline{T}$ cannot satisfy the "finite-determination property" (see Sect. 4 of the Mathematical Appendix), and therefore is not holonomic at threshold, unless $T_0 *_0 \dots *_0 T_0$ (R factors T_0) vanishes identically for some integer R. The latter condition is pathological and can be excluded physically.

Nonholonomic structure at β integer. It can be analyzed in two ways [BI2]:

(i) via orthogonal decompositions of T following from a diagonalization of the unitarity equation. This can be achieved at $m = 2$ via "partial wave analysis." T, expressed in terms of s and of the angle θ (between $\vec{p}_1$, $\vec{p}_3$ in a reference frame in which $\vec{p}_1 + \vec{p}_2 = \vec{p}_3 + \vec{p}_4 = 0$) is written in the form:

$$T(s, \theta) = \sum_{\ell=0}^{\infty} (2\ell + 1) f_\ell(s) P_\ell(\cos\theta) \qquad (1.83)$$

with conversely:

$$f_\ell(s) = \int_0^{2\pi} T(s, \theta) P_\ell(\cos\theta)\, \mathrm{d}\theta. \qquad (1.84)$$

The unitarity equation then gives for each ℓ:

$$f_\ell^{(0)}(s) - f_\ell^{(1)}(s) = a(s)f_\ell^{(0)}(s)f_\ell^{(1)}(s), \qquad (1.85)$$

where $a(s) = \text{cst } s^{-1/2}(s - 4\mu^2)^\beta$ is locally analytic in the neighborhood of $s = 4\mu^2$. Eq. (1.85) gives:

$$\left(1/f_\ell^{(1)}(s)\right) - \left(1/f_\ell^{(0)}(s)\right) = a(s)$$

and therefore:

$$f_\ell(s) = \left[a(s)\left(\frac{1}{2i\pi}\ell n\ \sigma\right) + b_\ell(s)\right]^{-1}, \qquad (1.86)$$

where $\sigma = s - 4\mu^2$ and b_ℓ is a locally analytic or uniform function. The function f_ℓ thus defined in (1.86) (involving the inverse of a logarithm) is not holonomic.

At $m > 2$, a similar result follows more generally from the further condition of hermitean analyticity. In fact, it can be shown that T can then be written in the form of the infinite sum (with convergence properties)

$$T(s; \Omega, \Omega') = \sum_i t_i(s)E_i(s, \Omega, \Omega'), \qquad (1.87)$$

where the projectors E_i are orthogonal ($E_i * E_j = E_i\delta_{i,j}$), the eigenvalues t_i are of the form:

$$t_i(s) = [(1/2i\pi)\ \ell n\ \sigma + b_i(s)]^{-1} \qquad (1.88)$$

with $\sigma = s - (m\mu)^2$, and the functions E_i, b_i are analytic at $\sigma = 0$ (or uniform around it).

Eqs. (1.83) and (1.87) provide a decomposition of T into elementary contributions which are nonholonomic but have a simple, well-specified form.

(ii) through the introduction of a suitable irreducible kernel U and a corresponding expansion of T in terms of holonomic contributions. Under weak conditions, this kernel can be introduced in both cases β half-integer or β integer via the equation:

$$\underline{T} = \underline{U} + \underline{T} \otimes \underline{U} = \underline{U} + \underline{U} \otimes \underline{T}, \qquad (1.89)$$

where

$$\otimes = \frac{1}{2} * \qquad\qquad \text{for } \beta \text{ half} - \text{integer,} \qquad (1.90)$$

$$\otimes = \left(\frac{i}{2\pi}\ell\text{n } \sigma\right) * \qquad \text{for } \beta \text{ integer.} \qquad (1.91)$$

In the case β half-integer, (1.89) is the standard equation defining the kernel called "K-matrix" in the 1960s at $m = 2$. The definition is modified at β integer in a way such that U will be, in either case, analytic at $\sigma = 0$ or uniform around it, as we now explain. First, via Fredholm theory (with complex parameters), or via the Neumann series expansion

$$\underline{U} = \underline{T} - \underline{T} \otimes \underline{T} + \underline{T} \otimes \underline{T} \otimes \underline{T} - \cdots, \qquad (1.92)$$

if the latter is convergent, $\underline{U}$ is well defined locally as an analytic or meromorphic function (with possible poles in s) in the physical sheet. The equality

$$U_0 = U_1, \qquad (1.93)$$

that is, the uniformity of U, is then a consequence of the unitarity equation $(T_0 - T_1 = T_0 *_0 T_1)$ and of the relation

$$\otimes_0 - \otimes_1 = *_0 \qquad (1.94)$$

checked by direct inspection in either parity case (in view of the properties of $*$ already mentioned: $*_1 = -*_0$ or $*_1 = *_0$ for β half-integer or integer respectively).

Proof of (1.93). Eq. (1.89) gives $T_0 = U_0 + U_0 \otimes_0 T_0$, $T_1 = U_1 + T_1 \otimes_1 U_1$. Lemma 1.1 is then applied to the sequence $U_0 = T_0 - U_0 \otimes_0 T_0$, $T_0 = T_1 + T_0 *_0 T_1$, $T_1 = U_1 + T_1 \otimes_1 U_1$. QED

We below restrict our attention for simplicity to the case $\beta > 0$ (β integer) and assume that U is analytic at $\sigma = 0$. This follows from the assumption that $\underline{T}$ is locally bounded in the physical sheet near $\sigma = 0$: the series (1.92) is then absolutely convergent in view of the factors σ^β in $*$. The analyticity of U will on the other hand be directly established in weakly coupled models of field theory (see Chapter IV).

Then the following expansion in terms of elementary holonomic contributions in $(\ell\text{n } \sigma)^n$, which is convergent near $\sigma = 0$, holds:

$$\underline{T} = \sum_{n=0}^{\infty} \underline{U}^{\hat{*}(n+1)} \left(\frac{i}{2\pi}\sigma^\beta \ell\text{n } \sigma\right)^n. \qquad (1.95)$$

In (1.95), $\hat{\ast}$ is obtained from $\ast$ after factorization of σ^β ($\ast = \sigma^\beta \hat{\ast}$), and all terms $U^{\hat{\ast}(n+1)}$ are, like U, analytic in a (common) neighborhood of $\sigma = 0$ where they are uniformly bounded in modulus by cst^n. The uniform absolute convergence of the series (1.95) follows in a neighborhood of $\sigma = 0$ (for any finite number of sheets).

Each term in $(\ell n \, \sigma)^n$ (and hence the sum of all terms up to n) satisfies, in view of the analyticity of the coefficients $(i/2\pi)\sigma^\beta U^{\hat{\ast}(n+1)}$, the finite determination property and is therefore holonomic. The nonholonomicity of T is linked from that viewpoint to the fact that the order of the finite determination property of $(\ell n \, \sigma)^n$ tends to infinity with n and that n is arbitrarily large, unless $U \ast \ldots \ast U \equiv 0$ for some number R of factors U, a condition that can be shown to be equivalent to $T_0 \ast \ldots \ast T_0 \equiv 0$. Rearrangements of terms of the infinite sum that would yield holonomicity for the sum, as in the case of the series $\sum (1/n!)(\ell n \, \sigma)^n$, equal to σ, can be excluded in the present situation.

The analogue of (1.95),, with possibly off-shell *external* energy-momenta, as well as related expansions in terms of Bethe-Salpeter type irreducible kernels, will be given in Chapter II. An interpretation in terms of refined ideas of macrocausal factorization will also be given there.

6.3. The nonsimplified theory: outlook and conjectures

As confirmed from the viewpoint of field theory, the S matrix is probably nonholonomic at its Landau singularities, apart from a particular class associated to graphs without sets of multiple lines or, in even dimension d, with at most sets of two lines between some vertices. The analysis of the simplified theory, as also further analysis in field theory, suggests on the other hand expansions in terms of an infinite number of regular holonomic contributions near any $+\alpha$-Landau point or in bounded parts of the physical region.

The existence of locally *convergent* expansions that would generalize the expansion (1.95) is an attractive conjecture [KS2], although there is so far no precise result to support it in the general case. The class of graphs to be considered is also unclear. Results of field theory, namely expansions in terms of holonomic Feynman-type contributions with irreducible kernels generalizing the Bethe-Salpeter kernel at each vertex, will be discussed in Chapter IV. They depend so far on the smallness of the coupling and involve a class of graphs that goes somewhat beyond those giving rise to causal diagrams in the sense of Sect. 3. It is not known whether analogues of the kernel U can be defined more generally.

Appendix: The multiparticle S matrix in two-dimensional space-time

Macrocausality and macrocausal factorization can be stated in the same way for theories in space-time dimension $d > 2$ or $d = 2$. However, specific features occur in the latter case (for on-shell initial and final energy-momenta. This is no longer true for off-shell Green functions in field theory). In particular:

(i) Two (nonparallel) trajectories *always meet each other* in space-time.

(ii) For a $2 \to 2$ process, the emc constraint $p_1 + p_2 = p_3 + p_4$ entails the conservation of the sets of (on-shell) initial and final momenta: $p_1 = p_3$, $p_2 = p_4$ or $p_1 = p_4$, $p_2 = p_3$. As a consequence, the momentum-space cluster decomposition property has no content and both $S_{2,2}$ and $S_{2,2}^c$ have the form:

$$S_{2,2}^{(c)}(p_1, \ldots, p_4) = \mathbb{1}_{2,2}(p_1, p_2; p_3, p_4) \, f^{(c)}(p_1, p_2), \qquad (1.96)$$

where $\mathbb{1}_{2,2}$ is the identity operator (and f or f^c depend on only one variable in view of Lorentz invariance).

(iii) Let us now consider a $3 \to 3$ process. As mentioned in Sect. 3.2 at $d \geq 2$, all physical region points of the submanifold

$$L = \{p = (p_1, \ldots, p_6) \, ; \; p_1 = p_6, \; p_2 = p_4, \; p_3 = p_5, \\ p_1, p_2, p_3 \text{ in a common plane}\}$$

lie on the eight $+\alpha$-Landau surfaces of two triangle graphs and of the six graphs with one internal line obtained by removing one vertex. These surfaces are tangent along L, which has a low dimension at $d > 2$, with two opposite physical sides. At $d = 2$, these surfaces now all coincide with L, which is a codimension-one surface of the physical region $M_{3,3}$: in view of the mass-shell and emc constraints, L can be defined there by a unique equation, such as $\vec{p}_1 = \vec{p}_6$.

Let P be a point of L that lies on no other $+\alpha$-Landau surface. As at $d > 2$, the scattering function $T_{3,3}$ is not the boundary value of an analytic function along L, but is instead locally a sum of two boundary values obtained from opposite plus $i\epsilon$ directions. Macrocausal factorization entails at $d = 2$ a more precise local structure, most easily expressed for the *nonconnected* S matrix. Let D_1, D_2 be the terms

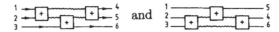 and

respectively. In view of (1.96), it is easily seen, in a theory with only one type of particle (see another case at the end), that

$$D_1 = D_2 = \mathbb{1}_{3,3}(p_1, \dots, p_6) \prod_{1 \leq i < j \leq 3} f(p_i, p_j), \qquad (1.97)$$

where the last product is locally analytic. Then, we show below that

$$s_{3,3}(p_1, \dots, p_6) \simeq d(p_1, \dots, p_6), \qquad (1.98)$$

where $d = d_1 = d_2$ and $\simeq$ means, as in Sect. 4, modulo a locally analytic background in the neighborhood of P. (As before, $s_{3,3}$ and d are obtained from $S_{3,3}$ and D after factorization of an overall δ-function. They still contain, as d, a partial emc δ-function, e.g. $\delta(\vec{p}_1 - \vec{p}_6)$.)

Proof of (1.98) The distribution d is concentrated on L ($d = 0$ locally outside L) and its singular directions at P are the two opposite causal directions $\hat{u}_1$, $\hat{u}_2$. Macrocausal factorization ensures that $\hat{u}_1 \notin \mathrm{ES}_P(s_{3,3} - d_1)$, $\hat{u}_2 \notin \mathrm{ES}_P(s_{3,3} - d_2)$. Since $d_1 = d_2 = d$, $\hat{u}_1$ and $\hat{u}_2$ are not singular directions of $s_{3,3} - d$ at P. Any other direction $\hat{u}$ is also absent from $\mathrm{ES}_P(d)$ and, by macrocausality, from $\mathrm{ES}_P(s_{3,3})$, hence from $\mathrm{ES}_P(s_{3,3} - d)$. The latter is thus empty and $s_{3,3} - d$ is locally analytic. QED

In view of (1.97), Eq. (1.98) is thus a factorization property of the momentum-space $3 \to 3$ S matrix into a product of $2 \to 2$ S matrices, modulo an analytic background. This is specific of the dimension $d = 2$. It is at the origin of an actual factorization *without background* ($S_{3,3} = D$) for a class of models enjoying the following properties (as is the case in field theory for "completely integrable" models):

(a) absence of particle production (or annihilation): $S_{m,n} = 0$ if $m \neq n$

(b) conservation of the sets of particle momenta (i.e. the S matrix vanishes unless each final momentum is equal to an initial momentum and conversely).

In fact, the background vanishes outside L from property (b) and hence, vanishes identically in view of its local analyticity. On the other hand, $+\alpha$-Landau surfaces of graphs with $m_v \to n_v$ vertices, $m_v \neq n_v$ are absent from property (a).

Properties (a) and (b), which would yield important inconsistencies at $d > 2$, can be satisfied (although they are not necessary) at $d = 2$.

For such models, the multiparticle S matrix $S_{m,m}$ satisfies a similar factorization property, again as a consequence of macrocausality and macrocausal factorization. Note that the analogue of the surface L, defined by the equality of the sets of initial and final momenta, has codimension > 1 at $m > 3$. In such models, previous exceptional $+\alpha$-Landau points are the actual relevant ones.

Factorization equations. For simplicity, we consider again a $3 \to 3$ process. Models of physical interest may include several types of particles with, for example, the same mass. There are in this case different possible attributions of intermediate particles for the internal lines of the diagrams involved. D_1 and D_2, which are sums of respective contributions, are no longer *a priori* equal. However, the analysis above still allows one to write locally $s_{3,3}$ as a sum of two boundary values a_1, a_2 of analytic functions (from opposite directions) with respective discontinuities d_1, d_2. If property (b) holds, $a_1 + a_2 = 0$ outside L. This implies, as easily seen by the edge-of-the-wedge theorem and as explained in a somewhat more explicit way below, that $d_1 = d_2$, hence, $s_{3,3} = d_1 = d_2$ as previously.

The equality $d_1 = d_2$ entails in this case "factorization equations" of interest between products of $2 \to 2$ scattering functions.

A more explicit derivation of the result above can be described as follows in the situation considered, in which discontinuities d_1, d_2 are concentrated on the surface L (and contain a corresponding δ-function). Let $q_1, q_2, \ldots$ be real analytic local coordinates in the neighborhood of a point P of L, such that L is represented locally by $q_1 = 0$ with $q_1 > 0$ representing one of the two opposite physical sides, associated with one of the sets of 4 graphs. Then, the general analysis shows that $s_{3,3}$ can be written locally in the form

$$s_{3,3} = d_1' \frac{1}{2i\pi} \frac{1}{q_1 + i\varepsilon} - d_2' \frac{1}{2i\pi} \frac{1}{q_1 - i\varepsilon} + a, \tag{1.99}$$

where a is locally analytic, $d_1 \equiv d_1' \delta(q_1)$, $d_2 \equiv d_2' \delta(q_1)$ and d_1', d_2' can be chosen independent of q_1. Eq. (1.99) can also be written as:

$$s_{3,3} = d_1' \delta(q_1) + \left[(d_1' - d_2') \frac{1}{2i\pi} \frac{1}{q_1 - i\varepsilon} + a\right]. \tag{1.100}$$

If $s_{3,3}$ vanishes outside L, i.e. at $q_1 \neq 0$, the bracket in the right-hand side, equal to $s_{3,3} - d_1' \delta(q_1)$, also vanishes there. Since it is the (minus $i\varepsilon$) boundary value of an analytic function, it vanishes identically also at $q_1 = 0$ so that $s_{3,3} = d_1' \delta(q_1) = d_1$. By a similar argument, $s_{3,3} = d_2' \delta(q_1) = d_2$ and thus $s_{3,3} = d_1 = d_2$. QED

CHAPTER II

Scattering Theory in Axiomatic Field Theory

INTRODUCTION AND GENERAL FORMALISM

§1. Introduction

1.1 Axiomatic field theory: general preliminaries

Let us first recall the traditional Wightman axioms in Minkowski space-time. This framework will be somewhat modified and completed for our purposes, but it can be considered as the fundamental starting point. Wightman axioms apply to field operators $A(x)$ acting on a Hilbert space $\mathcal{H}$ of physical states (to be later interpreted asymptotically in terms of particles in theories of interest); $x = (x_0, \vec{x})$ is a real space-time variable and each A is more precisely an operator-valued distribution which, to each test function f, associates an operator $A(f) \left(= \text{``} \int A(x)f(x)\mathrm{d}(x)\text{''} \right)$ in $\mathcal{H}$. (The fields A of interest here might be defined in terms of more fundamental underlying fields in theories with confinement.) The axioms include locality ($A(x)$ and $A(y)$ commute, or anticommute, if $x - y$ is space-like, depending on the bosonic or fermionic character of the field: see discussion in Sect. 2.1), the existence in $\mathcal{H}$ of a unitary representation of the Poincaré (inhomogeneous Lorentz) group and related relativistic transformation laws, positivity of the energy (the spectrum of the energy-momentum operators P_0, $\vec{P}$, generators of space-time translations, is assumed to be contained in the cone $\bar{V}_+ : p^2 \equiv p_0^2 - \vec{p}^2 \geq 0$, $p_0 \geq 0$), and the existence of a unique (up to scalar multiples) "vacuum vector" Ω in $\mathcal{H}$, invariant under Poincaré transformations, from which states in $\mathcal{H}$ can be generated by action of the fields. More precisely, it is assumed that all vectors in $\mathcal{H}$ can be obtained as linear combinations of vectors of the form $A(f_1)\ldots A(f_n)\Omega$ or as limits of sequences of such vectors. For most of our purposes, we shall consider for simplicity a theory with only one basic field, a "neutral" bosonic, scalar field A, from which $\mathcal{H}$ can be generated in that way. (Further fields, e.g. fields satisfying the locality condition, might be constructed from A, such as the product $:A(x)^n:$ from which "bound states," if there are any, might be defined in a more direct way by action on Ω.) Results to be described can be extended to other cases with only minor changes, at least for the class of theories in which we are mainly interested. General results of the 1960s such as TCP and the spin-statistics theorem, described in [SW], will be omitted.

The above axioms can be equivalently stated on the Wightman distributions $W_N(x_1, \ldots, x_N)$, $N \geq 2$, which are in the above framework vacuum expectation values of products of field operators $A(x_1), \ldots, A(x_N)$.

The Haag-Kastler (HK) axioms of local observables, also proposed in the 1960s (see [GJ, BLOT, Ha] and references therein) differ from the Wightman axioms in various technical and also conceptual respects, as will be mentioned in Sect. 2.1. Important developments (see [Fre, Ha] and references therein) on the possible structure of general theories have been carried out in that framework (charge superselection sectors, charges and statistics in low-dimensional field theories, particle or particle-like structure in theories with possible long-range forces and massless particles). However, the use of these axioms would introduce only minor changes for our purposes.

The Wightman or HK axioms take place in Minkowski space-time. On the other hand, interest in considering field theories in Euclidean space-time (imaginary times) became more and more apparent in the 1960s from various viewpoints, in particular for the definition of models in constructive theory. Euclidean axioms were then proposed in the first part of the 1970s[OS]. By analytic continuation from imaginary to real times, they allow one to reconstruct a theory satisfying the Wightman (or HK) axioms and are essentially equivalent to the latter. They will be briefly reviewed in Sect. 1 of Chapter III. The theory can be originally defined in this approach by a (probability) measure $d\mu(\varphi)$ on distributions φ in Euclidean space-time. Models of constructive theory will correspond to specific forms of this measure.

Finally, coming back to Minkowski space, it is useful (in particular in view of subsequent energy-momentum space analysis and scattering theory) to introduce "chronological" products of field operators and corresponding chronological functions: fields are ranged according to the order of time components of $x_1, \ldots, x_N$. The definition is only formal in view of problems at coincident points. However, properties of these products or of these functions can be derived formally from previous Wightman axioms. One may then start from an axiomatic framework in which relevant operators or functions are assumed to exist and to satisfy corresponding properties: see [EGS] and references therein. Euclidean axioms, supplemented by suitable conditions, allow one also to reconstruct a theory of this type in Minkowski space. (See [EE]).

Several other frameworks with their own interest have also been introduced, but we omit them for conciseness (see more details in [BLOT]).

The notion of particles is not *a priori* introduced in the Wightman framework (or in the alternative frameworks that have been mentioned). In the axiomatic framework, supplementary conditions are introduced to that purpose. For theories intended to describe systems of massive particles with short-range interactions, the first one is an assumption on the mass spectrum (spectrum of the mass operator $M = (P_0^2 - \vec{P}^2)^{1/2}$): existence of a "mass gap" between the origin (corresponding to the vacuum) and a first eigenvalue $\mu > 0$, and of a second mass gap between μ and the rest of the spectrum. As will be explained in Sect. 2, this condition allows one to define in $\mathcal{H}$ subspaces $\mathcal{H}_{\text{in}}$ and $\mathcal{H}_{\text{out}}$ of states, called "asymptotic states," which are naturally interpreted as free-particle states, corresponding to particles of mass μ, before or alternatively after interactions (Haag-Ruelle asymptotic theory). The existence of a partially isometric S matrix follows.

Apart from the origin and μ, the spectrum is expected to contain at least a continuum starting at 2μ corresponding to multiparticle states. It may also contain other eigenvalues μ', μ'', ..., for example, between μ and 2μ, corresponding to one-particle states of masses different from μ. In this case, the spaces $\mathcal{H}_{\text{in}}$, $\mathcal{H}_{\text{out}}$ of asymptotic states would be correspondingly enlarged. This may occur even if there is only one basic field A, from which "elementary" particles of mass μ but also "bound states" might be generated. For simplicity, we shall mainly restrict our attention, in this chapter, to a theory with only one type of (stable) particle, of mass μ.

The further condition of asymptotic completeness to be used in this chapter (see discussion below), which entails in particular that the S matrix is unitary, is the assertion that

$$\mathcal{H} = \mathcal{H}_{\text{in}} = \mathcal{H}_{\text{out}}. \tag{2.1}$$

Eq. (2.1) asserts that all states of $\mathcal{H}$ are asymptotically tangent both before and after interactions to free-particle states. As it stands, it asserts also that there are no other particles, different from the particles corresponding to the mass μ or possibly to the other masses of the spectrum. The situation may *a priori* be different. Let us consider, for example, a theory with only one basic field, a "neutral" field A. Neutral states that would correspond (asymptotically, before or alternatively after interactions) to a *pair* of "charged" particles, one of them with positive charge and the other one with the opposite charge, might *a priori* be created by action on the vacuum. (The "charge" involved here is not necessarily the "electric charge"; it may correspond to any suitable superselection rule.) Correspondingly, the (neutral) Hilbert space $\mathcal{H}$ would contain states corresponding to such pairs of charged particles. Being neutral, it would not contain states corresponding to *individual* charged particles, so that the latter would not appear as specified masses in the original mass spectrum. $\mathcal{H}_{\text{in}}$ and $\mathcal{H}_{\text{out}}$, which include only states corresponding to the original (neutral) particles, are in this case smaller than $\mathcal{H}$.

In such a situation, it is possible [Bu2] to *enlarge* the original space $\mathcal{H}$ by adding new "charged" superselections sectors: new charged fields, producing charged particles by action on the vacuum, can then be defined and Eq. (2.1) would then hold with larger spaces $\mathcal{H}_{\text{in}}$, $\mathcal{H}_{\text{out}}$ of asymptotic states. (The charged fields thus introduced are local in the class of theories describing massive particles with short-range interactions but not in more general cases: see [Bu2].)

1.2 Scattering theory: historical survey

We essentially consider theories of massive particles with short-range interactions and we start correspondingly with the special spectral condition with mass gap. For simplicity, moreover, we mainly consider a theory with only one particle of mass $\mu > 0$, corresponding to a mass spectrum composed of the origin, the mass μ and the continuum starting at 2μ. It is also assumed that one-particle states are generated by action of the basic field $A(x)$ itself on the vacuum: see Sect. 2.1. The case of theories with several types of particle is mentioned in the remarks that conclude this subsection.

We sometimes consider more specifically even theories, in which cases N-point functions vanish identically if N is odd. They will correspond in Chapters III and IV to models with even interactions, such as $P(\varphi)$ with even polynomial P.

Besides the Haag-Ruelle theory already mentioned, works on scattering theory have included the following steps, described below in a semihistorical order. Results described in paragraphs (i) to (v) are established in the "linear program" based on locality, the spectral condition and Lorentz invariance. They make use in particular of the interplay between support properties in x-space and in p-space provided by locality and the spectral condition, respectively. Results of the nonlinear program, based on the further condition of asymptotic completeness (and regularity conditions), are introduced in paragraphs (vi), (vii) and (viii).

(i) The link between the S matrix and underlying fields is made explicit by the LSZ "reduction formulae" [LSZ] which express (on-shell) scattering functions of $m \to n$ processes, $m + n = N$, as mass-shell restrictions of (connected, amputated) "chronological functions" $T_N(p_1, \ldots, p_N)$, defined in the space of real energy-momenta variables $p_1, \ldots, p_N$ satisfying energy-momentum conservation $(p_1 + \cdots + p_N = 0)$. For each N, T_N is obtained by Fourier transformation (and factorization of $\delta(p_1 + \cdots + p_N)$) from the connected, "amputated" vacuum expectation value of the time-ordered product of field operators $A(x_1), \ldots, A(x_N)$. The proof, first given in a rigorous form by K. Hepp (see [He]) and established so far away from M_0 points (some initial or some final p_k colinear), includes the proof that the "functions" T_N, which are as a matter of fact distributions, can be restricted to the mass-shell. This result is related to some of those described in (ii) and (v).

Scattering functions of crossed $m \to n$ processes where $m+n = N$ are restrictions, in their respective physical regions $(p_k^2 = \mu^2, k = 1, \ldots, N, (p_k)_0 < 0$ if k is initial, $(p_k)_0 > 0$ if k is final), of the same T_N.

(ii) Space-time cluster properties and more generally asymptotic causal properties of chronological functions and of the S matrix, also called macrocausal properties, have been established. First results in this domain were obtained in the 1960s in various particular situations (see in particular [AHR, He, O]). They have been more recently generalized in various respects in [BEG2] and in the related work [I8] in which rates of exponential fall-off arising from locality and the spectral condition, with a direct and simple physical content, have been established in general "non-causal" situations. We note that the results that can be derived in the linear program are, however, relatively weak, in general, in comparison with the macrocausal ideas presented in Chapter I for the S matrix: they express in particular the idea that energy-momentum can only propagate in future causal cones, up to well-specified exponential fall-off, but information on the content of the propagation in terms of particles remains at that stage very limited. Improved results will be obtained in the nonlinear program.

(iii) "N-points functions," analytic in a global "primitive domain" in the space of complex energy-momenta $p_1, \ldots, p_N$ (satisfying $p_1 + \cdots + p_N = 0$) are introduced. Each N-point function is analytic at real points in a neighborhood of the origin, as well as in a strip around Euclidean energy-momentum space (imaginary energies, real momenta), and can be analytically continued away from it at least in various "tubes" in complex space: $\text{Im}\,(p_1, \ldots, p_N)$ in a given open cone with apex at the origin, $\text{Re}\,(p_1, \ldots, p_N)$ arbitrary. It admits corresponding "cell" boundary values which are distributions in real p-space: as a matter of fact, these distributions are defined first, by various methods, and are shown to be boundary values of functions analytic in the above tubes. Boundary values from "adjacent" cells, moreover, coincide in corresponding real regions and the edge-of-the-wedge theorem (in its refined version: see Appendix) allows one to obtain the common analytic N-point function mentioned above.

For each N, T_N is shown to coincide, in the neighborhood of any real point $p = (p_1, \ldots, p_N)$, with at least one cell boundary value (depending on p) and is thus also a boundary value of the analytic N-point function, from directions which now depend on the real region considered. As a consequence, values of T_N in neighborhoods of physical regions of crossed processes are linked by analytic continuation in the primitive domain.

First results on the subject were obtained in the early 1960s[Ru1]. They have been improved and completed in various works of the 1960s and 1970s ([B2,EGS] and references therein). However, we note, at that stage, that although it admits real mass-shell physical regions on its boundary, the primitive domain has an empty intersection with the (real or complex) mass-shell, so that no result of interest is obtained on the S matrix on the mass-shell.

(iv) Results have been improved by noting that the primitive domain introduced previously is *not* a "natural domain" of holomorphy: any function analytic in this domain (and whose cell boundary values satisfy relevant algebraic relations) can be analytically continued in a larger domain called its holomorphy envelope. At $N \leq 4$, it has been shown already in the early sixties that the latter does intersect the complex mass-shell, and (on-shell) physical-sheet analyticity properties have then been established at $N = 4$, including hermitean analyticity, crossing and dispersion relations (in one variable) for $2 \rightarrow 2$ processes. Initial (and partial) results on the subject, due to [Bo], have been largely improved in subsequent works and in particular in [BEG1] (see further references therein). The results entail, as a particular aspect, that the $2 \rightarrow 2$ scattering function of, for example, the channel $(1, 2) \rightarrow (3, 4)$ is, in its physical region, the boundary value of the analytic 4-point function (restricted to the complex mass-shell) from the directions Im $s > 0$.

(v) Only more limited results can be expected at $N > 4$. We first discuss below the "local" situation near real physical points and will briefly come back to global problems in (viii). Let us consider a physical point P and the tube of analyticity that gives rise (by taking the boundary value) to the function T_N in a real neighborhood of P. If the holomorphy envelope has locally a nonempty intersection with the complex mass-shell, the *on-shell* scattering function $T_{m,n}$ will itself be locally the boundary value of an analytic function, namely the restriction of the analytic N-point function to the complex mass-shell. However, it is known from the analysis of Chapter I that there may exist some physical points in the neighborhood of which $T_{m,n}$ *cannot* be expected to be the boundary value of an analytic function. While these points are ultimately expected to be "exceptional" (see Chapter I), results on the holomorphy envelope are expected to be weaker. As a matter of fact, further use of the edge-of-the-wedge theorem allows one, as mentioned in [B4], to show, for any $m \to n$ process, "local analyticity" of $T_{m,n}$ in the neighborhood of some physical points in the sense above (i.e. the fact that $T_{m,n}$ is locally the boundary value of an analytic function) but there are important restrictions on the set of points thus obtained. Best results are obtained in [BEG2] at $N = 5$, that is, for a $2 \to 3$ or $3 \to 2$ process (a case in which the exceptional $+\alpha$-Landau points of Chapter I do not occur), but a full neighborhood of the 3-particle threshold is still excluded in spite of a refined use of Lorentz invariance. Partial results have also been given in [MP] for a $3 \to 3$ process. (More precisely, the latter results are obtained in the neighborhood of physical region points p such that the sets of initial and final energy-momenta coincide, and apply to the function obtained by subtracting pole terms: as explained in Chapter I, all points above are $+\alpha$-Landau points, where $T_{3,3}$ itself is not the boundary value of one analytic function.)

The best results to be expected in general in the linear program near real points (apart possibly from minor improvements) are probably those of [BEG2], reobtained in a more direct way in [I8] from the asymptotic causality properties discussed in paragraph (ii). They go (locally) beyond the holomorphy envelope and provide in fact well-specified local decompositions of the analytic N-point function, near any physical point, into (finite) *sums* of functions analytic in domains that now do intersect the complex mass-shell. Corresponding local decompositions of $T_{m,n}$ in the neighborhood of any physical point, as a sum of boundary values of analytic functions, follow. The decompositions obtained are different in various parts of the physical region. If, in a neighborhood of a given point, there are common directions of analyticity for all terms, the scattering function is there the boundary value of a unique analytic function.

The fact that more than one boundary value is needed at some points is not surprising, as mentioned above. However, the structure obtained at that stage is still remote from that ultimately desired. In particular, Landau singularities are not extricated. This is related to the remarks made in paragraph (ii) on the character of macrocausal properties in the linear program.

(vi) Exploitation of 2-particle unitarity or asymptotic completeness equations (and needed regularity conditions: see [Ma]) have led to further results already in the 1960s at $N = 4$, including in particular second-sheet analyticity (or more precisely meromorphy) across the cut associated to each $2 \rightarrow 2$ channel in the low-energy region (with possible poles in s). This result, first derived for the on-shell $2 \rightarrow 2$ S matrix (see Chapter I), has been reobtained and extended to the off-shell 4-point function in {B1}, via the use of a "2-particle irreducible" Bethe-Salpeter type kernel as outlined below, and more recently in [EGI] via a more direct method.

In the method of [B1], of interest also for later extensions, the BS-type kernel is defined (from the axiomatic viewpoint) in terms of the 4-point (connected, amputated) analytic function F via the possibly regularized Fredholm-type equation

$$F = G_\alpha + F \circ_\alpha G_\alpha \qquad (2.2)$$

first considered in Euclidean energy-momentum space. In (2.2), $\circ$ denotes a "Feynman-type" convolution with two internal lines and a 2-point function on each internal line, and $\circ_\alpha$ includes further suitable analytic "cut-off" factors, with sufficient decrease at infinity (in Euclidean directions), on these lines. The use of this regularization allows one to cover potentially more general theories for which divergence problems in integrals such as $F \circ G$ would otherwise occur. In contrast to G, which is from the perturbative viewpoint the usual Bethe-Salpeter kernel in super-renormalizable theories (i.e. the formal sum of "2-particle irreducible" Feynman graphs), G_α will not have a simple perturbative content, but its use is nevertheless satisfactory. By analytic continuation away from Euclidean space (based on Fredholm theory in complex space as developed by J. Bros) and use of 2-particle asymptotic completeness, G or G_α is shown to be analytic, up to possible poles, in a larger strip than F around Euclidean space, going up to $s = (3\mu)^2$, or $(4\mu)^2$ in an even theory, in Minkowski space: that is, the cut starting for F at $s = (2\mu)^2$ has been locally removed. This information, applied in the converse direction in Eq. (2.2) (from properties of G or G_α to properties of F), yields in turn, modulo some regularity assumptions on G or G_α, the 2-sheeted (d even) or multisheeted (d odd) structure of F around $s = 4\mu^2$.

(vii) Further works in the 1970s and 1980s have been based on a more general exploitation of asymptotic completeness. Given any N, the latter condition directly provides discontinuity formulae of interest that express differences between boundary values of the N-point function in terms of convolution integrals, over internal *on-shell* energy-momenta, involving specified boundary values of N'-point functions (with various values of N'). Besides discontinuity formulae for "absorptive parts" applying to adjacent cells and noted as useful below, let us mention, in particular, a class of formulae that express discontinuities in terms of multiple on-shell convolutions involving only "physical" boundary values, namely chronological functions and their mass-shell restrictions, i.e. S-matrix kernels. We note, however, that boundary values involved here are *a priori* obtained in general from off-shell directions. It is believed, but not proved so far in general, that results should extend in a suitable way to the complex mass-shell and should thus yield at least part of the discontinuity formulae, or "generalized optical theorems," needed for multiparticle dispersion relations (see Chapter I.5).

On the other hand, various improvements of previous analyticity properties have as a matter of fact been obtained, at least in some situations, with results on the complex mass-shell. As in the case $N = 4$ already mentioned in paragraph (vi), asymptotic completeness can be exploited in two different ways: either[EGI] by direct methods, or through the introduction of more general Bethe-Salpeter type irreducible kernels in the line of the ideas originally proposed by K. Symanzik [Sy1] and developed by J. Bros and collaborators see [B1,B3,B4]. The two methods have provided related results which do exhibit a refined structure "in terms of particles," in particular for 5- and 6-point functions and in turn for $2 \to 3$ and $3 \to 3$ scattering functions, in low-energy regions, namely below the 4-particle threshold in the $2 \rightarrow 3$ or $3 \rightarrow 3$ channel considered (or below the 5-particle threshold in an even theory). At $N = 6$, the following decomposition is, for example, essentially derived on the chronological function

$$\widetilde{\ominus} = \sum_{\substack{i=1,2,3 \\ j=4,5,6}} \widetilde{\ominus}^{\,j} + \sum_{\substack{i=1,2,3 \\ j=4,5,6}} \widetilde{\ominus\ominus}_{\,j} + R, \quad (2.3)$$

where explicit terms in the right-hand side are "Feynman-type" convolution integrals with 4-point functions at each vertex, and R is shown (after factorization of its emc δ-function) to be the boundary value of an analytic function from the directions Im $s >$ 0, in the region $(3\mu)^2 < s < (4\mu)^2$, or $(5\mu)^2$ in an even theory (and away from 2-particle thresholds in initial or final subenergies). It is, moreover, analytic if supplementary regularity assumptions are used. Eq. (2.3) then appears as a more precise version of the results stated on the S matrix in Sect. 4.3 of Chapter I, which are reobtained on-shell. An extension of the result is also obtained in the second method away from real points and then involves relevant analytic N-point functions.

The second method makes use, as at $N = 4$, of various general results on Fredholm theory in complex space (also due to J. Bros and collaborators) and on analyticity properties, in complex energy-momentum space, of convolution Feynman-type integrals. It is based at $N = 6$, in an even theory and at $s < (5\mu)^2$, on a suitable set of integral equations allowing one to define a further 3-particle irreducible kernel shown to satisfy suitable analyticity properties. While it requires stronger regularity assumptions in comparison with the "direct" method, it has provided more refined results in the above situation and seems potentially best adapted for many-particle structural analysis in more general energy regions. However, technical (and possibly conceptual) problems have blocked further progress. A heuristic analysis and further investigations and conjectures have been given in [I5]. Related developments in constructive or semi-axiomatic frameworks will be given in Chapter IV.

(viii) Other results of interest include some partial results on crossing properties (on the complex mass-shell) and dispersion relations at $N > 4$. In massive theories, such results are claimed in [LMPS] (and references therein) in the linear program for $3 \rightarrow 3$ amplitudes. However, the proof does not seem correct. In a more recent work [B5], J. Bros indicates that the question whether crossing properties can be established in the linear program at $N > 4$ "remains an unsolved and challenging problem" and he proves crossing properties at $N = 5$ through a combination of "global" methods generalizing those of [BEG1] in the case $N = 4$ and local results of the nonlinear program. As in [BEG1], "asymptotic crossing domains" are obtained (namely, here, cut-neighborhoods of infinity in suitable energy variables). Some steps towards extensions of the method to more general situations are also given. Another partial result of Bros, which involves a bounded crossing domain (joining physical points in the neighborhood of the forward scattering situation: same sets of initial and final energy-momenta) and also makes recourse to the nonlinear program for the local part, has been previously given in [B4] for $3 \to 3$ amplitudes.

This last topic will not be treated in this book.

We conclude with some remarks. As we shall see, the existence of a particle of mass μ will appear as a pole of the 2-point function at $s = \mu^2$. The nonamputated, connected N-point functions have poles (at $p_i^2 = \mu^2$) in each variable p_i, $i = 1, \dots, N$ in energy-momentum space. The analysis shows that they have also in general a pole at $s_I = \mu^2$ $(s_I = (\sum_{i \in I} p_i)^2)$ in the physical sheet, in any given channel, with a residue that factorizes into a product of N'-point functions. For example, the 4-point function has a pole in the channel $(1, 2; 3, 4)$ and is locally the product of this pole with two 3-point functions, if the theory is not even. In an even theory, this pole is absent in connection with the vanishing of 3-point functions.

For a theory with a spectrum composed of the origin, the mass μ, a new mass μ', $\mu < \mu' < 2\mu$, and the continuum, the analysis would yield further poles at μ'^2 in the physical sheet. If the mass μ' corresponds, for example, to a "2-particle bound state" generated by the action of $:A(x)^2:$ on the vacuum (see Sect. 1.1), and if the theory is even, this pole will not appear in the 2-point function itself (again in connection with the vanishing of 3-point functions), but will appear in the 4-point function. There is again locally a factorized form involving new 3-point functions.

As a matter of fact, in theories including several types of particles, various N-point functions (and in turn on-shell collision amplitudes of processes involving various initial and final particles) can be defined in terms of the field operators that directly create corresponding particles by action on the vacuum (either basic fields A_i or, for example, $B(x) = :A(x)^2:$ in the case mentioned above of 2-particle bound states).

Particles considered above are stable particles. Further poles (with again factorized residues) might occur, in all previous theories, in unphysical sheets (see Chapter I.5 and Sects. 5 and 6) with interpretation in terms of unstable particles.

1.3. Description of contents

The general formalism is presented in Sect. 2. The axiomatic framework is described in Sect. 2.1. The Haag-Ruelle asymptotic theory is outlined (with some minor changes) in Sect. 2.3 after a preliminary discussion in Sect. 2.2 of asymptotic localization properties in space-time which is intended to prepare also the later analysis of macrocausal properties.

Results on scattering theory are described in Sects. 3 to 7. Results of the linear program are treated first. We start, in Sect. 3, with the results of [BEG2,I8] mentioned in paragraphs (ii) and (v) of Sect. 1.2. Reduction formulae (paragraph (i)) are obtained as a byproduct in Sect. 3.2 (by methods which are essentially those of Hepp[He] with, however, some minor modifications and simplifications). With respect to that of paragraph (iii) on analytic N-point functions, the analysis of Sect. 3 is simpler from the algebraic viewpoint, although it involves, as a counterpart, elements of analytic essential support theory relying on a localized version of the Fourier transformation (see Appendix), whereas results of paragraph (iii) mainly rely on the usual Fourier-Laplace transformation. For conciseness, we shall not here specify the precise rates of exponential fall-off in noncausal situations or the corresponding domains of analyticity.

In Sect. 4, we present the complementary results of paragraph (iii), following the analysis of [EGS] which seems simplest, most satisfactory and best unified with that of Sect. 3. Relevant distributions, some of which will give rise to the cell boundary values (as already mentioned in Sect. 1.2), are defined in this approach in a way similar to chronological functions, except that more complicated combinations of products of time-ordered operators of fields are involved.

Improved results at $N \leq 4$ (paragraph (iv) of Sect. 1.2) are briefly outlined (without proofs) in Sect. 4.2.

We next come to the methods and results of the nonlinear program. Discontinuity formulae for "absorptive parts," following directly from asymptotic completeness and useful both in Sects. 5.2 to 5.4 and in Sect. 6, are first described in Sect. 5.1, where the generalized optical theorems mentioned in the first part of paragraph (vii) are also introduced. Sect. 5 then presents results on 4-, 5- and 6-point functions in low-energy regions obtained by direct methods. Results at $N = 4$ (analytic continuation of the 4-point function as a meromorphic function in a 2-sheeted or multisheeted domain around the 2-particle threshold, depending on the dimension d of space-time) are described in Sect. 5.2. The irreducible kernel U of Chapter I is also extended off-shell and expansions of interest of the 4-point function follow (under some conditions) locally. They provide a first type of "structural equation."

Local results at $N = 5$ and $N = 6$ are given in Sects. 5.3 and 5.4, respectively. They include the derivation (up to some technical limitations) of Eq. (2.3) which is a further example of a structure equation of interest. It applies, however, in a particularly simple situation (no singularity of graphs with sets of multiple lines and no "related" $+\alpha$-Landau surfaces). Further structure equations in terms of irreducible kernels will be discussed later.

Methods and results based on the introduction of irreducible Bethe-Salpeter type kernels are then outlined in Sect. 6. Some general preliminary theorems are first mentioned in Sect. 6.1. The 2-particle irreducible kernel G or G_α (see Sect. 1.2) is introduced in a more precise way in Sect. 6.2 and corresponding results are described. The treatment of the 6-point function below the 5-particle threshold, in an even theory, is then outlined in Sect. 6.3.

Further results and conjectures on macrocausal properties (macrocausality and macrocausal factorization), which complement those of Sect. 3 on macrocausality, are finally given in Sect. 7. They include off-shell extensions of the macrocausal properties described in Part I. A refined analysis of macrocausal factorization in the case of intermediate particles "travelling together" will also be given in simple situations (2-particle threshold or simplified theory of the m-particle threshold).

§2. General formalism

2.1. The axiomatic framework

In this subsection, we briefly complete the description of the Wightman axioms given in Sect. 1.1 (see [SW, Jo, GJ, BLOT] for more details) and introduce Wightman and chronological functions. As indicated in Sect. 1.1, the Wightman theory assumes the existence of a Hilbert space $\mathcal{H}$ of states and of operator-valued distributions A acting on $\mathcal{H}$ and satisfying the following axioms. We consider for simplicity the case of only one basic field A.

(i) Locality:

$$[A(x), A(y)]_\pm = 0 \qquad \text{if} \quad (x - y)^2 < 0, \qquad (2.4)$$

where $[A(x), A(y)]_\pm = A(x)A(y) \pm A(y)A(x)$ and $x^2 = x_{(0)}^2 - \vec{x}^2$.
There is commutation (sign $-$) in the "bosonic" case. Eq. (2.4)
then expresses the idea that $A(x)$ and $A(y)$ can be applied in any
order if $x - y$ is space-like, i.e. if there is no causal connection
between them. Mathematically, Eq. (2.4) entails corresponding
commutation rules between operators $A(f_1)$, $A(f_2)$ for functions
f_1, f_2 with space-like separated supports in space-time. For f
real, $A(f)$ might be considered as an "observable" associated to
the field A, and commutation of $A(f_1)$, $A(f_2)$ then means that
"these observables are independent." Comments on anticommutation will be given below.

(ii) Covariance properties. It is assumed that there exists a continuous unitary representation U in $\mathcal{H}$ of the Poincaré group which,
to each element $g = (a, \Lambda)$ of this group (where a is a space-time
translation and Λ is a Lorentz transformation), associates a unitary operator $U(g)$ in $\mathcal{H}$. Covariance properties of the fields are
assumed, in particular:

$$U(g)A(f)U^{-1}(g) = A(f_g), \qquad (2.5)$$

where $f_g(x) = f\left(g^{-1}(x)\right)$, $g(x) = \Lambda x + a$.

(iii) Positivity of the energy. It is assumed that the energy-momentum
operators $P_{(0)}, \vec{P}$ (generators of the subgroup of space-time translations: $U(a) = e^{i\, P.a}$, $P.a = P_{(0)}a_{(0)} - \vec{P}.\vec{a}$) have spectrum in
the forward cone $\bar{V}_+$ ($p^2 \geq 0$, $p_{(0)} \geq 0$).

(iv) Vacuum state. It is assumed that there is a unique "vacuum vector" Ω in $\mathcal{H}$, up to scalar multiples, invariant under the operators
$U(g)$, Ω belonging, moreover, to the domain of any polynomial
in the field operators $A(f)$. The subspace of states generated by
vectors $A(f_1) \ldots A(f_n)\Omega$ is assumed to be dense in $\mathcal{H}$.

Remarks.

(1) As already in the particular case of the "free field," operators
$A(f)$ are unbounded. They do not act on the full space $\mathcal{H}$, but
only on a domain of $\mathcal{H}$. It is a supplementary assumption to
consider that this domain is the (dense) domain generated by
the algebra of the operators $A(f)$ acting on the vacuum.

(2) In the case of anticommutation in (2.4), relevant fields are still
considered as "fundamental," and will create fermionic (or an-
tifermionic) particles, but the physical interpretation is less clear.
(Note, however, that suitable products of fields are again expected to satisfy commutation rules.)

In the theory of local observables, one introduces algebras of operators associated with bounded regions $\mathcal{O}$ in space-time (and replacing the operators $A(f)$, for f with support in $\mathcal{O}$). These observables should correspond to physical "detectors." Observables associated with regions $\mathcal{O}_1$ and $\mathcal{O}_2$ are then always assumed to *commute* if $\mathcal{O}_1$ and $\mathcal{O}_2$ are space-like separated. One of the aims of the theory is then to determine, from general arguments, what are the possible superselection sectors, corresponding, for example, to fermionic, and possibly other, "charges." It is then possible in that context to reconstruct "charged" field operators that will create charged particles, including fermionic particles, by action on the vacuum. In a class of theories (e.g. those that will correspond to systems of massive particles with short-range interactions), these operators should satisfy the commutation or anti-commutation rule (2.4), possibly in a somewhat weakened version, according to their fermionic charge. In other theories, these operators are no longer local but "string-like" ([Bu2]). These theories are not treated in this book, and we restrict our attention to the Wightman axioms.

Wightman functions. The Wightman "functions" $W_N = W_N(x_1, \ldots, x_N)$ are the well-defined (tempered) distributions

$$W_N(x_1, \ldots, x_N) = \langle \Omega \, | A(x_1) \ldots A(x_N) | \, \Omega \rangle . \qquad (2.6)$$

Because of translation invariance $W(x_1 + a, \ldots, x_N + a) = W(x_1, \ldots, x_N)$, so that they can be regarded as functions of $N - 1$ differences, for example, $x_j - x_1$, $j = 2, \ldots, N$, and their Fourier transforms contain an energy-momentum conservation δ−function $\delta(\sum_{k=1}^N p_k)$. Wightman axioms yield, and can be equivalently expressed as, corresponding properties of these functions. They entail relativistic transformation laws on these functions as also "positive definiteness conditions" arising from the positivity of the norm in Hilbert space. Locality entails that $W(x_1, \ldots, x_N)$ is unchanged (up to a sign in the case of anti-commutation) under permutations of space-like separated points, and the condition on the spectrum gives a related support property on the Fourier transforms $\tilde{W}_N(p_1, \ldots, p_N)$ in energy-momentum space. Finally, uniqueness of the vacuum entails a cluster property for space-like separation of subgroups of points ($W_N(x_1, \ldots, x_j, x_{j+1} + \lambda a, \ldots, x_N + \lambda a) - W(x_1, \ldots, x_j) W(x_{j+1}, \ldots, x_N) \to 0$ as $\lambda \to \infty$ if a is space-like). Conversely, if a set of distributions satisfies these properties, they can be considered as the vacuum expectation values of a field theory satisfying previous axioms (reconstruction theorem: see more precise statements in [SW]).

Chronological functions. The "chronological functions" $T_N(x_1, \ldots, x_N)$, which are also tempered distributions, are well-defined (if all points x_i, $i = 1, \ldots, N$, are different from each other) as the vacuum expectation values of chronological products $C(x_1, \ldots, x_N)$ of field operators $A(x_1), \ldots, A(x_N)$, that is, fields are ordered according to the time components of $x_1, \ldots, x_N$. $T_N(x_1, \ldots, x_N)$ is equal to $W_N(x_{i_1}, \ldots, x_{i_N})$ if the time components of $x_1, \ldots, x_N$ are in the chronological order $i_1, \ldots, i_N$ or equivalently:

$$
T_N(x_1, \ldots, x_N) = \sum_\pi \theta\left(x_{i_1,0} - x_{i_2,0}\right) \ldots \theta\left(x_{i_{N-1},0} - x_{i_N,0}\right)
$$
$$
\times W_N\left(x_{i_1}, \ldots, x_{i_N}\right), \tag{2.7}
$$

where $\theta(u) = 0$ if $u < 0$, $\theta(u) = 1$ if $u > 0$ and the sum $\sum$ runs over permutations π of $(1, \ldots, N)$ with, for each π, $i_k = \pi(k)$.

If the time components of some points x_i are equal, but not their space components, there is no difficulty to define chronological products or functions; the order is then irrelevant in view of locality. On the other hand, the definition is *a priori* ambiguous if some of the points x_i coincide (in which case the multiplication of θ-functions with the distribution W_N in (2.7) is *a priori* not defined). For some of our purposes, "regularized" functions can be introduced by replacing the θ-function with smooth functions different from θ in a small neighborhood of the origin. The use of such functions introduces some complications but results are essentially unchanged. However, this procedure gives more basic problems for a number of other topics. As mentioned in Sect. 1.1, an alternative axiomatic procedure is the following: from their "sharp" formal definition, chronological operators and functions are shown formally to satisfy various properties such as those presented in Sect. 3.1. One may then consider an axiomatic framework in which relevant operators or functions are the basic quantities assumed to exist and to satisfy these properties.

The functions T_N are symmetric, i.e. are unchanged by permutation of $x_1, \ldots, x_N$. Their Fourier transforms $\tilde{T}(p_1, \ldots, p_N)$ also contain an energy-momentum conservation δ–function $\delta(p_1 + \cdots + p_N)$. For later purposes, we shall also define amputated functions $\tilde{T}_{\mathrm{amp}}$, in a theory with a specified mass μ (see Sect. 2.2), by the relation

$$\tilde{T}_{\text{amp}}(p_1, \ldots, p_N) = \left[\prod_{k=1}^{N} (p_k^2 - \mu^2) \right] \tilde{T}(p_1, \ldots, p_N). \qquad (2.8)$$

As we shall see in Sect. 3.2, multiplication by the factors $(p_k^2 - \mu^2)$ removes poles $(p_k^2 - \mu^2 + i\varepsilon)$ present in $\tilde{T}$.

Connected functions. Given a set of functions $f_1(x)$, $f_2(x_1, x_2), \ldots,$ $f_N(x_1, \ldots, x_N), \ldots,$ connected functions $(f_N)_c$ are defined inductively by the formulae:

$$f_1(x) = (f_1)_c(x), \qquad (2.9)$$

$$f_2(x_1, x_2) = (f_2)_c(x_1, x_2) + (f_1)_c(x_1)(f_1)_c(x_2), \qquad (2.10)$$

$$\begin{aligned} f_3(x_1, x_2, x_3) = {} & (f_3)_c(x_1, x_2, x_3) + (f_2)_c(x_1, x_2)(f_1)_c(x_3) \\ & + (f_2)_c(x_1, x_3)(f_1)_c(x_2) + (f_2)_c(x_2, x_3)(f_1)_c(x_1) \\ & + (f_1)_c(x_1)(f_1)_c(x_2)(f_1)_c(x_3), \end{aligned} \qquad (2.11)$$

and more generally:

$$f_N(x_1, \ldots, x_N) = \sum_{\substack{\text{partitions of } 1,\ldots,N \\ \text{into subsets } \pi_1,\ldots,\pi_k; \\ k=1,\ldots,N}} \prod_{j=1}^{k} f_c(x_{\pi_j}), \qquad (2.12)$$

where x_{π_j} is the subset of points x_i in π_j. (The index $|\pi_j|$ on the functions f_c in the right-hand side, which denotes the number of points in π_j, has been left implicit.) If the functions f_i are not symmetric, the order of points is everywhere the natural order.

This definition was already that used in Chapter I for the definition of connected S matrices (some terms were absent there because S matrices with zero initial or zero final particles do not appear, as also $m \to 1$ or $1 \to n$ S matrices with $m > 1$ or $n > 1$).

Connected Wightman or chronological functions in energy-momentum space also contain an energy-momentum conservation δ-function (as follows by induction on N).

2.2 Fields and particles: preliminary discussion

The Wightman theory is not by itself a theory of particles. As mentioned in Sect. 1, we wish to consider theories in which all states are asymptotically tangent to free-particle states, corresponding to physical massive (stable) particles, before and after interactions. Then, the spectrum of the energy-momentum operator should be that of the associated Fock space. More precisely, it should be composed of the origin (corresponding to the vacuum: $P_\alpha \mid \Omega >= 0$, $\alpha = 0, \ldots, d-1$ since Ω is invariant under operators $U(a) = e^{iP \cdot a}$), isolated hyperboloids $H_+(\mu_i)$ $(p^2 = \mu_i^2$, $p_0 > 0$) corresponding to free one-particle states of masses $\mu_1, \mu_2, \ldots$ and continuous regions corresponding to multiparticle states, for example, $p^2 \geq 4\mu^2$ in a theory with one mass μ.

An example is shown in Figure 2.1 in a theory with two masses μ, μ' $(0 < \mu < \mu' < 2\mu)$, the hyperboloid $H_+(\mu')$ being removed if there is only one mass μ.

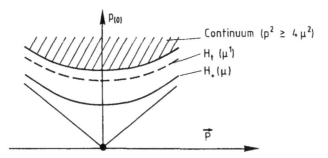

Figure 2.1 Spectrum of the energy-momentum operator in a theory with a basic particle of mass μ and a bound state of mass μ', $\mu < \mu' < 2\mu$.

The Haag-Ruelle theory will start from the Wightman axioms complemented by the following further condition.

Spectral condition. The spectrum is contained in the union of the origin, an hyperboloid $H_+(\mu)$ $(\mu > 0)$ and $\bar{V}_+(M) = \{p; p^2 \geq M^2,\ p_{(0)} > 0\}$, with $M > \mu$.

In a theory with only one mass μ, one may also assume that $M = 2\mu$, following comments made above, but part of the Haag-Ruelle results does not require this value of M. We denote below by $\mathcal{H}_1$ the subspace of $\mathcal{H}$ corresponding to the eigenvalue μ. $\mathcal{H}$ is Lorentz-invariant and the representation of the inhomogeneous Lorentz group reduces in $\mathcal{H}_1$ to a representation $U_1(a, \Lambda)$ which will be assumed to be irreducible (with spin zero in cases we consider below). $\mathcal{H}_1$ is then interpreted (in view of Wigner's analysis) as the space $\mathcal{F}_1$ of one-particle states (of mass μ) already described in Chapter I.

One-particle states. Let f be a function in x-space whose Fourier transform $\tilde{f}$ is a C^∞ function of $p = (p_{(0)}, \vec{p})$ with support in a neighborhood of a point P of $H_+(\mu)$ that does not intersect the region $\bar{V}_+(M)$ (and stays within the cone $\bar{V}_+$). The spectral condition entails, as can be checked, that $A(f)\Omega$ depends only on the restriction $\hat{f}$ of $\tilde{f}$ to the mass-shell hyperboloid $H_+(\mu)$. If $A(f)\Omega$ is nonzero, it is naturally interpreted as the free one-particle state $\mid \hat{f} >$ with wave function $\hat{f}$. More precisely:

$$A(f)\Omega = \text{cst} \mid \hat{f} > . \qquad (2.13)$$

Previous assumptions do not prevent A from having vanishing matrix elements between Ω and $\mathcal{H}_1$, in which case one-particle states of $\mathcal{H}_1$ would have to be generated by more complicated products of basic field operators. The Haag-Ruelle theory would still work in such cases. In "simple" theories, the constant in (2.13) is nonzero. It can be chosen equal to one with suitable conventions.

Localization properties in energy-momentum space. As we have seen, localization properties of $\tilde{f}$ in energy-momentum space give information on localization properties of the state $A(f)\Omega$ (again in energy-momentum space). This is in fact true also in more general cases in the following sense. Let $\Psi = A(f_1)\ldots A(f_n)\Omega$. Then, if $\tilde{f}_1, \ldots, \tilde{f}_n$ have supports around points $P_1, \ldots, P_n$, it can be shown (see [Jo]) that the mean value $\langle \Psi, P\ \Psi \rangle / \langle \Psi / \Psi \rangle$ of the energy-momentum operator P is close to $P_1 + \cdots + P_n$. This property is analogous to that of free n-particle states.

Asymptotic localization properties in space-time (one-particle states). Let us next study localization properties in space-time. As a general rule, exact or approximate localization of f or $f_1, \ldots, f_n$ in given regions R or $R_1, \ldots, R_n$ in space-time cannot be expected to give precise information of interest on space-time localization of the states $A(f)\Omega$ or $A(f_1)\ldots A(f_n)\Omega$. On the other hand, information on asymptotic localization in space-time of f, or $f_1, \ldots, f_n$, will provide information on asymptotic localization of corresponding states, as we now explain.

Let us consider functions f such that $\tilde{f}$ is C^∞ with compact support. From the viewpoint of asymptotic localization, f is localized around the origin: it has a rapid decrease in all directions under the above conditions on $\tilde{f}$. If $\tilde{f}$ has support around a point P of $H_+(\mu)$, $A(f)\Omega = \text{cst} \mid \hat{f} >$ is asymptotically localized in the velocity cone $V(\hat{f})$, up to rapid fall-off outside $V(\hat{f})$: see Chapter I. This is the particular case $u = 0$ of Figure 2.2a. It will be of interest to consider also classes of functions f_τ depending on a parameter τ that will tend to infinity. We shall more precisely consider classes of functions of one of the following forms (a), (b), or (c). Class (c) is that of direct interest in the Haag-Ruelle asymptotic theory (following the presentation of K. Hepp[He]). Classes (a) and (b) will be of interest in later sections in the discussion of macrocausal properties. Functions of class (a) are simply obtained by the space-time translation τu; functions of class (b) rather than (a) will be needed if one wishes to get results in the sense of *exponential* (rather than rapid) fall-off. In fact, if $\tilde{f}$ in (2.14) has a compact support, it has at least C^∞ singularities (hence is not analytic) on the boundary of its support, so that f or f_τ cannot decay exponentially. If $\tilde{f}$ does not have a compact support and is, for example, a fixed gaussian centered around P, the presence of points p at a possibly large, but fixed distance of P in the support would spoil the asymptotic causality properties (in the $\tau \to \infty$ limit) discussed in Sect. 3 for given values of the energy-momenta. The choice indicated in (b) of gaussian-type functions with width shrinking with τ allows one to avoid these problems.

(a) $$\tilde{f}_\tau(p) = \tilde{f}(p) e^{i(p.u)\tau}, \tag{2.14}$$

where $\tilde{f}$ is C^∞ with support around a point P of $H_+(\mu)$ and u is a given space-time translation vector. Correspondingly, $f_\tau(x) = f(x - \tau u)$. Respective asymptotic localizations of f_τ and of $A(f_\tau)\Omega$ in the $\tau \to \infty$ limit are then the point τu and the velocity cone $V_{\tau u}(\hat{f})$ with apex at τu, as shown in Figure 2.2a, in the following sense: $f_\tau(\tau v)$ has a rapid decay when $\tau \to \infty$ for any given $v \neq u$ (i.e. $\tau v \neq \tau u$), and similarly the space-time wave function associated to $\tilde{f}_\tau$ (Fourier transform of $\tilde{f}_\tau \delta_+ (p^2 - \mu^2)$: see Chapter I) has a rapid decay as $\tau \to \infty$ if $v \notin V_u(\hat{f})$ (i.e. $\tau v \notin V_{\tau u}(\hat{f})$).

(b) $$\tilde{f}_\tau(p) = \chi(p) e^{i(pu)\tau} e^{-\gamma\tau|p-P|^2}, \tag{2.15}$$

where $P \in H_+(\mu)$, χ is analytic at P and $\gamma > 0$; $|p - P|^2 = (p_{(0)} - P_{(0)})^2 + (\vec{p} - \vec{P})^2$.

The function $\tilde{f}_\tau$ is more and more localized around P, up to exponential fall-off in τ as $\tau \to \infty$ (with width of the order of $1/\sqrt{\gamma\tau}$). Its Fourier transform f_τ is on the other hand asymptotically localized around the point τu: given any point v, $v \neq u$, $f_\tau(\tau v)$ now decays exponentially as $\tau \to \infty$. Rates of exponential fall-off depend on v, γ and the analyticity domain of χ and can be specified. If, for example, $\chi \equiv 1$, f_τ is, up to a normalization factor and to a phase, the gaussian $\exp -(|x - \tau u|^2/4\gamma\tau)$

centered at τu and with width of the order of $\sqrt{\gamma \tau}$ (small compared to τ as $\tau \to \infty$), so that a decay in $\exp -(|v-u|^2/4\gamma)\tau$ is obtained. It may be lower if the analyticity domain of χ around P is small. On the other hand, the state $A(f_\tau)\Omega$ is now localized along the trajectory $(P, \tau u)$ shown in Figure 2.2b, parallel to P and passing through τu, again up to exponential fall-off for all points v outside the trajectory (P, u).

Results above follow from those of the Appendix since χ is analytic at P. The case of the state $|\hat{f}_\tau\rangle$ has already been mentioned in Sect. 4.1 of Chapter I (in a slightly different form).

(c) $$\tilde{f}_\tau(p) = \tilde{f}(p)e^{i[p_0 - \omega(\vec{p})]\tau u_0}, \qquad (2.16)$$

where $\tilde{f}$ is C^∞ with compact support around a point P of $H_+(\mu)$, u_0 is fixed and $\omega(\vec{p}) = (\vec{p}^2 + \mu^2)^{1/2}$.

The asymptotic localization of f_τ, up to rapid fall-off, is the region R_τ shown in Figure 2.2c, namely the section of the velocity cone $V(\hat{f})$ at time τu_0. We show this below where, for example, $\hat{f}$ is taken of the form $g(\vec{p})h(p_0 - \omega(\vec{p}))$, g is C^∞ with compact support around $\vec{P}$ and h is C^∞ with compact support around the origin where it is equal to one. In this case, one obtains easily:

$$f_\tau(x) = \int g(\vec{p})e^{-i[\omega(\vec{p})x_0 - \vec{p}.\vec{x}]}d\vec{p}$$

$$\times \int h(p_0 - \omega(\vec{p}))\, e^{i(p_0 - \omega(\vec{p}))(\tau u_0 - x_0)}dp_0. \qquad (2.17)$$

The first integral has a rapid decrease outside the velocity cone $V(g)$ (see Sect. 4.1 of Chapter I), while the second one has a rapid decrease as $\tau u_0 - x_0 \to \infty$. QED

Since $\hat{f}_\tau(p) \equiv \hat{f}(p)$ (independently of τ), the asymptotic localization of the state $A(f_\tau)\Omega \equiv A(f)\Omega$ is again $V(\hat{f})$ (up to rapid fall-off).

More general space-time localization properties—preliminary heuristic remarks. In cases described above, states $A(f)\Omega$ or $A(f_\tau)\Omega$ are free one-particle states "sub specie aeterninatis" (i.e. with all their evolution). Let us next consider states $A(f_1) \ldots A(f_n)\Omega$, $n > 1$. This is in general an "interacting state" for which there is a priori no reason to expect any simple particle interpretation, even "before interactions" or "after interactions." On the other hand, let us consider states $\phi_\tau = A(f_{1,\tau}) \ldots A(f_n, \tau)\Omega$ where $f_{1,\tau}, \ldots, f_{n,\tau}$ all belong to one of the classes (a), (b), or (c) (with respective points $P_k \in H_+(\mu)$, u_k and functions χ_k) and let us first consider cases such that the following conditions are satisfied:

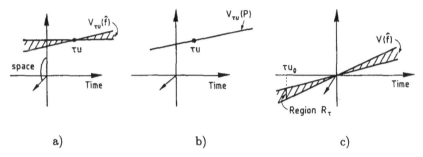

Figure 2.2 Asymptotic space-time localization of functions f_τ and states $A(f_\tau)\Omega$ for functions of classes (a), (b), or (c) in the sense of rapid fall-off (cases a c) or exponential fall-off (case b). The asymptotic localization of f_τ is the point τu (cases a b) or the region R_τ (case c). The asymptotic localization of $A(f_\tau)\Omega$ is the velocity cone $V_{\tau u}(\hat{f})$ (case a), the trajectory $V_{\tau u}(P)$ (case b) or the velocity cone $V(\hat{f})$ (case c).

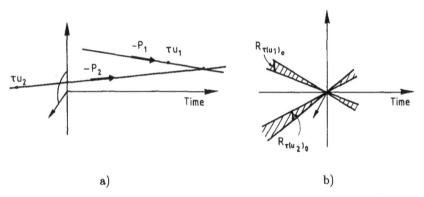

Figure 2.3 Examples of trajectories or velocity cones which meet each other and are such that conditions (i) and (ii) stated in the last part of Sect. 2.2 are satisfied. (Functions $\tilde{f}_\tau$ of class (b) and (c) respectively).

(i) the asymptotic localizations of the functions $f_{1,\tau}, \ldots, f_{n,\tau}$ are more and more remote from each other as $\tau \to \infty$.

(ii) the velocity cones (or trajectories) along which individual states $A(f_{i,\tau})\Omega$ would be asymptotically localized either do not intersect each other or their intersections all lie (strictly) in the future, respectively in the past, of all localization regions of the functions $f_{1,\tau}, \ldots, f_{n,\tau}$.

Examples are shown in Figures 2.3a and 2.3b for functions $f_{1,\tau}, f_{2,\tau}$ in classes (b) and (c) respectively. We note that these conditions are always satisfied for functions $f_{1,\tau}, \ldots, f_{n,\tau}$ of class (c) if all $(u_i)_0$ are < 0 or all > 0 respectively and if the functions $\tilde{f}_i$ have nonoverlapping velocities.

Then, it seems reasonable to believe that ϕ_τ is close for large τ (in some suitable sense) to a state ψ_τ asymptotically tangent before, respectively after, interactions to the free n-particle state composed of the n individual states $A(f_{1,\tau})\Omega, \ldots, A(f_{n,\tau})\Omega$ with wave functions $\hat{f}_{1,\tau}, \ldots, \hat{f}_{n,\tau}$. ($\psi_\tau$ should e.g. be tangent before interactions to a free 2-particle state in the cases of Figure 2.3.)

For functions of class (c), states $A(f_{1,\tau})\Omega, \ldots, A(f_{n,\tau})\Omega$ are independent of τ, so that ψ_τ should itself be a state ψ independent of τ and ϕ_τ should converge to ψ in the $\tau \to \infty$ limit. This idea is confirmed by the Haag-Ruelle theory now outlined in Sect. 2.3.

2.3. Asymptotic states and S matrix (Haag-Ruelle theory)

We consider, in Theorem 2.1 below, functions $\tilde{f}_{i,\tau}$, $i = 1, \ldots, n$, or $\tilde{f}'_{i,\tau}$, of the form (2.16) with C^∞ functions $\tilde{f}_i$, or $\tilde{f}'_i$, having supports around points P_i, or P'_i, of $H_+(\mu)$ and a common nonzero u_0 (independent of i). As in Sect. 2.2, the supports of the functions $\tilde{f}_i$, or $\tilde{f}'_i$, are assumed to be contained in $\bar{V}_+$ and to have an empty intersection with the continuum part of the spectrum. Extensions of Theorem 2.1 will be indicated in remarks that follow the proof of the theorem.

Theorem 2.1. Let $\phi_\tau = A(f_{1,\tau}) \ldots A(f_{n,\tau})\Omega$.

(i) There exists a state ψ in $\mathcal{H}$ such that $\|\phi_\tau - \psi\| \to 0$ as $\tau \to \infty$. The decrease of $\|\phi_\tau - \psi\|$ with τ is rapid if the velocity cones $V(\tilde{f}_i)$ do not overlap. Otherwise, $\|\phi_\tau - \psi\|$ decreases at least like $\tau^{-\frac{d-3}{2}}$.

(ii) The limit state ψ depends only on the mass-shell restrictions $\hat{f}_1, \ldots, \hat{f}_n$ of $\tilde{f}_1, \ldots, \tilde{f}_n$.

The scalar product in $\mathcal{H}$ of states $\phi_\tau = A(f_{1,\tau}) \ldots A(f_{n,\tau})\Omega$ and $\phi'_\tau = A\left(f'_{1,\tau}\right) \ldots A\left(f'_{n',\tau}\right)\Omega$ converges to the scalar product in Fock space of the free-particle states composed of the one-particle states with wave functions $\hat{f}_1, \ldots, \hat{f}_n$ and $\hat{f}'_1, \ldots, \hat{f}'_{n'}$ respectively (i.e. to zero if $n \neq n'$).

Proof (outline). In the nonoverlapping case, results are based on the one hand, on the localization properties (up to rapid fall-off) of the functions $f_{i,\tau}$ in the space-like separated regions $R_{i,\tau}$ (whose distance tends to infinity with τ), and on the other hand, on results on the spatial behaviour of connected Wightman functions $W_c(x_1, \ldots, x_n) =$

$\langle \Omega \,|\, A(x_1) \ldots A(x_n) \,|\, \Omega \rangle_c$. These functions are shown to fall off exponentially for large *space-like* separations of subgroups of the points $x_1, \ldots, x_n$. (The rate of exponential fall-off is related to the mass μ. Rapid fall-off would be sufficient for present purposes.) We omit here the direct proof of this result[AHR,He], which is a particular aspect (up to some technical details) of more general macrocausal properties of chronological functions established in Sect. 3.1. (Note that chronological and Wightman functions coincide for space-like separations of the arguments.)

Strong convergence of ϕ_τ is obtained as follows. The vector $\mathrm{d}\phi_\tau/\mathrm{d}\tau$ is a sum of terms in which one of the functions $f_{i,\tau}$ is replaced by its derivative. The norm squared $\|\mathrm{d}\phi_\tau/\mathrm{d}\tau\|^2 = \langle \mathrm{d}\phi_\tau/\mathrm{d}\tau \,|\, \mathrm{d}\phi_\tau/\mathrm{d}\tau \rangle$ of this vector can thus be written as a sum of integrals of the form $\int W(x_1, \ldots, x_r) \prod_k g_{k,\tau}(x_k)\, \mathrm{d}x_k$ where $g_{k,\tau}$ is one of the functions $f_{k,\tau}$ or its derivative. From the definition of connected functions, it is in turn expressed as a sum of products of integrals involving connected Wightman functions. There are terms involving exclusively 2-point functions. At least one factor in such a term involves at least one derivative. It vanishes since $A(f_\tau) \,|\, \Omega >$ is independent of τ. Other terms involve at least 3-point functions. If velocity cones do not overlap, at least two of the (asymptotic) localization regions of the functions $f_{k,\tau}$ (or their derivatives) involved have a space separation that tends to infinity with τ. It can then be proved that such a term has a rapid decrease with τ. From the inequality $\|\phi(\tau_2) - \phi(\tau_1)\| \leq \int_{\tau_1}^{\tau_2} \|\mathrm{d}\phi_\tau/\mathrm{d}\tau\|\, \mathrm{d}\tau$, one shows easily in turn the existence of ψ such that $\|\phi_\tau - \psi\|$ tends rapidly to zero as $\tau \to \infty$.

If some velocity cones overlap, a similar proof can be carried out. It now uses the fact that, given points x in the region R obtained at $\tau = 1$, the values of f_τ at the points τx of R_τ have a decrease in $1/\tau^{(d-1)/2}$. (This result follows e.g. from Eq. (2.17) and from the results mentioned in Chapter I on the decrease of functions $\int g(p)\mathrm{e}^{ipx}\delta_+\left(p^2 - \mu^2\right) dp$ inside the velocity cones.) Then, $\|\mathrm{d}\phi_\tau/\mathrm{d}\tau\|$ does not have a rapid decrease with τ but it is shown to decrease at least like $\tau^{-(d-1)/2}$. This yields again the existence of ψ and a decrease of $\|\phi_\tau - \psi\|$ at least like $\tau^{-(d-3)/2}$ if $d \geq 4$. This result can be, as a matter of fact, improved to some extent and will then apply for $d \geq 2$ [Bu3]. The existence of ψ (with no information on the decrease) follows in more general cases from density arguments.

Results on scalar products are proved in a similar way, from expansions of $\langle \phi_\tau' \,|\, \phi_\tau \rangle$ as a sum of integrals involving connected Wightman functions. All terms vanish as $\tau \to \infty$ by the same arguments as above, except those involving only 2-point functions (which now do not vanish since no derivative is involved). The result then follows from those on vectors $A(f_\tau)\Omega$ (which are states with wave functions $\hat{f}$).

Remarks.

(1) Previous results apply equally in a theory in which the strict locality condition (2.4) is weakened, for example, is replaced by a rapid decrease for space-like separations as $|x - y| \to \infty$. In this case, connected Wightman functions do not decay exponentially for space-like separations of the arguments but still have a rapid decrease which is sufficient for present purposes.

(2) A supplementary argument[He] shows that vectors of the form $A(f_{1,\tau_1})A(f_{2,\tau_2})\ldots A(f_{n,\tau_n})\Omega$ still converge (in norm) to (the same) ψ when all $\tau_i, i = 1, \ldots, n$, tend to infinity but are not necessarily equal. This is also the case if we consider the same τ but different $(u_i)_0$ with a common sign (as in Figure 2.3b). This result is not needed, in our presentation, for our later purposes and we thus omit the proof.

The S matrix. Subspaces of $\mathcal{H}$ generated by $\tau \to \infty$ limit vectors of previous vectors ϕ_τ where $u_0 > 0$ or $u_0 < 0$ are denoted respectively $\mathcal{H}_{\text{in}}$ and $\mathcal{H}_{\text{out}}$. Vectors in these subspaces are naturally interpreted as representing states of $\mathcal{H}$ that are asymptotically tangent, before or after interactions respectively, to corresponding free-particle states. Although $\mathcal{H}_{\text{in}}$ and $\mathcal{H}_{\text{out}}$ can both be put in a 1-1 correspondence with the same Fock space $\mathcal{F}$, there is *a priori* no guarantee that they are the same subspace of $\mathcal{H}$.

An operator S' from $\mathcal{H}_{\text{out}}$ to $\mathcal{H}_{\text{in}}$ can be defined as follows: to each vector of $\mathcal{H}_{\text{out}}$ asymptotically tangent after interactions to a given free-particle state, it associates the vector of $\mathcal{H}_{\text{in}}$ which is asymptotically tangent before interactions to the same free-particle state. If we consider states $\phi_{\tau,+}$ and $\phi_{\tau,-}$ obtained from the same functions $f_{i,\tau}$ except that u_0 is chosen equal to $+1$ and to -1 respectively, and their $\tau \to +\infty$ limits ψ_+ and ψ_-, one thus has:

$$S'\psi_+ = \psi_-. \tag{2.18}$$

This operator conserves scalar products and is thus (partially) isometric.

If asymptotic completeness is moreover assumed ($\mathcal{H} = \mathcal{H}_{\text{in}} = \mathcal{H}_{\text{out}}$), S' becomes a unitary operator from $\mathcal{H}$ to $\mathcal{H}$, hence, equally from $\mathcal{F}$ to $\mathcal{F}$ if $\mathcal{H}$ is put in a (1-1) correspondence with the Fock space $\mathcal{F}$ (e.g. by associating, to each vector of $\mathcal{H}$, the free-particle state with which it coincides after interactions). As easily seen, S' then coincides with the scattering operator S of Chapter I which, to each vector of $\mathcal{F}$ representing a physical state before interactions, associates the vector of $\mathcal{F}$ that represents the same physical state after interactions. In view of the equality $\mathcal{H}_{\text{in}} = \mathcal{H}_{\text{out}}$, S is (as in Chapter I) a unitary operator.

As we have seen above, the Haag-Ruelle analysis can be carried out in the Wightman axiomatic framework. The further introduction of chronological products will allow one to get more explicit expressions of the S matrix that will be easier to handle: connected S-matrix kernels will be shown in Sect. 3.2 to be mass-shell restrictions of the connected, amputated chronological functions in energy-momentum space introduced in Sect. 2.1. Related expressions of the free-particle states with which any given vector coincides before or after interactions follow by analogous methods, but the expressions obtained are more complicated in general. We only give below a simple example. Let $\tilde{f}_1$, $\tilde{f}_2$ be C^∞ functions with supports around on-shell points P_1, P_2 and, for simplicity, such that $(p_1 + p_2)^2 < 9\mu^2$ when p_1 and p_2 belong to the respective supports, and let us consider the two vectors $\mathcal{C}\left(A\left(f_1\right)A\left(f_2\right)\right)\Omega$ and $\bar{\mathcal{C}}\left(A\left(f_1\right)A\left(f_2\right)\right)\Omega$ in which $A\left(x_1\right)A\left(x_2\right)$ has been replaced by its chronological or antichronological product. The free-particle states obtained belong to the 2-particle Fock space $\mathcal{F}_2$ in view of the condition $(p_1 + p_2)^2 < 9\mu^2$. In the case of the chronological product, the 2-particle wave function *after* interactions is the sum of two contributions: (i) the (symmetrized) product of the on-shell restrictions $\hat{f}_1$, $\hat{f}_2$ of $\tilde{f}_1$, $\tilde{f}_2$ and (ii) the on-shell restriction (with respect to p_1, p_2) of the integral

$$(p_1^2 - \mu^2)(p_2^2 - \mu^2) \int \tilde{T}_c(p_1, p_2, p_1', p_2')\tilde{f}(p_1')\tilde{f}(p_2')\mathrm{d}p_1' \; \mathrm{d}p_2',$$

where $\tilde{T}_c$ is the connected, nonamputated chronological function. In the antichronological case, the 2-particle wave function *before* interactions is similar except that $\tilde{T}_c$ is replaced in (ii) by the "antichronological" function. (Multiplication by $\left(p_1^2 - \mu^2\right)\left(p_2^2 - \mu^2\right)$ amounts to amputation with respect to p_1, p_2 and removes corresponding poles. Factors $\left(p_i'^2 - \mu^2 + i\varepsilon\right)^{-1}$, $i = 1, 2$, in $\tilde{T}_c$, or $\left(p_i'^2 - \mu^2 - i\varepsilon\right)^{-1}$ in the antichronological function, are still present.) The 2-particle wave function before interactions in the first case, or after interactions in the second one, are obtained by action of S^{-1}, or S, respectively. By using the unitarity equation with possibly off-shell energy-momenta introduced later, results are similar to those above with, however, some complementary terms.

Alternatively, one may start with functions f_1, f_2 with space-like separated supports in space-time, in which case the two previous vectors coincide with the unique vector $A(f_2)A(f_1)\Omega \equiv A(f_1)A(f_2)\Omega$, but $\tilde{f}_1$, $\tilde{f}_2$ do not have compact supports.

CAUSALITY AND ANALYTICITY IN THE LINEAR PROGRAM

§3. Causality and local analyticity

3.1. Asymptotic causality properties of chronological N-point functions

Throughout this section, we consider for simplicity a theory with one mass $\mu > 0$. The energy-momentum spectrum is correspondingly assumed to be contained in $\{0\} \cup H_+(\mu) \cup \bar{V}_+(2\mu)$ where $H_+(\mu) = \{p;\ p^2 = \mu^2,\ p_0 > 0\}$ and $\bar{V}_+(2\mu) = \{p;\ p^2 \geq 4\mu^2, p_0 > 0\}$. The locality condition (2.4) at the microscopic space-time level is not by itself a causality condition in the sense to be used in this section: it does not distinguish past and future. However, we note for later purposes that it directly yields the following microcausal factorization property on chronological products of field operators in view of the very definition of the latter (see discussion in Sect. 2.1 on the status of these products and of related results):

$$C(x_1, \ldots, x_N) = C(x(I))C(x(J)) \quad \text{if} \quad x(I) \gtrsim x(J), \qquad (2.19)$$

where I, J are two complementary nonempty subsets of indices among $(1, \ldots, N)$; $x(I)$ is the set of points x_i such that $i \in I$; and $x(I) \gtrsim x(J)$ means that there is no point x_j, $j \in J$ in the closed causal future of $x(I)$, which is the union of the closed future cones of each x_i, $i \in I$ (see example in Figure 2.4). In other words, each x_j, $j \in J$ is either in the past of, or is space-like with respect to, each x_i, $i \in I$.

Eq. (2.19) will be of interest because of the link, established later, between chronological functions and the S matrix (reduction formulae). However, we note that (2.19) does *not* yield at that stage a factorization property of chronological functions. It yields in fact

$$\mathcal{T}(x_1, \ldots, x_N) = \mathcal{T}_I(x_1, \ldots, x_N) \quad \text{if} \quad x(I) \gtrsim x(J), \qquad (2.20)$$

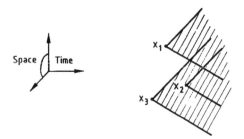

Figure 2.4 The closed causal future of a set of points.

where

$$T_I\,(x_1,\ldots,x_N) = \langle\Omega|\mathcal{C}(x(I))\mathcal{C}(x(J))|\Omega\rangle \tag{2.21}$$

is in general different from the product $T(x(I))T(x(J))$.

Before pursuing the discussion of causality properties, we give some definitions and notations. We shall later consider connected functions $(T_I)_c$. To define them, we need to define functions $T_I(x(K))$ for subsets K of $(1,\ldots,N)$: they are defined as vacuum expectation values, either of the product $\mathcal{C}(x(I\cap K))\mathcal{C}(x(J\cap K))$ if both $J\cap K$ and $I\cap K$ are nonempty, or of $\mathcal{C}(x(K))$ itself otherwise.

Functions R_I and $(R_I)_c$ are defined as:

$$(R_I)_{(c)}(x) = T_{(c)}(x) - (T_I)_{(c)}(x)\ ,\quad x = (x_1,\ldots,x_N). \tag{2.22}$$

Eq. (2.20), also valid for connected functions, can be written in the form:

$$\text{Support of } (R_I)_c \subset \Sigma_I = \bigcup_{\substack{(i,j)\\ i\in I, j\in J}}\ \{x;\ x_j - x_i \in \bar{V}_+\}, \tag{2.23}$$

where $\bar{V}_+$ is the closed future cone issued from the origin.

Either Wightman or chronological functions cannot be expected to satisfy by themselves strict causal properties. They will at best satisfy suitable fall-off or factorization properties in the limit of large separations of points or subgroups of points in space-time. The simplest result[AHR] of this type (already mentioned in Sect. 2.3) is, in a massive theory, the exponential fall-off of connected Wightman functions $W_c\,(x_1,\ldots,x_N)$ for large *space-like* separations of subgroups of points (with a rate depending on the relative configuration and on the mass μ). (A factorization property of nonconnected functions follows, up to exponential fall-off, again for space-like separations.) We do not prove this result here, but the same result will be established on connected chronological functions $\mathcal{T}_c(x_1, \ldots, x_n)$ as a byproduct of more general properties: see Theorem 2.2. (Results on W and $\mathcal{T}$ are closely related for space-like separations.)

In order to get actual causality properties, in which past and future are distinguished, one has (i) to consider the (connected) *chronological* functions rather than Wightman functions and (ii) to introduce localization properties in energy-momentum space. To that purpose, one may consider the action of $\mathcal{T}_c$ on test functions $f_{i,\tau}$ of the form (2.14) or (2.15) for each $i = 1, \ldots, N$, with corresponding points u_i, P_i, and functions $\tilde{f}_i$ in (2.14) with supports around respective points P_i. The latter do not necessarily belong to $H_+(\mu)$ so far. As already mentioned in Sect. 2.2, the choice (2.15) is that for which exponential (rather than rapid) fall-off properties will be obtained. For most purposes below, we can restrict our attention to the case $\chi \equiv 1$ in (2.15) for which best results on rates of exponential fall-off can be established in this subsection. We note that $\mathcal{T}_c(\{f_{i,\tau}\})$ can then be written in the form:

$$\mathcal{T}_c(\{f_{i,\tau}\}) = \int \tilde{\mathcal{T}}_c(p_1, \ldots, p_N) \, e^{i \sum_{i=1}^{N} (p_i \cdot \tau u_i) - \gamma\tau \sum_{i=1}^{N} |p_i - P_i|^2} \mathrm{d}p_1 \ldots \mathrm{d}p_N, \tag{2.24}$$

that is, is the generalized Fourier transform of $\tilde{\mathcal{T}}_c$ at $P = (P_1, \ldots, P_N)$ in the sense of Eq. (A.1) of the Appendix, with $x = (x_1, \ldots, x_N) = \tau u$, $u = (u_1, \ldots, u_N)$ and γ replaced by $\gamma/|u|$. It will be useful to rewrite (2.24) as the convolution product

$$\mathcal{T}_c(\{f_{i,\tau}\}) = \mathrm{cst}(\gamma\tau)^{-Nd/2} \int \mathcal{T}_c(x') \, e^{-|x' - \tau u|^2 / (4\gamma\tau)} e^{iP \cdot (x' - \tau u)} \mathrm{d}x', \tag{2.25}$$

where x', u, P are defined as above (e.g. $x' = (x'_1, \ldots, x'_N)$).

Results on causality properties in the $\tau \to \infty$ limit will be based on Eqs. (2.20) and (2.23) in x-space, which apply equally to connected functions $\mathcal{T}_c$ and $(\mathcal{T}_I)_c$ (or to the corresponding amputated quantities defined in (2.8)), and on the following support property of $(\tilde{\mathcal{T}}_I)_c$ in p-space which is a consequence of the spectral condition:

Support of $(\tilde{\mathcal{T}}_I)_c \quad \subset \{p = (p_1, \ldots, p_N); p_1 + \cdots + p_N = 0,$
$$p_I \in \bar{V}_+(2\mu) \cup H_+(\mu)\} \tag{2.26}$$

Support of $(\tilde{\mathcal{T}}_I)_{c,\mathrm{amp}} \subset \{p = (p_1, \ldots, p_N); p_1 + \cdots + p_N = 0,$
$$p_I \in \bar{V}_+(2\mu)\} \quad \text{if } |I| = 1 \text{ or } |I| = N - 1, \tag{2.27}$$

where $p_I = \sum_{i \in I} p_i$. The result (2.26) is obtained by inserting a complete set of intermediate states (of $\mathcal{H}$) between the operators $\mathcal{C}(x(J))$ and $\mathcal{C}(x(I))$ and by taking into account the assumed spectral condition on the energy-momentum operators. Because we consider connected functions, it is easily seen (by induction) that the vacuum state Ω does not contribute. If we next consider amputated functions obtained, as in (2.8), by multiplication in p-space by the factors $p_k^2 - \mu^2$, $k = 1, \ldots, N$, the improved result (2.27) at $|I| = 1$ or $N - 1$ is due to the fact that, because of the amputation, the one-particle state does not contribute when $\mathcal{C}(x(J))$ or $\mathcal{C}(x(I))$ consists of a single field operator.

Before we analyze more general consequences of the support properties (2.23), (2.26) and (2.27), we first show:

Theorem 2.2. The chronological function $\mathcal{T}_c(x_1, \ldots, x_N)$ decays exponentially, in a massive theory ($\mu > 0$), for large space-like separations of subgroups of points.

Proof (outline). We consider, for example, a situation in which the set of points $(x_1, \ldots, x_N)$ can be divided into two subsets $x(I_1), x(I_2)$ such that all differences $x_{i_1} - x_{i_2}$ are space-like $\forall \, (i_1, i_2), i_1 \in I_1, i_2 \in I_2$ and will tend to infinity under dilation of the configuration $(x_1, \ldots, x_N)$, i.e. in the $\tau \to \infty$ limit if we put $x_i = \tau u_i$, $i = 1, \ldots, N$. Then $(R_{I_1})_c(x)$ and $(R_{I_2})_c(x)$ both vanish as a particular consequence of (2.20). On the other hand, given any real point $P = (P_1, \ldots, P_N)$ ($\sum P_k = 0$), at least one of the functions $(\tilde{T}_{I_1})_c$ or $(\tilde{T}_{I_2})_c$ vanishes in the neighborhood of P (for $\mu > 0$) in view of (2.26), since $p_{I_1} = -p_{I_2}$. If, for example, $(\tilde{T}_{I_1})_c$ vanishes locally, we use the relation

$$\tilde{T}_c = (\tilde{R}_{I_1})_c + (\tilde{T}_{I_1})_c. \tag{2.28}$$

Since $(R_{I_1})_c(x) = 0$, the generalized Fourier transform of $(\tilde{R}_{I_1})_c$ at any real point, hence at the point P considered, decays exponentially in the direction of $x = \tau u$ as $\tau \to \infty$: this follows from the formula analogous to (2.25) with $\mathcal{T}_c$ replaced by $(R_{I_1})_c$. Since $(\tilde{T}_{I_1})_c(p) = 0$ locally, the generalized Fourier transform of $(\tilde{T}_{I_1})_c$ at P (defined in a way analogous to (2.24)) decreases exponentially in all directions, as easily seen. From this information, the generalized Fourier transform of $\tilde{T}_c$ at P decays exponentially in the direction of x. The same result is obtained similarly if the function that vanishes locally is $(\tilde{T}_{I_2})_c$, and the more detailed analysis shows that there is a uniform strictly positive lower bound (independent of P) on the rate of exponential fall-off. Hence, the ordinary Fourier transform $\mathcal{T}_c(x_1, \ldots, x_N)$ of $\tilde{T}_c$ also decays exponentially in this direction: to see this, one may use formulae analogous to

Eqs. (A.2) and (A.3) of the Appendix that express (for a suitable given > 0 value of γ) $T_c(x)$ as an integral, over all values of $P = (P_1, \ldots, P_N)$, of the generalized Fourier transform of $\tilde{T}_c$ at P. (For a rigorous version of this argument, see [I8].) QED

Remark. The same analysis yields the dependence of the rate of exponential fall-off on the relative configuration of points and on the mass $\mu > 0$. If $\mu = 0$, no result is obtained here because it is no longer true that either $(\tilde{T}_{I_1})_c$ or $(\tilde{T}_{I_2})_c$ vanishes locally in the neighborhood of the origin.

By the same methods, we establish below properties of exponential fall-off of $T_c(\{f_{i,\tau}\})$ in more general situations. As already mentioned, we shall not here specify the precise rates of exponential fall-off that come out[18] from locality and the spectral condition and results will thus be conveniently stated in terms of the "microlocal" (analytic) essential support (as defined in Sect. 1 of the Appendix).

Theorem 2.3. Given test functions of the form (2.15), $T_c(\{f_{i,\tau}\})$ decays exponentially as $\tau \to \infty$ (for any $\gamma > 0$), apart from possibly "causal" configurations (P, u), $P = (P_1, \ldots, P_N)$, $u = (u_1, \ldots, u_N)$ of the set Σ' defined as follows: $P_1 + \cdots + P_N = 0$ and, given any $I \subset (1, \ldots, N)$ such that $u(I)$ has no other point u_j in its closed causal future, then:

$$P_I \in H_+(\mu) \cup \bar{V}_+(2\mu). \qquad (2.29)$$

The same result applies to $T_{c,\text{amp}}$ with the set Σ' replaced by a (smaller) set Σ: the condition $P_I \in H_+(\mu) \cup \bar{V}_+(2\mu)$ is replaced by $P_I \in \bar{V}_+(2\mu)$ if $|I| = 1$ or $N - 1$.

Condition (2.29) can be equivalently replaced by

$$P_I \in H_-(\mu) \cup \bar{V}_-(2\mu), \qquad (2.30)$$

if $u(I)$ has no other point in its closed past and $P_I \in \bar{V}_-(2\mu)$ in the case of $\tilde{T}_{c,\text{amp}}$ and $|I| = 1, N - 1$. ($H_-(\mu)$ and $V_-(2\mu)$ are opposite to $H_+(\mu)$ and $\bar{V}_+(2\mu)$ respectively.)

In terms of microsupports:

$$\text{ES}_P(\tilde{T}_c) \subset \Sigma'_P = \bigcap_{I;I \in S'_P} \Sigma_I, \qquad (2.31)$$

$$\text{ES}_P(\tilde{T}_{c,\text{amp}}) \subset \Sigma_P = \bigcap_{I;I \in S_P} \Sigma_I \qquad (2.32)$$

at any real point P, where Σ'_P, Σ_P are the restrictions at P of Σ', Σ, Σ_I is the set defined in the right-hand side of (2.23) and S'_P, S_P denote respectively the classes of sets I such that $(\tilde{T}_I)_c$ and $(\tilde{T}_I)_{c,\text{amp}}$ vanish locally in the neighborhood of P.

Proof. Let P be a real point *outside* the support of $(\tilde{T}_I)_c$ for a given I. The argument already used in the proof of Theorem 2.2, starting from Eq. (2.28) (with I_1 replaced by I), yields the exponential fall-off of $T_c(\{f_{i,\tau}\})$ if u does not belong to Σ_I. The result then follows from the consideration of the various possible sets I. Properties (2.31) and (2.32) are a straightforward consequence of the definition of the microsupport given in the Appendix and of the facts, easily checked, that rates of exponential fall-off obtained in noncausal situations satisfy continuity properties and are at least proportional to γ at small γ. The equalities in the right-hand sides of (2.31) and (2.32) directly follow from the various definitions.

Remark. The result (2.32) will yield an improved result on $\text{ES}_P(\tilde{T}_c)$: see Theorem 2.5.

We give below simple upper bounds on Σ_P, Σ'_P at physical region points P. They will show in a more transparent way how Eqs. (2.31) and (2.32) do express the idea of (asymptotic) causality. (The bound on Σ_P will also be sufficient to show in Sect. 3.3 that $\tilde{T}_{c,\text{amp}}$ can be restricted to the mass-shell at non M_0 points P.) Results derived from (2.31) and (2.32) on local analyticity properties in p-space are given in Sect. 3.2.

Theorem 2.4. Let P be a point in the physical region of a $m \to n$ process with $(1,\ldots,m)$ initial, $(m+1,\ldots,m+n)$ final $(P_i \in H_-(\mu)$, $i = 1,\ldots,m$, $P_j \in H_+(\mu)$, $j = m+1,\ldots,m+n)$. Then:

(i)
$$\Sigma'_P \subset \hat{\Sigma}'_P, \tag{2.33}$$

where $\hat{\Sigma}'_P$ is the set of points $u = (u_1,\ldots,u_N)$ such that each "final" point u_j, $j = m+1,\ldots,m+n$, is in the future cone of at least one "initial" point u_i, $i = 1,\ldots,m$, and each initial point is in the past cone of at least one final point.

(ii)
$$\Sigma_P \subset \hat{\Sigma}_P, \tag{2.34}$$

where $\hat{\Sigma}_P$ is the subset of points of $\hat{\Sigma}'_P$ such that there is no isolated initial point u_i in the past of, or space-like separated with respect to, all other initial points, and similarly no final point u_j in the future of, or space-like separated with respect to, all other final points. That is, any initial singleton u_i lies in the future cone of at least one pair of initial points that lie at the same space-time position, and the analogous condition holds for final singletons.

An example of a point $(y_1, \ldots, y_8)$ in $\hat{\Sigma}_P$ is shown in Figure 2.5. Other (simpler) examples will be given at the end of Sect. 3.3.

The (simple) proof of Theorem 2.4 is given after the following comments.

(1) A better bound on $\mathrm{ES}_P(\tilde{T}_c)$ will follow from Theorem 2.5 (ii), namely:

$$\mathrm{ES}_P(\tilde{T}_c) \subset \hat{\Sigma}''_P, \qquad (2.35)$$

where $\hat{\Sigma}''_P$ is the set of points $(u_1, \ldots, u_N)$ for which there exists a point $(v_1, \ldots, v_N)$ in $\hat{\Sigma}_P$ and real scalars $\lambda_1, \ldots, \lambda_N$, $\lambda_i \leq 0$ $i = 1, \ldots, m$, $\lambda_j \geq 0$, $j = m+1, \ldots, m+n$ such that $u_k - v_k = \lambda_k P_k$, $k = 1, \ldots, N$. Each u_k lies on the trajectory passing through v_k and parallel to P_k and is in the past of v_k on this trajectory if k is initial, or in the future of v_k if k is final. An example of a point $(x_1, \ldots, x_8)$ in $\hat{\Sigma}''_P$ is also shown in Figure 2.5.

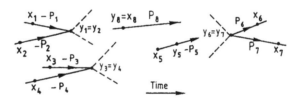

Figure 2.5 Possible causal configurations $(x_1, \ldots, x_8)$ and $(y_1, \ldots, y_8)$ relative to the connected nonamputated and amputated 8-point chronological function at a physical point $P = (P_1, \ldots, P_8)$, $1, \ldots, 5$ inital, 6, 7, 8, final.

(2) The interpretation of the property $\mathrm{ES}_P(\tilde{T}_c) \subset \hat{\Sigma}'_P$ in terms of causality is simple: the only possible causal configurations $(u_1, \ldots, u_N)$ are those for which energy-momentum incoming at initial points can be outgoing at final points in their future, and conversely. On the other hand, the condition on $\hat{\Sigma}_P$ in (ii) will correspond to the fact that at least two initial particles have to interact first, if there are interactions at all (with a similar condition on final particles), and the condition $\mathrm{ES}_P(\tilde{T}_{c,\mathrm{amp}}) \subset \hat{\Sigma}_P$ means that corresponding initial points must lie at the position of

the interaction vertex. The condition (2.35) relative to the non-amputated function means that they must lie in the past of this vertex, on the corresponding trajectories.

(3) For more general conjectures on macrocausality properties in terms of intermediate particles, which cannot be established at this stage, see Sect. 7.2. They will be partly established in the nonlinear program.

(4) Note that, given points $(u_1, \ldots, u_N)$ in $\mathrm{ES}_P(\tilde{\mathcal{T}}_c)$ or $\mathrm{ES}_P(\tilde{\mathcal{T}}_{c,\mathrm{amp}})$, some final points can be in the past of some initial points (e.g. x_8 is in the past of x_5 in Figure 2.5). This will still be the case after improvements of the results in Sects. 5 and 6 and in the general conjectures of Sect. 7.2, as was the case for vertices involving initial and final particles in the classical diagrams $\mathcal{D}_+$ of Chapter I.

Proof of Theorem 2.4.

(i) Consider, for example, a configuration with one or more initial points in the future of, or space-like with respect to, all final points. Let I denote the set of indices of these initial points. Then $u(I)$ has no other point in its closed future, but $p_I \notin H_+(\mu) \cup V_+(2\mu)$ (since it belongs to $H_-(\mu) \cup V_-(2\mu)$).

(ii) We consider, for example, a configuration $(x_1, \ldots, x_N)$ in Σ'_P, such that there is at least one single final point x_i in the future of, or space-like with respect to, all other final points. In view of (i), the only possible initial points in the closed future of x_i lie at the same space-time position. If there are none, let I be composed of the unique index i. Then, p_I cannot belong to $\bar{V}_+(2\mu)$ since it belongs to $H_+(\mu)$. If there are, let I be the union of i and of relevant initial indices. Then p_I, the sum of $p_i \in H_+(\mu)$ and of points in $H_-(\mu)$, cannot belong to $H_+(\mu) \cup V_+(2\mu)$. QED

We next state

Theorem 2.5.

(i) $$\tilde{\mathcal{T}}_c(p_1, \ldots, p_N) = \left[\prod_{k=1}^{N} \left[p_k^2 - \mu^2 + i\epsilon \right]^{-1} \right] \tilde{\mathcal{T}}_{c,\mathrm{amp}}(p_1, \ldots, p_N),$$
$$(2.36)$$

where the right-hand side of (2.36) is a well-defined product of the distributions $(p_k^2 - \mu^2 + i\epsilon)^{-1}$ and $\tilde{\mathcal{T}}_{c,\mathrm{amp}}$.

(ii) If P is a physical point of a $m \to n$ process (with $1, \ldots, m$ initial, $m+1, \ldots, m+n$ final), then:

$$\mathrm{ES}_P(\tilde{T}_c) \subset \{u = (u_1, \ldots, u_N); u_k = v_k + \lambda_k P_k, k = 1, \ldots, N$$
$$\text{with} \quad v = (v_1, \ldots, v_N) \in \mathrm{ES}_P(\tilde{T}_{c,\mathrm{amp}})$$
$$\lambda_i \leq 0, i = 1, \ldots, m, \lambda_j \geq 0, j = m+1, \ldots, m+n\}.$$

$$(2.37)$$

Proof. We restrict below our attention to physical region points P ($P_k^2 = \mu^2, \forall\, k, (P_k)_0 < 0$ if k is initial, $(P_k)_0 > 0$ if k is final). In fact, if $P_k^2 \neq \mu^2$, $\forall\, k$, Eq. (2.36) holds trivially in the neighborhood of P and intermediate cases ($P_k^2 = \mu^2$ for some indices k) are treated in a way similar to that indicated below.

By direct inspection $\mathrm{ES}_P(\prod_{k=1}^N (p_k^2 - \mu^2 + i\epsilon)^{-1})$ is the set of points of the form $(\lambda_1 P_1, \ldots, \lambda_N P_N)$, $\lambda_k \leq 0$ if k is initial, $\lambda_k \geq 0$ if k is final. On the other hand, $\mathrm{ES}_P(\tilde{T}_{c,\mathrm{amp}})$ contains no (nonzero) point opposite to previous ones in view of its upper bound $\hat{\Sigma}_P \subset \hat{\Sigma}'_P$ and of the definition of $\hat{\Sigma}'_P$. By part (a) of Sect. 3 of the Appendix, the product in the right-hand side of (2.36) is thus a well-defined distribution $\tilde{T}'_c$ and

$$\mathrm{ES}_P(\tilde{T}'_c) \subset \text{right-hand side of (2.37)}. \qquad (2.38)$$

The equality (2.36) can then be checked by noting that

$$\prod_k (p_k^2 - \mu^2) \left[\tilde{T}_c - \tilde{T}'_c\right] = 0, \qquad (2.39)$$

since $\Pi\, (p_k^2 - \mu^2)\, \tilde{T}_c = \Pi\, (p_k^2 - \mu^2)\, \tilde{T}'_c = \tilde{T}_{c,\mathrm{amp}}$. The difference $\tilde{T}_c - \tilde{T}'_c$ is thus at most a sum of terms containing δ–functions $\delta\, (p_k^2 - \mu^2)$. In view of essential supports of $\tilde{T}_c$ and $\tilde{T}'_c$, hence of $\tilde{T}_c - \tilde{T}'_c$, it can be checked that such terms cannot occur, hence $\tilde{T}_c = \tilde{T}'_c$. QED

3.2 Momentum-space analyticity and local decompositions

We first give an alternative expression of the cone Σ_P involved in Eq. (2.32), from which local analyticity properties of $T(p_1, \ldots, p_N) = \tilde{T}_{c,\mathrm{amp}}(p_1, \ldots, p_N)/\delta(\mathrm{emc})$ will in turn follow.

Lemma 2.1.

$$\Sigma_P \subset \bigcup_{(k,k')} C_{k,k';P}, \qquad (2.40)$$

where the union runs over pairs of "choices" k, k', a choice k is a map which, to each proper subset I of $(1, \ldots, N)$, associates an index $k(I)$ contained in I, and

$$C_{k,k';P} = \bigcap_{I;I \in S_P} \{x = (x_1, \ldots, x_N) ; x_{k(J)} - x_{k'(I)} \in \bar{V}_+\} \qquad (2.41)$$

(where J is the complement of I).

The interest of Lemma 2.1 is that the cones $C_{k,k';P}$, which (as earlier ones) are invariant under global space-time translations of all x_k by a common space-time vector, yield closed convex salient cones with apex at the origin in the space of points $x = (x_1, \ldots, x_N)$ defined modulo global space-time translation. (By abuse, we shall use the same notation for a translation invariant cone and the corresponding cone obtained in the latter space.) As a matter of fact, a more detailed investigation ([BEG2]) shows that some cones of this family are contained in others and that there are uniquely determined maximal elements C_β (depending on P) such that:

$$\Sigma_P \subset \bigcup_\beta C_\beta. \qquad (2.42)$$

The decomposition theorem of Sect. 2 of the Appendix then yields

Theorem 2.6. Given any real point $P = (P_1, \ldots, P_N) \, (\Sigma P_k = 0)$, there exist distributions f_β, each of which is locally the boundary value of an analytic function $\underline{f}_\beta$ from the (imaginary) directions of the (open) dual cone Γ_β of C_β such that:

$$T(p_1, \ldots, p_N) = \sum_\beta f_\beta \qquad (2.43)$$

in the neighborhood of P. (Here $\underline{f}_\beta$ is analytic in a domain of the space of complex energy-momenta satisfying $p_1 + \cdots + p_N = 0$.)

Remarks.

(1) It can be seen that the convex envelope of Σ_P is itself a closed convex salient cone C_P (although this is not true for individual Σ_I in (2.32)). Correspondingly, all functions f_β have common directions of analyticity locally, namely (at P) those of the dual cone Γ_P of C_P, and T is thus also locally the boundary value of a unique analytic function from these directions. However, the

decomposition (2.43) gives more information: the analyticity domains of each f_β are much larger than the domain of T and will
in fact intersect the complex mass-shell as shown in Sect. 3.3.

(2) The terms f_β and the cones C_β depend on P. The analysis, carried
out in a slightly more careful way, shows that they are indepen-
dent of P in various open real regions.

On the other hand, each $\underline{f}_\beta$ can be shown to be analytic at
least in a "truncated" local tube: $\mathrm{Re}\, p$ in a given region, $\mathrm{Im}\, p \in
\Gamma_\beta \cap \{\mathrm{Im}\, p \,|< \epsilon\}$ with ϵ independent of the direction in Γ_β.

3.3. Reduction formulae and results on the S matrix

We first present some general results and then describe somewhat improved results for $2 \to 2$ and $2 \to 3$ processes. As in Chapter I, $M_{m,n}$
will denote the physical region of a $m \to n$ process with, for definiteness, $1, \ldots, m$ initial and $m + 1, \ldots, m + n$ final, and $m + n = N$.
By convention, $M_{m,n}$ is here the set of $p = (p_1, \ldots, p_N)$ such that
$p_k^2 = \mu^2$, $\forall k = 1, \ldots, N$, $(p_k)_0 < 0$ if k is initial, $(p_k)_0 > 0$ if k is final and $p_1 + \cdots + p_N = 0$. M_0 is the set of points of $M_{m,n}$ such that
some initial or some final p_k are colinear. Finally, $T_{m,n}(-p_1, \ldots, -p_m;
p_{m+1}, \ldots, p_{m+n})$ will denote the scattering function of the $m \to n$ pro-
cess as defined in Chapter I, namely the connected kernel of the S matrix after factorization of its overall energy-momentum conservation δ-
function.

Theorem 2.7.

(i) The distribution $T_N(p_1, \ldots, p_N)$, $N > 2$, can be restricted in a
well-defined sense to $M_{m,n}$, at least away from M_0 points.

(ii) Its restriction satisfies the relation:

$$T_{m,n}(-p_1, \ldots, -p_m; p_{m+1}, \ldots, p_{m+n}) = T_N(p_1, \ldots, p_N)|_{M_{m,n}}.$$
(2.44)

(iii) For any (non M_0) point P of $M_{m,n}$:

$$\mathrm{ES}_P(T_{m,n}) \subset (\Sigma_P)_{\mathrm{ext}},$$
(2.45)

where $(\Sigma_P)_{\mathrm{ext}}$ is the set of (relative) configurations of external
trajectories that can be constructed from (relative) configura-
tions of points $(u_1, \ldots, u_N)$ in Σ_P by replacing each u_k by the
trajectory (P_k, u_k) (passing through u_k and parallel to P_k).

Remarks.

(1) Eq. (2.45) provides exponential fall-off properties of connected collision amplitudes between *on-shell* initial and final wave functions of the form (2.15), apart from causal configurations in $(\Sigma_p)_{\text{ext}}$. Rates of exponential fall-off can be specified (see [18]). Local decompositions of $T_{m,n}$ into (sums of) boundary values of analytic functions are given in Theorem 2.8.

(2) It is desirable and probably possible to extend these results to M_0 points. However, the standard results on restrictions of distributions at "$u \neq 0$ points" (see Appendix) are not sufficient, "$u = 0$" results are needed. The set $(\Sigma_p)_{\text{ext}}$ should probably be defined in a more complicated way at M_0 points.

(3) Results analogous to the above follow from Theorem 2.7 for nonconnected functions, but some care is needed in the statements because of the occurrence of 2-point functions. (The emc δ-function $\delta(p_1 + p_2)$ cannot be restricted to the mass-shell since $\vec{p}_1 + \vec{p}_2 = 0$ implies $(p_1)_0 + (p_2)_0 = 0$ on-shell.)

Proof of Theorem 2.7. We first show in (i) that T_N can be restricted to $M_{m,n}$ and that $\text{ES}_P(T_N \,|\, M_{m,n})$ is contained in $(\Sigma_P)_{\text{ext}}$. The relation (2.44) will be derived in (ii).

(i) Given any non M_0 point P of $M_{m,n}$, the conormal set N_P at P to the manifold $M_{m,n}$ is the set of (relative) configurations $(u_1, \ldots, u_N)$ of the form $(\lambda_1 P_1, \ldots, \lambda_N P_N)$ where the $\lambda'_k s$ are real (positive or negative) scalars. The relation

$$\text{ES}_P(T_N) \cap N_P = \{0\} \qquad (2.46)$$

then follows from the bound $\hat{\Sigma}_P$ (see (2.34)) on $\text{ES}_P(T_N)$: given a point of the form $(\lambda_1 P_1, \ldots, \lambda_N P_N)$ in $\hat{\Sigma}_P$, initial and final pairs must lie at the origin. (E.g. $\lambda_{i_1} P_{i_1} = \lambda_{i_2} P_{i_2} \rightarrow \lambda_{i_1} = \lambda_{i_2} = 0$ since P_{i_1} and P_{i_2} are not colinear.) Other points must then also lie at the origin since $\hat{\Sigma}_P \subset \hat{\Sigma}'_P$.

Announced results then directly follow from part b of Sect. 3 of the Appendix.

(ii) Given free-particle states φ, ψ (of the Fock space $\mathcal{F}$) composed of one-particle states with (smooth) wave functions $\hat{f}_1, \ldots, \hat{f}_m$ and $\hat{f}_{m+1}, \ldots, \hat{f}_{m+n}$ respectively, the analysis of Sect. 2.3 allows one to express $\langle \psi | S | \varphi \rangle$ as the scalar product $\langle \Psi \mid \Phi \rangle$ where

$$\Phi = \lim_{\tau \to -\infty} \Phi_\tau \quad \text{with} \quad \Phi_\tau = \prod_{i=1}^{m} A(f_{i,\tau})\Omega, \qquad (2.47)$$

$$\Psi = \lim_{\tau \to +\infty} \Psi_\tau \quad \text{with} \quad \Psi_\tau = \prod_{j=m+1}^{m+n} A(f_{j,\tau})\Omega, \quad (2.48)$$

and each $f_{k,\tau}$ is a function of the form (2.16) where $u_{(0)} = 1$, support around an on-shell point P_k and with on-shell restriction $\hat{f}_k$. From the convergence of Φ_τ, Ψ_τ to Φ, Ψ as $\tau \to -\infty$ and $\tau \to +\infty$ respectively, $\langle \psi|S|\varphi\rangle$ is also equal to $\lim_{\tau \to +\infty} \langle \Psi_\tau, \Phi_{-\tau}\rangle$. In view of the asymptotic localization of the functions $f_{i,\tau}(x)$ and $f_{j,\tau}(x)$ in regions in past and future cones respectively, it can be checked that the product $A(x_1)\dots A(x_N)$ in the integral that defines $\langle \Psi_\tau, \Phi_{-\tau}\rangle$, can be replaced in the $\tau \to \infty$ limit by the chronological product $\mathcal{C}(x_1,\dots,x_N)$. Hence:

$$\langle \psi|S|\varphi\rangle = \lim_{\tau \to \infty} \int \mathcal{T}(x_1,\dots,x_N) \prod_{k=1}^{m} f_{i,-\tau}(x)$$

$$\prod_{k=m+1}^{m+n} f_{j,\tau}(x)\mathrm{d}x_1 \dots \mathrm{d}x_N \quad (2.49)$$

or by Fourier transformation:

$$\langle \psi|S|\varphi\rangle = \lim_{\tau \to \infty} \int \mathrm{d}p_1 \dots \mathrm{d}p_N \; \tilde{\mathcal{T}}(p_1,\dots,p_N)$$

$$\times \prod_{i=1}^{m} \tilde{f}_i(-p_i)\mathrm{e}^{i\tau[(p_i)_0 + \omega(\vec{p}_i)]} \prod_{j=m+1}^{m+n} (\tilde{f}_j)(p_j)\mathrm{e}^{i\tau[(p_j)_0 - \omega(\vec{p}_j)]}.$$

$$(2.50)$$

The relation (2.44) is then easily obtained (by passage to connected functions): to see it, one may use the expression (2.36) of $\tilde{\mathcal{T}}_c$ and replace each propagator $(p_k^2 - \mu^2 + i\varepsilon)^{-1}$ by a contribution in $\delta(p_k^2 - \mu^2)$ and a contribution in $(p_k^2 - \mu^2 - i\varepsilon)^{-1}$. From essential support properties of $\tilde{\mathcal{T}}_{c,\mathrm{amp}}$, its multiplication either by mass-shell δ-functions or "minus $i\varepsilon$" propagators is allowed. Moreover, results on the essential support of terms containing minus $i\varepsilon$ propagators, obtained by part a of Sect. 3 of the Appendix, show that corresponding integrals vanish in the $\tau \to \infty$ limit. (Rapid fall-off is obtained as the particular case $\gamma = 0$ of results mentioned in the remark that concludes Sect. 1 of the Appendix, if the functions $\tilde{f}_k$ are C^∞ with sufficiently small supports: a simple change of variables can be used to that purpose.) The only remaining term, with mass-shell δ-functions is independent of τ and gives the desired result. QED

We conclude generalities with

Theorem 2.8.

(i) Given any non M_0 point P of $M_{m,n}$, the analyticity domains of the function $\underline{f}_\beta$ involved in Theorem 2.6 have locally a nonempty intersection with the complex mass-shell $\underline{M}$, and the boundary values of their restrictions to $\underline{M}$ (from the directions determined by those of the cones Γ_β) define locally distributions $\hat{f}_\beta$ which are restrictions of f_β to $M_{m,n}$. The decomposition

$$T_{m,n} = \sum_\beta \hat{f}_\beta \qquad (2.51)$$

of $T_{m,n}$ as the sum of the boundary values $\hat{f}_\beta$ then holds in the neighborhood of P.

(ii) For any $m \to n$ process, there is (at $d > 2$) a nonempty open set of points P of $M_{m,n}$ such that the analyticity domain of $\underline{f} = \sum_\beta \underline{f}_\beta$ has itself (locally) a nonempty intersection with $\underline{M}$. Then $T_{m,n}$ is locally the boundary value of the restriction of $\underline{f}$ to $\underline{M}$.

Proof.

(i) The same methods as above and as in the Appendix. The fact that the analyticity domain of $\underline{f}_\beta$ has locally a nonempty intersection with $\underline{M}$ follows from the fact that $\mathrm{ES}_P(f_\beta) \cap N_P = \{0\}$. This shows that the trace of $\underline{M}$ (in Im p-space) for real points p in the neighborhood of P lies (locally) inside the cone Γ_β.

(ii) The proof that the set of points P under consideration is nonempty is omitted here. Examples will be mentioned in Sect. 4.1 in the part devoted to the partial results on the holomorphy envelope of the primitive domain. The remaining part of the statement follows as before. Directions of analyticity from which the boundary value is obtained are those determined in $\underline{M}$ by the directions of the cone $\Gamma = \bigcap_\beta \Gamma_\beta$.

Examples. We consider below $2 \to 2$ and $2 \to 3$ processes. (The case of a $3 \to 3$ process will be presented in Sect. 5.) The set Σ_P then coincides as easily seen with the upper bound $\hat{\Sigma}_P$ and is moreover, independent of P. There is only one cone $C_\beta = C$ in the $2 \to 2$ case and there are three cones C_β in the $2 \to 3$ case (shown below).

(a) $N = 4$, $2 \to 2$ process

The microsupports of T_4 and of the scattering function $T_{2,2}$ that follow from locality and the spectral condition at any physical point P, or any non M_0 physical-region point P respectively, are shown in Figure 2.6a. The improved result obtained in this case from Lorentz invariance is shown in Figure 2.6b.

The dual cone Γ of $C = \{u; u_1 = u_2,\ u_3 = u_4,\ u_3 - u_1 \in \bar{V}_+\}$ (where points $u = (u_1, \ldots, u_4)$ are defined here modulo global space-time translation) is the set:

$$\Gamma = \{\operatorname{Im} p = \operatorname{Im}(p_1, \ldots, p_4);$$
$$\operatorname{Im} p_1 + \cdots + \operatorname{Im} p_4 = 0,\ \operatorname{Im} p_3 + \operatorname{Im} p_4 \in V_+\}.$$
$$(2.52)$$

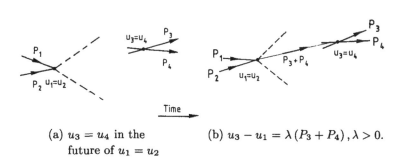

Time →

(a) $u_3 = u_4$ in the   (b) $u_3 - u_1 = \lambda (P_3 + P_4),\ \lambda > 0.$
future of $u_1 = u_2$

Figure 2.6 Microsupports of T_4 and $T_{2,2}$ obtained from locality and spectrum (Fig. a) and improved results obtained from Lorentz invariance (Fig. b). The microsupports are relative configurations of points $(u_1, \ldots, u_4)$ and of trajectories $(P_1, u_1), \ldots, (P_4, u_4)$, respectively.

If Lorentz invariance is used, the scattering function $T_{2,2}$ is shown in this case to be at all (non M_0) points the boundary value of an analytic function $\underline{T}_{2,2}$ from the direction $\operatorname{Im} s > 0$. M_0 points are those of the 2-particle threshold $s = 4\mu^2$.

(b) $N = 5,\ 2 \rightarrow 3$ process

The microsupport of T_5 at any physical point P is the union of three cones C_3, C_4, C_5 shown in Figure 2.7:

$$ES_P(T_S) \subset C_3 \cup C_4 \cup C_5 \qquad (2.53)$$

The common intersection of $\Gamma_3, \Gamma_4, \Gamma_5$ is the cone:

$$\Gamma = \ \{\operatorname{Im} p;\ \operatorname{Im} p_1 + \cdots + \operatorname{Im} p_5 = 0,\ \operatorname{Im} p_3 + \operatorname{Im} p_4 \in V_+,$$
$$\operatorname{Im} p_3 + \operatorname{Im} p_5 \in V_+, \operatorname{Im} p_4 + \operatorname{Im} p_5 \in V_+\},$$
$$(2.54)$$

Figure 2.7 The cone $C_\ell, \ell = 3, 4, 5$ ($u_m - u_\ell \in \bar{V}_+$, $u_\ell - u_1 \in \bar{V}_+$, $u_1 = u_2$, $u_m = u_n$, with e.g. $m = 4, n = 5$ if $\ell = 3$).

and T_5 is the boundary value of a unique analytic function from the directions of Γ. This is still the case for $T_{2,3}$ if the intersection with the complex mass-shell is nonempty. In the $2 \to 3$ case under consideration, it is possible to show that $T_{2,3}$ is the boundary value of a single analytic function (from directions of a relatively small cone depending on P) away from a limited part of the physical region.

The use of Lorentz invariance yields some improvements which, however, are more limited than in the case $N = 4$. It is shown (see [BEG2]) that $T_{2,3}$ is the boundary value of a unique analytic function in the part of the physical region such that

$$s > (4, 8\ \mu)^2, \tag{2.55}$$

where $s = (p_1 + p_2)^2 = (p_3 + p_4 + p_5)^2$. The condition $s \geq (3\mu)^2$ is satisfied in the physical region (in view of the mass-shell constraints on p_3, p_4, p_5) so that only a relatively small neighborhood of the threshold $s = (3\mu)^2$ is excluded. Results are more limited for more general processes.

§4. The analytic N-point functions

4.1 General primitive results of the linear program

The following definitions will be useful.

A *paracell* S of $(1, \dots, N)$ is a class of proper subsets I of indices such that, if $I_1 \in S$ and $I_2 \in S$, then $I_1 \cup I_2 \in S$ or $I_1 \cap I_2 \in S$. (A proper subset is a set I different from $(1, \dots, N)$ and nonempty.)

A *precell* S of $(1, \dots, N)$ is a paracell such that, given any pair (I_1, I_2) of proper complementary subsets $(I_1 \cup I_2 = (1, \dots, N), I_1 \cap I_2$ empty), either I_1 or I_2 belongs to S. (Only one of them can belong to S in view of the definition of paracells.)

A *cell* S of $(1, \ldots, N)$ is a class of proper subsets I of indices such that there exist real scalars $s_1, \ldots, s_N$ satisfying

$$\sum_{k=1}^{N} s_k = 0, \tag{2.56}$$

$$\sum_{k \in I} s_k > 0, \quad \forall \ I \in S, \tag{2.57}$$

$$\sum_{k \in J} s_k < 0, \quad \forall \text{ proper subset } J \notin S. \tag{2.58}$$

In other words, a cell is a consistent set of signs $+$ or $-$ attributed to each proper subset of indices of $(1, \ldots, N)$. As easily seen, a cell is a precell such that there exist $s_1, \ldots, s_N$ satisfying (2.56) and (2.57). There is one cell of $(1, \ldots, N)$ associated to each one of the (convex, polyhedral) cones determined (in the real space of the variables $s_1, \ldots, s_N$) by the hyperplanes $s_I \left(= \sum_{i \in I} s_i \right) = 0$ obtained for all proper subsets I of $(1, \ldots, N)$ (with one hyperplane for two complementary subsets). Each one of these cones is called the *geometrical cell S*.

Remark. A paracell can be composed of a *unique* proper subset I: then, it is clearly not a precell, and thus not a cell, unless $N = 2$. All paracells are cells at $N = 2$. All precells are cells at $N \leq 4$, but some precells are not cells at $N > 4$.

The *paracell operator* $\mathcal{R}_S (x_1, \ldots, x_N)$ associated to a paracell S is defined as

$$\mathcal{R}_S (x_1, \ldots, x_N) = \sum_{\nu=1}^{N} (-1)^{\nu-1} \sideset{}{'}\sum_{(J_1, \ldots, J_\nu)} \mathcal{C}(x(J_1))\mathcal{C}(x(J_2)) \ldots \mathcal{C}(x(J_\nu)), \tag{2.59}$$

where the sum Σ' runs over all ordered partitions $(J_1, \ldots, J_\nu)$ of $(1, \ldots, N)$ such that, for each $\nu \geq 2$ and $r \leq \nu - 1$, the complement of $J_1 \cup \ldots \cup J_r$ in $(1, \ldots, N)$ belongs to S:

$$(1, \ldots, N) \backslash (J_1 \cup J_2 \ldots \cup J_r) \in S, \quad \forall \ r \leq \nu - 1. \tag{2.60}$$

The term $\nu = 1$ in (2.59) is simply $\mathcal{C} (x_1, \ldots, x_N)$.

The *paracell function* $R_S (x_1, \ldots, x_N)$ is the distribution

$$R_S (x_1, \ldots, x_N) = \langle \Omega | \mathcal{R}_S (x_1, \ldots, x_N) | \Omega \rangle. \tag{2.61}$$

It coincides with the term $R_I (x_1, \ldots, x_N)$ defined in (2.22) if the paracell S is composed of the unique set I.

The causal factorization property (2.19) of C-products has the following consequence for any paracell S (see [EGS]):

Theorem 2.9.

$$\text{Support of } R_S \subset \bigcap_{I \in S} \Sigma_I, \tag{2.62}$$

where Σ_I has been defined in Sect. 3.1 (Eq. (2.23)), or equivalently:

$$\text{Support of } R_S \subset \bigcup_k \bigcap_{I \in S} \left\{ x = (x_1, \ldots, x_N), x_{k(J)} - x_{k(I)} \in \bar{V}_+ \right\}$$

$$\equiv \bigcup_k C_{S,k}, \tag{2.63}$$

where $J = (1, \ldots, N) \backslash I$ in (2.63), the union runs over choices k as defined in Lemma 2.1 in Sect. 3.1, and each $C_{S,k}$ is a closed convex salient cone (with apex at the origin).

The following Theorems 2.10, 2.11, 2.12 can be established *if S is a cell*:

Theorem 2.10.

$$R_S (x_1, \ldots, x_N) = (R_S)_c (x_1, \ldots, x_N). \tag{2.64}$$

In order to define $(R_S)_c$ in (2.64) (for a cell S of $(1, \ldots, N)$), functions $R_S(x(K))$ are defined, for subsets K of indices of $(1, \ldots, N)$, by replacing sets J_ℓ by $J_\ell \cap K$ for nonempty sets $J_\ell \cap K$. Theorem 2.10 can be checked by induction.

Theorem 2.11. If S is a cell, the right-hand side of (2.62) and (2.63) is contained in a closed convex salient cone (with apex at the origin) C_S. The (nonempty, open) dual cone of C_S is the cone

$$\Gamma_S = \left\{ q = (q_1, \ldots, q_N), \sum_{k=1}^N q_k = 0, q_I \in V^+, \quad \forall \ I \in S \right\}, \tag{2.65}$$

where q_I is the sum of the energy-momentum vectors q_i, $i \in I$.

The Fourier-Laplace transform theorem thus yields:

Theorem 2.12. For every cell S of $(1, \ldots, N)$, the distribution $\tilde{r}_S(p_1, \ldots, p_N)$ $(= (\tilde{r}_S)_c (p_1, \ldots, p_N))$ obtained from $(\tilde{R}_S)_{(c)}$ after factorization of its emc δ-function is at all real points the boundary value (from the directions of Γ_S) of a function $\underline{r}_S(p_1,\ldots,p_N)$ analytic in the tube Re p arbitrary, Im $p \in \Gamma_S$ (in the space of complex $p = (p_1,\ldots,p_N)$ satisfying $p_1 + \cdots + p_N = 0$).

Remark. In the derivation of Theorem 2.12, only part of the information in (2.62) and (2.63) has been used. More complete information (valid also for paracells) follows from the fact that the right-hand side of (2.63) is a union of closed convex salient cones $C_{S,k}$. Namely, $\tilde{r}_S$ is at all real points a sum (over k) of boundary values of functions analytic in the tubes Re p arbitrary, Im p in the dual cone $\Gamma_{S,k}$ of $C_{S,k}$. These functions depend on S. Steinmann relations stated below between some cell functions yield some global (i.e. valid in all real p-space) decompositions of these cell functions in terms of common boundary values. (This fact is used for the enlargement of the primitive domain of analyticity introduced below.)

Ruelle identity and Steinmann relations. The following identity, useful for several purposes, has been established for cell operators in [Ru2]. It is also easily derived[EGS] for general paracell operators if one starts from the definition (2.59). It gives an expression for the difference $\mathcal{R}_{S_+} - \mathcal{R}_{S_-}$ when S_+, S_- are two (para) cells of $(1,\ldots,N)$ "adjacent" along a common "(pseudo)-face" (I_1, I_2). For simplicity, we only give below the definition of cells adjacent along a common face. For the (simple) more general definition of paracells adjacent along a common pseudoface, see [EGS]. Two cells S_+, S_- are said to be adjacent along the face (I_1, I_2), where I_1, I_2 are complementary subsets of $(1,\ldots,N)$, if the geometrical cells associated to S_+, S_- are adjacent along the face determined by the hyperplane $s_{I_1} = s_{I_2} = 0$. These two cells are then composed of the same sets I except that one of the sets I_1, I_2 belongs to S_+ (hence not to S_-) and the other one to S_- (hence not to S_+). Let $\hat{S}_1$ be the set of all subsets of $(1,\ldots,N)$ that belong to S_+ (or S_-) and are strictly contained in I_1. As easily seen, $\hat{S}_1$ is a cell of I_1. $\hat{S}_2$ is defined similarly and is a cell of I_2. (Cells of a set I different from $(1,\ldots,N)$ are defined in the same way as above.)

We then state:

Theorem 2.13. Given two (para) cells S_+, S_- adjacent along the (pseudo) face $I_1, I_2, I_1 \in S_+, I_2 \in S_-$,

$$\mathcal{R}_{S_+} - \mathcal{R}_{S_-} = \left[\mathcal{R}_{\hat{S}_1}, \mathcal{R}_{\hat{S}_2}\right], \tag{2.66}$$

where the right-hand side is the commutator of $\mathcal{R}_{\hat{S}_1}$ and $\mathcal{R}_{\hat{S}_2}$.

This result is checked by inspection. A particular consequence is a set of algebraic relations (Steinmann relations) which we state below for cells. (A similar statement applies to paracells.) We consider four proper subsets I_1, I_2, I_3, I_4 of $(1, \ldots, N)$ such that I_1, I_2 are complementary subsets, as also I_3, I_4, and such that none of them is contained in any other. Let $\mathcal{S}_{++}$, $\mathcal{S}_{+-}$, $\mathcal{S}_{-+}$, $\mathcal{S}_{--}$ be four cells of $(1, \ldots N)$ composed of the same subsets I apart from $I_1, \ldots, I_4$ and of two of the latter: $I_1, I_3 \in \mathcal{S}_{++}$, $I_2, I_3 \in \mathcal{S}_{-+}$, $I_1, I_4 \in \mathcal{S}_{+-}$, $I_2, I_4 \in \mathcal{S}_{--}$. Cells $\mathcal{S}_{++}$, $\mathcal{S}_{-+}$, as also $\mathcal{S}_{+-}$, $\mathcal{S}_{--}$ are adjacent along (I_1, I_2) while $\mathcal{S}_{++}$, $\mathcal{S}_{+-}$, as also $\mathcal{S}_{-+}$, $\mathcal{S}_{--}$ are adjacent along (I_3, I_4). Then:

$$\mathcal{R}_{\mathcal{S}_{++}} - \mathcal{R}_{\mathcal{S}_{+-}} + \mathcal{R}_{\mathcal{S}_{--}} - \mathcal{R}_{\mathcal{S}_{-+}} = 0. \qquad (2.67)$$

Eq. (2.67) follows from Eq. (2.66), since $\mathcal{R}_{\mathcal{S}_{++}} - \mathcal{R}_{\mathcal{S}_{-+}}$ and $\mathcal{R}_{\mathcal{S}_{+-}} - \mathcal{R}_{\mathcal{S}_{--}}$ are equal to a common commutator.

Results derived from the spectral condition. Results below are now derived from the spectral condition. By definition, $\tilde{r}_{\mathcal{S}}$ and $(\tilde{r}_{\mathcal{S}})_{\text{amp}}$ are Fourier transforms of $R_{\mathcal{S}}$ after factorization of $\delta(\Sigma p_k)$, and in the case of $(\tilde{r}_{\mathcal{S}})_{\text{amp}}$, multiplication by $\Pi(p_k^2 - \mu^2)$.

Theorem 2.14. Let $\mathcal{S}$ be a cell. Then

$$(\tilde{r}_{\mathcal{S}})_{\text{amp}}(p_1, \ldots, p_N) = T_N(p_1, \ldots, p_N) \qquad (2.68)$$

in the region:

$$\Omega_{\mathcal{S}} = \{p = (p_1, \ldots, p_N), \ \Sigma p_k = 0, \ p_I \notin \bar{V}_-(2\mu), \ \forall I \in \mathcal{S}\}. \qquad (2.69)$$

This result is completed by:

Lemma 2.2. Given any real point P, there always exist one or more cells such that $P \in \Omega_{\mathcal{S}}$.

We next state:

Theorem 2.15. Let $\mathcal{S}_1, \mathcal{S}_2$ be two cells of $(1, \ldots, N)$ adjacent along a face (I_1, I_2). Then

$$\tilde{r}_{\mathcal{S}_1}(p_1, \ldots, p_N) = \tilde{r}_{\mathcal{S}_2}(p_1, \ldots, p_N) \qquad (2.70)$$

in the region:

$$\Omega_{I_1, I_2} = \left\{p = (p_1, \ldots, p_N), \ \Sigma p_k = 0, \ p_{I_1}^2 (= p_{I_2}^2) < 4\mu^2, \ p_{I_1}^2 \neq \mu^2\right\}. \qquad (2.71)$$

This result is a consequence of the Ruelle identity (2.66) and of (an improved version of) the following result derived from the spectral condition:

Theorem 2.16. The Fourier transforms of $\langle \Omega | \mathcal{R}_{\hat{S}_1} \mathcal{R}_{\hat{S}_2} | \Omega \rangle_c$ and $\langle \Omega | \mathcal{R}_{\hat{S}_2} \mathcal{R}_{\hat{S}_1} | \Omega \rangle_c$ vanish for $p_{I_1} \notin \bar{V}^-(\mu)$ and $p_{I_2}(= -p_{I_1}) \notin \bar{V}^-(\mu)$ respectively.

This result yields (2.70) for $p_{I_1}^2 < \mu^2$. A further analysis leads us to replace $\bar{V}^-(\mu)$ in Theorem 2.16 with $\bar{V}^-(2\mu) \cup H^-(\mu)$ and Theorem 2.15 follows.

Remark. It can also be shown that $(p_{I_1}^2 - \mu^2)(\tilde{R}_{S_1} - \tilde{R}_{S_2}) = 0$ for $p_{I_1}^2 < 4\mu^2$.

Theorem 2.15 is completed by:

Lemma 2.3. Given two cells S_1, S_2 of $(1, \ldots, N)$, there exists at least one (and in general several) sequence $S_1', S_2', \ldots, S_\ell'$, $S_1' = S_1$, $S_\ell' = S_2$ of cells such that S_i, S_{i+1} are adjacent along a common face (depending on i), $i = 1, \ldots, \ell - 1$.

Primitive domain and analytic N-point functions. The set of previous results and an (almost straightforward) application of the edge-of-the-wedge theorem (in its refined version: see Appendix) yield the existence, for each N, of a well-defined, analytic function H_N with the following properties:

(i) H_N is analytic in the union D_N of tubes with bases Γ_S ($\mathrm{Im}\, p \in \Gamma_S$) associated to all cells of $(1, \ldots, N)$, and of domains of the form $\mathcal{N}_{S_1, S_2} \cap \{$tube with basis $(\Gamma_{S_1} \cup \Gamma_{S_2})^{\mathrm{conv.}}\}$ associated with all pairs of adjacent cells (S_1, S_2) of $(1, \ldots, N)$.

In this statement, for each pair S_1, S_2 of cells adjacent along a face (I_1, I_2), $\mathcal{N}_{S_1, S_2}$ is an open complex neighborhood of the set of real points such that $p_{I_1}^2 < \mu^2$, and $(\Gamma_{S_1} \cup \Gamma_{S_2})^{\mathrm{conv.}}$ is the convex envelope of Γ_{S_1} and Γ_{S_2}. (A tube with basis Γ is the set of complex $p = (p_1, \ldots, p_N)$, $\Sigma\, p_k = 0$, such that $\mathrm{Re}\, p$ is arbitrary, $\mathrm{Im}\, p \in \Gamma$.)

(ii) Each cell function $\tilde{r}_S$ is the boundary value of H_N from the directions $\mathrm{Im}\, p \in \Gamma_S$.

The domain D_N contains Euclidean points (real momenta $\vec{p}_k$, imaginary energy components $(p_k)_0$) in any given Lorentz frame. For later purposes, we also note that D_N contains a neighborhood of the set $D_N^{(0)}$, defined (in the space of complex $p_1, \ldots, p_N$ satisfying $p_1 + \cdots + p_N = 0$) by the conditions:

$$D_N^{(0)} = \left\{ p; \vec{p}_1, \ldots, \vec{p}_N \text{ real}, \ (p_{1,0}, \ldots, p_{N,0}) \notin \bigcup_I (\omega_I \cup \sigma_I) \right\}, \quad (2.72)$$

where I denotes an arbitrary proper subset of $(1, \ldots, N)$ and ω_I, σ_I denote, respectively, a pole and a cut with equations:

$$\omega_I : p_I^2 = \mu^2, \qquad\qquad\qquad (2.73)$$
$$\sigma_I : p_I^2 = (2\mu)^2 + \rho, \quad \rho \geq 0. \qquad (2.74)$$

On the other hand, it is easily seen that the intersection of D_N with the complex mass-shell is empty.

Holomorphy envelope of the primitive domain: some results. D_N is not a "natural" domain: any function analytic in D_N and whose boundary values satisfy the Steinmann relations can be analytically continued in a larger domain D'_N. The latter satisfies the following properties:

- invariance under complex Lorentz transformations

  This result is independent of the covariance properties of the field: the relativistic form of the axiom of locality and spectral condition yields by itself the invariance of the analyticity domain[BEG3].

  For a scalar field, the covariance properties of the field entail that the N-point function H_N can be expressed as a function of (complex) invariants $\zeta_k = p_k^2$ and $\zeta_{k_1,k_2} = p_{k_1} . p_{k_2}$ (for complex values of $p_1, \ldots, p_N$), linked by a system of algebraic relations.

- D'_N is contained in $\mathbf{C}^{d(N-1)}$ (i.e. it is not a multisheeted domain) so that H_N is uniformly defined. It is more precisely contained in a cut-domain D_N^{max} obtained by subtracting from $\mathbf{C}^{d(N-1)}$ all hypersurfaces with equations of the form $(p_{i_1} + \cdots + p_{i_q})^2 = 4\mu^2 + \rho$, $\rho \geq 0$. (It is strictly smaller than D_N^{max} at $N > 2$.) These results, obtained by H. Epstein, will be found in his course in [ChD].

- Given any real point P in the neighborhood of which more than two cell boundary values coincide, a further straightforward application of the edge-of-the-wedge theorem already yields a local enlargement of the analyticity domain: analyticity in the intersection of a complex neighborhood of P and of the tube whose basis is now obtained by convex envelopes of the cones associated to the various cells involved.

The following local result given in the framework of Sect. 3 can then be equally reobtained: given any $m \to n$ process, there exist points P, such that the analyticity domain does have a nonempty intersection (locally) with the complex mass-shell. Results given at the end of Sect. 3.2 at $N = 4$ and $N = 5$ are also reobtained. (There is in these cases no difference between the two treatments, at the local level.) Examples, for general processes, are some (but not all) points such that the squared center-of-mass energy s of the process is sufficiently large (depending on m, n). More precisely, let us consider a reference frame in which the sum of initial (or final) momenta $\vec{P}_k$ is equal to zero, hence, in which s is the sum of initial (or final) energies $(P_k)_0$ squared; examples are then points P such that all initial energies $|(P_k)_0|$ are equal and sufficiently large, as also all final energies, with physical momenta $\varepsilon_k \vec{P}_k$ ($\varepsilon_k = -1$ if k is initial, $\varepsilon_k = +1$ if k is final) not too close to each other. These examples include some $+\alpha$-Landau points, for example, some points of the surfaces $s = (r\mu)^2$ for sufficiently large values of r. They remain sufficiently far from other $+\alpha$-Landau surfaces. Conditions that have been mentioned exclude in particular $+\alpha$-Landau points such as those described in Chapter I in a $3 \to 3$ process, where $T_{m,n}$ is certainly not expected to be locally the boundary value of an analytic function. (The conditions can be somewhat weakened if one considers "1-particle irreducible" functions, introduced later, obtained by subtracting suitable pole terms.)

Absorptive parts. Given two cells $\mathcal{S}_1, \mathcal{S}_2$ adjacent along (I_1, I_2), we have seen that $\tilde{r}_{\mathcal{S}_1} - \tilde{r}_{\mathcal{S}_2}$ vanishes for $p_{I_1}^2 \neq \mu^2$, $p_{I_1}^2 < 4\mu^2$, that is, it has its support in the union of the disconnected regions $p_{I_1} \in H_+(\mu) \cup \bar{V}_+(2\mu)$ and $p_{I_2} = -p_{I_1} \in H_+(\mu) \cup \bar{V}_+(2\mu)$. The absorptive part $\Delta^{I_1}_{\hat{\mathcal{S}}_1 \hat{\mathcal{S}}_2} H_N$ is the value of $\tilde{r}_{\mathcal{S}_1} - \tilde{r}_{\mathcal{S}_2}$ for $p_{I_1} \in H_+(\mu) \cup \bar{V}_+(2\mu)$ and zero outside. Hence:

$$\tilde{r}_{\mathcal{S}_1} - \tilde{r}_{\mathcal{S}_2} = \Delta^{I_1}_{\hat{\mathcal{S}}_1 \hat{\mathcal{S}}_2} H_N - \Delta^{I_2}_{\hat{\mathcal{S}}_2 \hat{\mathcal{S}}_1} H_N. \tag{2.75}$$

From Ruelle identity, $\Delta^{I_1}_{\hat{\mathcal{S}}_1 \hat{\mathcal{S}}_2} H_N$ is the Fourier transform (after factorization of $\delta(\Sigma p_k)$) of $\langle \Omega | \mathcal{R}_{\hat{\mathcal{S}}_1} \mathcal{R}_{\hat{\mathcal{S}}_2} | \Omega \rangle_c$.

It can be shown that, given (I_1, I_2) and the various pairs of cells of $(1, \ldots, N)$ adjacent along (I_1, I_2), there is a common analytic function $\Delta^{I_1} H_N$ from which the absorptive parts $\Delta^{I_1}_{\hat{\mathcal{S}}_1 \hat{\mathcal{S}}_2} H_N$ are various boundary values. More precisely, $\Delta^{I_1} H_N$ is a distribution in the (real) variable p_{I_1} with support in $H_+(\mu) \cup \bar{V}_+(2\mu)$, which depends analytically on the $d(N-2)$ remaining independent energy-momenta variables, with analyticity inside the union of all "flat" tubes with basis defined by the condition:

$$\text{Im } p_{I_1} = 0, \quad \text{Im } p_{J_1} \in V_+, \quad \forall J_1 \in \hat{S}_1$$
$$\text{Im } p_{J_2} \in V_+, \quad \forall J_2 \in \hat{S}_2 \tag{2.76}$$

and of appropriate real regions connecting these flat tubes. The boundary value in the real region $H_+(\mu) \cup \bar{V}_+(2\mu)$ from directions of $\text{Im} p$ defined in (2.76) is the absorptive part $\Delta^{I_1}_{\hat{S}_1 \hat{S}_2} H_N$.

The following choice of variables will be convenient in $\mathbf{C}^{d(N-1)}$:

$$k = -p_{I_1} = p_{I_2}, \tag{2.77}$$

$$z = \{z_i\}_{i \in I_1} \quad \text{with} \quad \sum_{i \in I_1} z_i = 0, \quad z_i = -p_i - (k/|I_1|), \tag{2.78}$$

$$z' = \{z'_j\}_{j \in I_2} \quad \text{with} \quad \sum_{j \in I_2} z'_j = 0, \quad z'_j = +p_j - (k/|I_2|). \tag{2.79}$$

Then, it can be seen that:

$$\Delta^{I_1} H_N(k, z, z') =$$
$$\lim_{\substack{\varepsilon \to 0 \\ \varepsilon > 0}} \left[H_N(k_{(o)} + i\varepsilon, \vec{k}, z, z') - H_N(k_{(o)} - i\varepsilon, \vec{k}, z, z') \right]. \tag{2.80}$$

Remark. Steinmann's relations can be reexpressed as follows: given I_1, I_3 and their complements I_2, I_4 as above, the double discontinuity of H_N around $\sigma_{I_1} \cap \sigma_{I_3}$ (σ_I defined in (2.74)) vanishes:

$$\Delta^{I_1}(\Delta^{I_3} H_N) = 0. \tag{2.81}$$

As a matter of fact, Steinmann's relations can be reobtained from the knowledge of the analyticity domain of $\Delta^{I_3} H_N$ (described in (2.76)).

One-particle singularities. As a consequence of previous analysis, N-point functions have in general singularities at $p_I^2 = \mu^2$, $I \subset (1, \ldots, N)$, which are poles with factorized residues. This will be explained in Sect. 5.1 (Remark 4) as a particular aspect of more general results. Examples in particular situations will also be given later.

4.2. 2-point and 4-point functions

2-point function (N=2). There are two paracells composed respectively of the subset {1} or {2} of (1,2). The cell functions, $\langle \Omega IC(x_1, x_2) -$

$C(x_1)C(x_2)I\Omega\rangle$ and $\langle \Omega I C(x_1, x_2) - C(x_2)C(x_1)I\Omega\rangle$ where $C(x_i) \equiv A(x_i)$, can also be written:

$$R(x_1 - x_2) = \theta\big((x_1)_0 - (x_2)_0\big) \langle \Omega |\, [A(x_1), A(x_2)]\, |\Omega\rangle, \qquad (2.82)$$

$$A(x_1 - x_2) = -\theta\big((x_2)_0 - (x_1)_0\big) \langle \Omega |\, [A(x_1), A(x_2)]\, |\Omega\rangle. \qquad (2.83)$$

The support property (2.63) reduces to:

$$\text{Support of } R \subset \text{ future cone } \bar{V}_+ \qquad (2.84)$$

$$\text{Support of } A \subset \text{ past cone } \bar{V}_-\,, \qquad (2.85)$$

so that $\tilde{r}$ and $\tilde{a}$ (which depend on one energy-momentum variable p; for example, $p = p_1 = -p_2$) are respectively boundary values of functions analytic in tubes Im $p \in V_+$ and Im $p \in V_-$ respectively, from the directions of V_+ or V_- respectively. The spectral condition yields:

$$\tilde{A}(p)|\Omega\rangle = 0 \quad \text{if} \quad p \notin H_-(\mu) \cup \bar{V}_-(2\mu), \qquad (2.86)$$

since this vector represents (in distribution sense) a "state" of energy-momentum $-p$ (*the minus sign is due to the relation* $A(f) = \int A(x)f(x)$ $\mathrm{d}x = \int \tilde{A}(-p)f(p)\mathrm{d}p$). In turn,

$$\tilde{r}(p) = \tilde{a}(p) \quad \text{if} \quad p^2 \neq \mu^2, \quad p^2 < 4\mu^2. \qquad (2.87)$$

Amputated functions are defined at $N = 2$ by multiplication with only one factor $p^2 - \mu^2$. With this definition,

$$\tilde{r}_{\mathrm{amp}}(p) = \tilde{a}_{\mathrm{amp}}(p) \quad \text{if} \quad p^2 < 4\mu^2. \qquad (2.88)$$

The primitive domain is the union of the tubes with bases V_+ and V_- respectively, and of a complex neighborhood of the real region defined in (2.87) (or in (2.88) in the case of amputated functions). The holomorphy envelope is the complex space $\mathbf{C}^d$ minus the set $p^2 = \mu^2$, $p^2 = 4\mu^2 + \rho$, $\rho \geq 0$ (or $p^2 = 4\mu^2 + \rho$ in the amputated case), as a geometrical result.

From relativistic invariance, H_2 depends in turn only on the complex variable p^2 and is analytic in the cut-plane ($p^2 \neq \mu^2$, $p^2 \neq 4\mu^2 + \rho$), a result which can be established in a simpler way, if relativistic invariance is assumed from the outset.

As easily seen, H_2 has a pole at $p^2 = \mu^2$.

4-point function. As easily seen, there are 32 precells, all of which are cells in the case $N = 4$:

- the four cells S_j^+, which include in particular three subsets of one index. S_j^+ does not include the set $\{j\}$ but includes the three sets with only one index $\neq j$ among $(1, \ldots, 4)$. As easily seen, it then also includes the sets $\{k, m\}$, $\{k, n\}$, $\{m, n\}$, $\{k, m, n\}$ and no others, where $\{j, k, m, n\} = \{1, 2, 3, 4\}$.
- the four cells S_j^-, opposite to previous ones (i.e. composed of the sets complementary to those of S_j^+).
- 24 cells which include two sets of one index. If the indices that define these two sets are m, n, the possible cells under consideration must include, as easily seen, the subsets $\{m, n\}$, $\{m, n, j\}$, and $\{m, n, k\}$. Other subsets included are either $\{m, j\}$ or $\{n, k\}$, and either $\{m, k\}$ or $\{n, j\}$.

The four cells thus determined in the latter case for given m, n are those giving rise to Steinmann relations as indicated below. The cell denoted below as S_{jk}^+ includes $\{m\}$, $\{n\}$ as also $\{m, k\}$ and $\{n, k\}$. (Note that $S_{jk}^+ \neq S_{kj}^+$.) The cell S_{jk}^- is opposite to S_{jk}^+. (It includes $\{j\}$, $\{k\}$ as also $\{n, j\}$ and $\{m, j\}$.)

Cell functions associated with S_j^+ and S_j^- are usually denoted $\tilde{a}_j(p)$ and $\tilde{r}_j(p)$ respectively ($p = (p_1, \ldots, p_4)$, $p_1 + \cdots + p_4 = 0$). Functions associated to S_{jk}^+ and S_{jk}^- are denoted $\tilde{a}_{jk}$ and $\tilde{r}_{jk}$ respectively. Bases Γ_j^+, Γ_j^- associated with S_j^+, S_j^- and bases Γ_{jk}^+, Γ_{jk}^- associated with S_{jk}^+, S_{jk}^- can be defined by the relations:

$$\Gamma_j^+ = -\Gamma_j^- = \{\mathrm{Im}\, p = (\mathrm{Im}\, p_1, \ldots, \mathrm{Im}\, p_4),\ \mathrm{Im}\, p_1 + \cdots + \mathrm{Im}\, p_4 = 0,$$
$$\mathrm{Im}\, p_k \in V^+,\ \mathrm{Im}\, p_m \in V^+,\ \mathrm{Im}\, p_n \in V^+\} \tag{2.89}$$

$$\Gamma_{jk}^+ = -\Gamma_{jk}^- = \{\mathrm{Im}\, p = (\mathrm{Im}\, p_1, \ldots, \mathrm{Im}\, p_4),\ \mathrm{Im}\, p_1 + \cdots + \mathrm{Im}\, p_4 = 0,$$
$$\mathrm{Im}\, p_k \in V^-,\ \mathrm{Im}\, p_m + \mathrm{Im}\, p_k \in V^+,\ \mathrm{Im}\, p_n + \mathrm{Im}\, p_k \in V^+\}.$$
$$\tag{2.90}$$

There are six Steinmann relations, of the form:

$$\tilde{a}_{j,k} + \tilde{a}_{k,j} = \tilde{r}_{m,n} + \tilde{r}_{n,m}. \tag{2.91}$$

The following link between the (connected, amputated) chronological function $T(p_1, \ldots, p_4)$ and the functions $(\tilde{r}_S)_{\mathrm{amp}}$ follows from the spectral condition. Given the real domain:

$$\Omega_{1,2} = \big\{ p = (p_1, \ldots p_4),\ p_1 + \cdots + p_4 = 0,$$
$$p_k^2 < 4\mu^2,\ k = 1, \ldots, 4$$
$$p_3 + p_4 \in \text{complement of } \bar{V}_-(2\mu)$$
$$(p_1 + p_3)^2 < \mu^2, (p_2 + p_3)^2 < \mu^2 \big\}, \tag{2.92}$$

which contains the physical region of the process $(1,2) \to (3,4)$, one has:

$$T(p_1, \ldots, p_4) = (\tilde{r}_S)_{\mathrm{amp}}(p_1, \ldots, p_4) \tag{2.93}$$

in $\Omega_{1,2}$ for 16 cells S (functions $\tilde{r}_S$ coincide in $\Omega_{1,2}$ for these cells). Similar results hold for other $2 \to 2$ channels.

The primitive domain is determined from this information in the way indicated in the general case in Sect. 4.1. In the present situation, a refined use of the spectral condition, techniques of holomorphy envelope and Lorentz invariance provide in turn the more precise (physical-sheet) results mentioned in Sect. 1.2 on the complex mass-shell. First results had been obtained already at the end of the 1950s[Bo], by nongeometrical methods (namely, analytic continuation—with respect to the mass variables—of relevant Cauchy-type integral representations) in a more restricted formulation of the analytic framework of Quantum Field Theory, namely, analyticity in a single complex vector. These results include analyticity in a cut-plane of the type described in Sect. 5.1 of Chapter I (with poles at $s = \mu^2$ and $u = \mu^2$) in the complex variable $s = (p_1 + p_2)^2$, for fixed $t = (p_1 + p_3)^2$ in a region of the form $t_o < t < 0$, with related results on crossing and (fixed t) dispersion relations. Results apply, in particular, in a theory with only one type of particle of mass $\mu > 0$ (in which case $t_o = -28\mu^2$), and also in some more general cases such as pion-pion scattering in a theory including other particles (e.g. nucleons) with much higher masses. The deeper and more complete analysis made in the first part of the 1960s ([BEG1] and references therein) has led to various improvements in more general cases (e.g. nucleon-pion scattering) and for more general values of t.

Remarks.

  (1) The linear program does not provide, in general, analyticity in a cut-plane for general t: a local region is excluded (this does not obstruct the proof of crossing properties). The exclusion of such local regions cannot be avoided in this framework: further improvements require the use of the nonlinear program (unitarity equations).

  (2) Local results of Sect. 3.1 and 3.2 are in particular recovered at $N = 4$. It is interesting[BEG2] to note that, even at the local level, the results obtained are somewhat better. In Sect. 3.2, the scattering function $T_{2,2}$ was shown to be the boundary value of an analytic function $\underline{T}_{2,2}$ from the directions Im $s > 0$, but the analyticity domain of $\underline{T}_{2,2}$ (in the complex mass-shell) might *a priori* shrink to zero when the threshold $s = 4\mu^2$ is approached.

Knowing, as mentioned above, that the holomorphy envelope is invariant under complex Lorentz transformations, allows one to show that the analyticity domain will extend to a domain including the region $\{\text{Im } s > 0\} \cap \mathcal{N}$, where $\mathcal{N}$ is a complex open set containing the real region $\Omega_{1,2}$. Since, at $s < 4\mu^2$ all boundary values coincide (by the spectral condition), the only singularity of the 4-point function in a complex neighborhood of the real mass-shell (of the $2 \to 2$ process considered) is thus the s-cut.

The proof of previous information on the holomorphy envelope, not to speak of global results such as crossing, requires global methods of analytic completion in which analyticity near the complex infinity within the tubes plays an essential role.

PARTICLE ANALYSIS IN THE NONLINEAR PROGRAM

§5. The nonlinear program-direct methods

5.1 Discontinuity formulae for absorptive parts and a class of generalized optical theorems

Asymptotic completeness equations are obtained by inserting the identity operator between some operators involved in the relations of the linear program. More precisely, the identity operator is written, as a consequence of asymptotic completeness, in the form $E^{\text{in}} = \sum_\ell E_\ell^{\text{in}}$ or $E^{\text{out}} = \sum_\ell E_\ell^{\text{out}}$ where E_ℓ^{in} or E_ℓ^{out} are projectors onto the space of ℓ incoming or outgoing free-particle states. Finally, appropriate versions of reduction formulae are used. It may be convenient, as a formal guide, to write E_ℓ^{in} or E_ℓ^{out} in the usual form:

$$E_\ell^{\text{in}} = \int |k_1, \ldots, k_\ell \text{in}\rangle \langle k_1, \ldots, k_\ell \text{in}| \prod_{\alpha=1}^{\ell} \delta_+(k_\alpha^2 - \mu^2) \mathrm{d}k_\alpha \qquad (2.94)$$

$$E_\ell^{\text{out}} = \int |k_1, \ldots, k_\ell \text{out}\rangle \langle k_1, \ldots, k_\ell \text{out}| \prod_{\alpha=1}^{\ell} \delta_+(k_\alpha^2 - \mu^2) \mathrm{d}k_\alpha. \qquad (2.95)$$

However, there is no need for our purposes to give an actual definition of "states" $|k_1, \ldots, k_\ell \text{in (out)}\rangle$ with sharp values of the momenta, which do not belong to $\mathcal{H}_{\text{in}}$ or $\mathcal{H}_{\text{out}}$: results can be derived through the use of wave functions.

This procedure yields various formulae, including off-shell versions of the unitarity equations of Chapter I (external energy-momenta are no longer restricted to the mass-shell) and further results. We first describe formulae for absorptive parts that are useful later, and then a class of "generalized optical theorems" with the limitations already mentioned in paragraph (vii) of Sect. 1.2.

In the case of a $2 \to 2$ process, both types of formulae reduce (in a neighborhood of the mass-shell) to the discontinuity formula (1.21) (with possibly off-shell external energy-momenta) relative to the difference between the plus $i\varepsilon$ and minus $i\varepsilon$ boundary values of the 4-point function from above and below the cut $s > 4\mu^2$.

Discontinuity formulae for absorptive parts. Absorptive parts were introduced at the end of Sect. 4.1. The procedure described above is applied to the product $\mathcal{R}_{\hat{S}_1} \mathcal{R}_{\hat{S}_2}$ in the term $\langle \Omega | \mathcal{R}_{\hat{S}_1} \mathcal{R}_{\hat{S}_2} | \Omega \rangle_c$ from which $\Delta^{I_1}_{\hat{S}_1 \hat{S}_2} H_N$ is defined as indicated below Eq. (2.75). The following result is obtained (see simple examples, with a more transparent diagrammatical notation in Sect. 5.2 to 5.4):

$$\Delta^{I_1}_{\hat{S}_1 \hat{S}_2} H_N = \sum_{\ell \geq 1} \frac{1}{\ell!} \hat{r}^{(N_1 + \ell)}_{\Sigma_{1,\ell}} *_{(\ell)} \hat{r}^{(N_2 + \ell)}_{\Sigma_{2,\ell}}$$

$$= \sum_{\ell \geq 1} \frac{1}{\ell!} \hat{r}^{(N_1 + \ell)}_{\Sigma'_{1,\ell}} *_{(\ell)} \hat{r}^{(N_2 + \ell)}_{\Sigma'_{2,\ell}}, \tag{2.96}$$

where $N_1 = |I_1|$, $N_2 = |I_2|$, $*_{(\ell)}$ denotes on-mass-shell convolution over ℓ intermediate real energy-momenta variables $p_{N+1}, \ldots, p_{N+\ell}$, with the measure $[\prod_{r=1}^{\ell} \delta_+ (p^2_{N+r} - \mu^2) \, dp_r] \delta(p_{I_1} + \sum_{r=1}^{\ell} p_{N+r})$ and the functions $\hat{r}$ are defined as follows. They depend on the energy-momenta variables $\{p_i\}$, $i \in I_1$, $p_{N+1}, \ldots, p_{N+\ell}$ and $-p_{N+1}, \ldots, -p_{N+\ell}$, $\{p_j\}$, $j \in I_2$, respectively, and are equal, up to amputation in $p_{N+1}, \ldots, p_{N+\ell}$ (i.e. multiplication by $\prod_r (p^2_{N+r} - \mu^2)$), to the cell boundary values associated to cells $\Sigma_{1,\ell}$ and $\Sigma_{2,\ell}$ (first expression), or $\Sigma'_{1,\ell}$ and $\Sigma'_{2,\ell}$ (second expression), of $I_1 \cup \{N+1, \ldots, N+\ell\}$ and $I_2 \cup \{N+1, \ldots, N+\ell\}$, respectively. $\Sigma_{\alpha,\ell}$, $\alpha = 1, 2$, is composed of all proper subsets J of indices such that $J \subset (N+1, \ldots, N+\ell)$ or $J \cap I_\alpha \in \hat{S}_\alpha$. Similarly, $\Sigma'_{\alpha,\ell}$ is composed of all proper subsets J such that $J \supset I_\alpha$ or $J \cap I_\alpha \in \hat{S}_\alpha$.

The δ-function $\delta(p_{I_1} + \Sigma p_{N+r})$ is included in the integration since all energy-momentum conservation δ-functions have been factored out in all terms involved. Alternatively, individual terms in the convolution contain the respective factors $\delta\left(p_{I_1} + \Sigma p_{N+r}\right)$ and $\delta\left(-p_{I_2} + \Sigma p_{N+r}\right)$ and the convolution product contains the overall factor $\delta\left(p_{I_1} + p_{I_2}\right)$. If all terms in the right-hand sides are amputated both in the internal and external variables, a corresponding formula is obtained with H_N replaced by $(H_N)_{\mathrm{amp}}$.

Remarks.

(1) Expressions in the right-hand side of (2.96) involve restrictions to the mass-shell of the internal variables $p_{N+1}, \ldots, p_{N+\ell}$. By the methods of Sect. 3.3, these restrictions can be shown to be well defined, at least away from M_0 points. The latter points occur in integration domains. Corresponding problems are not discussed here.

(2) The energy-momentum conservation and mass-shell constraints entail, as in the unitarity equations in Chapter I, that terms of order ℓ or more vanish for $p_{I_1}^2 < (\ell\mu)^2$, so that there is only a finite number of terms in any bounded region.

(3) The absorptive part $\Delta^{I_1} H_N$, defined at the end of Sect. 4.1 as an analytic function in variables other than $p_{I_1} = -p_{I_2}$ (and from which the functions $\Delta^{I_1}_{\hat{S}_1 \hat{S}_2} H_N$ are various boundary values), satisfies:

$$\Delta^{I_1} H_N(k, z, z') = \sum_{\ell \geq 1} \frac{1}{\ell!} \left(\hat{H}_+^{(N_1+\ell)} *_{(\ell)_+} \hat{H}_-^{(N_2+\ell)} \right) (k, z, z')$$

$$= \sum_{\ell \geq 1} \frac{1}{\ell!} \left(\hat{H}_-^{(N_1+\ell)} *_{(\ell)_+} \hat{H}_+^{(N_2+\ell)} \right) (k, z, z'),$$

$$(2.97)$$

where the subscript $+$ or $-$ in $\hat{H}_{+,-}$ denotes the half-space Im $k_{(0)} > 0$ or Im $k_{(0)} < 0$ from which the corresponding boundary values of the analytic functions $H^{(N_1+\ell)}$ and $H^{(N_2+\ell)}$ are taken. $\hat{H}_\epsilon^{(N_1+\ell)}$ and $\hat{H}_\epsilon^{(N_2+\ell)}$ depend respectively on $k, z, p_{N+1}, \ldots, p_{N+\ell}$, and $k, z', -p_{N+1}, \ldots, -p_{N+\ell}$, and the right-hand side of (2.97) is the $\epsilon \to 0$ ($\epsilon > 0$) limit of the (on-mass-shell) convolution product of $\hat{H}^{(N_1+\ell)}(k_{(0)} + i\epsilon, \vec{k}, \ldots)$ and $\hat{H}^{(N_2+\ell)}(k_{(0)} - i\epsilon, \vec{k}, \ldots)$ (first expression), or of $\hat{H}^{(N_1+\ell)}(k_{(0)} - i\epsilon, \ldots)$ and $\hat{H}^{(N_2+\ell)}(k_{(0)} + i\epsilon, \ldots)$ (second expression).

(4) One-particle singularities

The contribution $\ell = 1$ to the right-hand side of (2.96) and (2.97) is the product of the two functions $\hat{r}$ or $\hat{H}$, restricted at $k^2 = \mu^2$, and of the δ-function $\delta\left(k^2 - \mu^2\right)$, where $k = p_{I_1}$. The existence of corresponding pole singularities, with a factorized residue, of N-point functions follows.

As a matter of fact, this result can be obtained without recourse to asymptotic completeness but as a consequence of the previously known information on one-particle states associated with the part $H_+(\mu)$ of the spectrum. (Asymptotic completeness is used in (2.96) to analyze the continuous part of the spectrum.)

(5) For the purposes of Sect. 6, it will be convenient to define amputation in p-space by multiplication with $\prod_{k=1}^{N} \left[H_2(p_k)\right]^{-1}$ instead of $\prod_{k=1}^{N} \left(p_k^2 - \mu^2\right)$. As a matter of fact, it is known from previous results that H_2 has a pole at $p^2 = \mu^2$ with a residue Z, which is in general (and is assumed here to be) a nonzero constant. Previous formulae then apply equally with this definition of amputation, up to multiplicative constants equal to one if conventions are such that $Z = 1$.

In Sects. 5.2 to 5.4, amputation can be defined by either procedure without change in the results. For definiteness, functions F_N will denote the analytic N-point functions H_N after amputation with this new procedure:

$$F_N\left(p_1, \ldots, p_N\right) = \prod_{k=1}^{N} \left[H_2(p_k)\right]^{-1} H_N\left(p_1, \ldots, p_N\right). \qquad (2.98)$$

Generalized optical theorems. As already mentioned in Chapter I, relevant discontinuities in multiparticle dispersion relations involve differences of boundary values of various analytic continuations of the S matrix (see Chapter I). Generalized optical theorems should express these discontinuities in terms only of *physical S* matrices (or their hermitean conjugates), rather than more complicated boundary values.

It is believed in general that cell functions provide some of the boundary values needed, and that they will ultimately be shown to be, on-shell, (almost everywhere) boundary values of functions analytic in domains of the complex mass-shell. As already discussed in the case of chronological functions, this is not established in the linear program. This problem is not treated here, but we now explain how (on-shell) expressions of the functions $\tilde{r}_S$, as sums of on-mass-shell convolution integrals involving physical S-matrix kernels only (or their hermitean conjugates), can be obtained. It is convenient to start from the expression (2.59) of the operators $\mathcal{R}_S$. Products $E_{\text{in}} E_{\text{out}}$ are then inserted between each pair of successive operators $\tilde{C}(p(J_\alpha))$, $\tilde{C}(p(J_{\alpha+1}))$ (obtained after Fourier transformation). Using, for example, the expressions [(2.94) and (2.95)] of $E_{\ell,\text{in}}$ and $E_{\ell',\text{out}}$ as a (formal) guide, the desired expressions follow from the fact that $\langle k_1, \ldots, k_\ell, \text{in} \mid k_1', \ldots, k_{\ell'}', \text{out} \rangle$ is the corresponding S-matrix kernel, while

$$\langle k_1, \ldots, k_\ell, \text{in} \mid \tilde{C}(p(J)) k_1', \ldots, k_{\ell'}', \text{out} \rangle$$

can be shown to be a partially connected S-matrix kernel between initial energy-momenta $k_1, \ldots, k_\ell, (p(J))_{\text{in}}$ and final energy-momenta $k_1', \ldots, k_{\ell'}', (p(J))_{\text{out}}$, where $(p(J))_{\text{in}}$ and $(p(J))_{\text{out}}$ denote sets of energy-momenta $p_j, j \in J$ such that $(p_j)_0 < 0$ and $(p_j)_0 > 0$ respectively. Partially connected here means connected with respect to the initial and final variables of the sets $(p(J))_{\text{in}}$ and $(p(J))_{\text{out}}$ only.

5.2. 4-point function and 2-particle threshold

The amputated analytic 4-point function $F_{(4)}$ is known from the results of the linear program (Sect. 4.2) to be analytic in a complex domain including, for example, the real set $\Omega_{1,2}$ defined in (2.92), which contains the physical region of the $2 \to 2$ process $(1, 2) \to (3, 4)$, minus the cut $s = 4\mu^2 + \rho$, $\rho \geq 0$, and the pole at $s = \mu^2$ if the theory is not even. There are only two different boundary values at $s \geq 4\mu^2$, denoted below as F_+ and F_- (leaving the index $N = 4$ implicit), obtained respectively from the directions $\text{Im } s > 0$ and $\text{Im } s < 0$ of the physical sheet. The discontinuity formula on absorptive parts of Sect. 4.3 reduces here at $s < (3\mu)^2$, or $s < (4\mu)^2$ in an even theory, to the off-shell unitarity equation

$$F_+ - F_- = F_+ *_+ F_-, \tag{2.99}$$

where $*_+$ is on-mass-shell convolution over two internal energy-momenta. Eq. (2.99) reduces to the standard on-shell unitarity equation $T_+ - T_- = T_+ *_+ T_-$ when external energy-momenta are restricted to the mass-shell. The following result then holds (with $M < 3\mu$, or 4μ in an even theory):

Theorem 2.17. If the on-shell functions T_+ and T_- are assumed to be continuous in the region $4\mu^2 < s < M^2$, F_+ and F_- are analytic in the region $4\mu^2 < s < M^2$ and F is a meromorphic function (with possible poles in s) in a 2-sheeted (d even) or multisheeted (d odd) domain around $s = 4\mu^2$.

Proof. This result has already been established on-shell in Sect. 5.2 (and 6.2) of Chapter I, starting from the on-shell unitarity equation.

The argument used there does not directly extend to F: Eq. (2.99) is no longer a Fredholm-type equation. However, the result can be easily extended to F as follows. It is first extended to the case when, for example, initial energy-momenta are allowed to vary off-shell, while final energy-momenta are still restricted to the mass-shell. The corresponding restriction of F will be denoted $F_|$. From Eq. (2.99), $(F_-)_|$ is equal to

$$(F_-)_| = (F_+)_| \quad - \quad (F_+) *_+ (F_-)_|$$
$$= (F_+)_| \quad - \quad (F_+) *_+ T_-, \qquad (2.100)$$

where the second equality follows from the fact that $(F_-)_|$ in the convolution product $F_+ *_+ (F_-)_|$ is restricted to the mass-shell on both sides.

In view of the analyticity of T_- already established on-shell for s real, $4\mu^2 < s < M^2$, the second expression of $(F_-)_|$ in (2.100) shows that it is the *plus iε* boundary value of an analytic function. Since it is both a plus iε and a minus iε boundary value of (*a priori* different) analytic functions, the edge-of-the-wedge theorem ensures that these analytic functions are analytic continuations of each other and that $(F_-)_|$ is analytic at s real, $4\mu^2 < s < M^2$. A simple extension of the argument yields also the meromorphy of $(F_-)_|$ in a 2-sheeted or multisheeted domain. A second application of the procedure, using the relation

$$F_- = F_+ \quad - \quad F_- *_+ F_+$$
$$= F_+ \quad - \quad (F_-)_| *_+ F_+ \qquad (2.101)$$

allows one to conclude the analysis.

Remark. The analyticity of F_+, hence of T at any physical-region point P, such that $4\mu^2 < s < M^2$ is equivalent to the assertion that $\mathrm{ES}_P(T)$ is empty (apart from the origin), that is:

$$\mathrm{ES}_P(T) = \{x = (x_1, \ldots, x_4), x_1 = x_2 = x_3 = x_4\}, \qquad (2.102)$$

instead of the result shown in Figure 2.6a in the linear program.

The following theorem is an off-shell extension of the results of Sect. 6.2 of Chapter I involving the irreducible kernel U. As was already the case in Chapter I, we state it either at $m = 2$ or more generally in a simplified theory, at $m > 2$, in which the 2m-point function satisfies analyticity properties analogous to above (analyticity in a cut-domain, with the cut at $s \geq (m\mu)^2$) and the discontinuity formula (2.99) with

$*_+$ denoting on-mass-shell convolution over m internal energy-momenta. As in Chapter I, by definition:

$$\otimes \; = (1/2)* \qquad (m = 2, \;\; d \text{ even}, \;\; \text{or} \;\; \beta \text{ half} - \text{integer})$$
$$\otimes \; = (i/2\pi)\ell n \; \sigma \quad (m = 2, \;\; d \text{ odd}, \;\; \text{or} \;\; \beta \text{ integer}),$$

where $*$ denotes analytic continuation in s of $*_+$ (in the same sense as in Chapter I), $\sigma = s - (m\mu)^2$ and $\beta = [(m-1)d - m - 1]/2$.

Theorem 2.18. There is a well-defined kernel U, analytic or meromorphic in a complex domain containing the m-particle threshold $s = (m\mu)^2$ such that:

$$F = U + (F \otimes U) = U + (U \otimes F). \tag{2.103}$$

If $\beta > 0$ and if U is analytic at $s = (m\mu)^2$, F is equal to its (locally absolutely convergent) Neumann series in terms of U:

$$F = U + (U \otimes U) + (U \otimes U \otimes U) + \cdots \tag{2.104}$$

in the neighborhood of $s = (m\mu)^2$.

Remarks.

(1) U is analytic at $s = 4\mu^2$, in particular, if F is locally bounded in the physical sheet near $s = 4\mu^2$ (in which case the series $U = F - F \otimes F + F \otimes F \otimes F - \cdots$ of U in terms of F is locally absolutely convergent).

(2) As in Chapter I, Eq. (2.104) is, at β integer, a local expansion of the (non-holonomic) function F in terms of holonomic contributions of the form $a_n(p)(\ell n \; \sigma)^n$, a_n locally analytic.

Proof of Theorem 2.18. The kernel U can be defined by the relation:

$$U = F - (F \otimes F) + (F \otimes U \otimes F), \tag{2.105}$$

where U is involved in the right-hand side only by its *on-shell* values already defined in terms of T in Chapter I. Therefore, Eq. (2.105) does define U off-shell in a suitable complex domain. Eq. (2.103) follows in turn. The same algebraic argument as in Chapter I, starting from Eq. (2.103), shows that $U_+ = U_-$ at $s \geq (m\mu)^2$, that is, the uniformity of U around $s = (m\mu)^2$. The expansion (2.104) under the conditions mentioned in the theorem follows in the same way as in Chapter I.

5.3. The 5-point function in the 3-particle region

In this subsection, we consider for definiteness a $2 \to 3$ process $(1, 2)$ $\to (3, 4, 5)$ ($3 \to 2$ processes are treated similarly). The distribution T_5 (denoted below T), considered in a real neighborhood of the physical region $M_{2,3}$, is already known from the results of Sect. 4.1 to be the boundary value of the analytic (amputated) 6-point function from the directions

$$\begin{aligned} \mathrm{Im}\,(p_3 + p_4) \in V^+, &\quad \mathrm{Im}\,(p_3 + p_5) \in V^+, \\ \mathrm{Im}\,(p_4 + p_5) \in V^+, &\quad \mathrm{Im}\,(p_1 + p_2) \in V^- \end{aligned} \qquad (2.106)$$

and satisfies more refined local properties (see end of Sect. 3) near each physical point P, in particular:

$$\mathrm{ES}_P(T) \subset C_3 \cup C_4 \cup C_5. \qquad (2.107)$$

The aim of this section is to show how the latter can be improved at non M_0 points in the low-energy region $(3\mu)^2 < s < (4\mu)^2$. As a matter of fact, if one believes in ideas of macrocausality in terms of intermediate particles, the essential support $\mathrm{ES}_P(T)$ should be empty at these points $(x_1 = x_2 = x_3 = x_4 = x_5)$, and T should equivalently be analytic in that region. (M_0 points in that region are those such that two final energy-momenta are colinear, i.e. equal).

The following weaker result has been established[EGI], as outlined below, from the direct exploitation of 2-particle asymptotic completeness equations and from the analyticity of the 4-point function T_4 at physical points in the 2-particle region $4\mu^2 < s < 9\mu^2$, established in Sect. 5.2 (and equivalent to the essential support property (2.102) at these points).

Theorem 2.19.

$$\mathrm{ES}_P(T) \subset C = \left\{ x = (x_1, \ldots, x_5), x_1 = x_2, x_3 = x_4 = x_5, x_3 - x_1 \in \bar{V}_+ \right\} \qquad (2.108)$$

at any physical point P such that $(3\mu)^2 < s < (4\mu)^2$, and outside the submanifolds $(P_i + P_j - P_k)^2 = \mu^2$, where (i, j, k) is any permutation of $(3, 4, 5)$.

Equivalently, T is (at the points P above) the boundary value of an analytic function from the directions

$$\mathrm{Im}\,(p_3 + p_4 + p_5) \in V^+ \quad \left(\mathrm{Im}\,(p_1 + p_2) \in V^- \right). \qquad (2.109)$$

Remarks.

(1) If Lorentz invariance is used, it is moreover possible to show that

$\mathrm{ES}_P(T) \subset$
$$\{x; x_1 = x_2, x_3 = x_4 = x_5, x_3 - x_1 = \lambda (P_3 + P_4 + P_5), \ \lambda \geq 0\}$$
$$(2.110)$$

at the same points P as in Theorem 2.19; T is correspondingly the boundary value of an analytic function from the directions $\mathrm{Im}\ s > 0$ $(s = (p_3 + p_4 + p_5)^2)$.

(2) Related results follow for the on-shell scattering function $T_{2,3}$, which is shown to be the boundary value (at points P covered by Theorem 2.19) of a function, analytic in a domain of the complex mass-shell, from the directions dual to (2.108) or (2.110), that is, the directions $\mathrm{Im}\ s > 0$ in the latter case.

(3) The submanifolds $(P_i + P_j - P_k)^2 = \mu^2$ contain the M_0 points (two final energy-momenta colinear) where the results are not expected to be valid. The exclusion of other points of these submanifolds is due to technical reasons: results are expected to hold at all non M_0 points.

(4) As already mentioned, better results are expected (analyticity at points P considered) but their derivation would require exploitation of 3-particle asymptotic completeness equations and further regularity conditions.

Proof of Theorem 2.19.

(a) Ingredients to be used: besides T, three other boundary values $f_k^{(-)}$ of the (amputated) 5-point function will be used, namely, those obtained for each $k = 3, 4, 5$ from the respective directions:

$$\mathrm{Im}\ (p_3 + p_4 + p_5) \in V_+, \ \mathrm{Im}\ (p_i + p_j) \in V_-, \ \mathrm{Im}\ (p_i + p_k) \in V_+,$$
$$\mathrm{Im}\ (p_j + p_k) \in V_+, \qquad (2.111)$$

where (i, j, k) is a permutation of (3, 4, 5). Besides (2.107), the following essential support properties can be derived at any physical region point by methods analogous to those of Sect. 3.1:

$$\mathrm{ES}_P(f_k^{(-)}) \subset C_i \cup C_j \cup C_k^{(-)}, \qquad (2.112)$$

where cones C_i, C_j are those defined in Sect. 3.1 (see Figure 2.7) and

$$C_k^{(-)} = (u; \ u_1 = u_2, \ u_i = u_j, \ u_k - u_i \in \bar{V}_+, \ u_k - u_1 \in \bar{V}_+). \qquad (2.113)$$

On the other hand, Eq. (2.96) provides the following formulæ on absorptive parts for $I_1 = (1, 2, k)$, $I_2 = (i, j)$ in the region $(3\mu^2) < s < (4\mu)^2$:

$$\begin{array}{c}\text{1}\underbrace{}_{\text{2}}\hspace{-1em}\bigodot\hspace{-1em}\underbrace{}_{\text{5}}^{\text{3}}\, - \, \text{1}\underbrace{}_{\text{2}}\bigotimes\underbrace{}_{\text{k}}^{\text{i}}_{\text{j}} \, = \, \text{1}\underbrace{}_{\text{2}}\bigoplus\bigoplus\underbrace{}_{\text{k}}^{\text{i}}_{\text{j}}\, ,\end{array} \qquad (2.114)$$

where $\bigodot$ and $\bigoplus$ denote corresponding functions T_4 and

$T = T_5$, and $\bigotimes_k$ denotes similarly $f_k^{(-)}$ times the global

relevant emc δ-function. As in Chapter I, lines ∿ denote on-mass-shell convolution. (Line k has been put on the same side as the other final lines i, j.) In (2.114), the term $\ell = 1$ is absent because $s > \mu^2$ and terms with $\ell > 2$ are absent because $s < (4\mu)^2$.

(b) The theorems of Sect. 3 of the Appendix on products and integrals will yield, as explained below, information on the essential support of the integral in the right-hand side of Eq. (2.114). The equality (2.114) will yield in turn improved information on the essential supports of both individual terms in the left-hand side and in particular of T_5. The result obtained at that stage is still weaker than (2.108), which will be established by an iteration of the procedure.

We now give some details.

(i) Information on the integral of the right-hand side of (2.119).

The theorems of the Appendix directly yield the following result. If $P = (P_1, \ldots, P_5)$ is not a "$u = 0$ point" for this integral, the only possible configurations $(x_1, \ldots, x_5)$ in its essential support at P are those for which there exist real energy-momenta P_i', P_j' in the integration domain ($P_i' + P_j' = P_i + P_j$, $P_i', P_j' \in H_+(\mu)$), space-time points x_i', x_j', x_i'', x_j'' and real scalars λ_i, λ_j such that:

$$\left(x_1, x_2, x_k, x_i', x_j'\right) \in \mathrm{ES}_{(P_1, P_2, P_k, P_i', P_j')} \bigotimes_k, \qquad (2.115)$$

$$\left(x_i'', x_j'', x_i, x_j\right) \in \mathrm{ES}_{(P_i', P_j', P_i, P_j)} \bigodot, \qquad (2.116)$$

$$x_i'' - x_i' = \lambda_i P_i', \qquad (2.117)$$

$$x_j'' - x_j' = \lambda_j P_j'. \qquad (2.118)$$

Eqs. (2.117) and (2.118) indicate that x_i', x_i'' and x_j', x_j'' must lie on (internal) trajectories parallel to P_i' and P_j' respectively.

By definition, P is a $u = 0$ point if Eqs. (2.115)–(2.118) can be satisfied with $x_1 = x_2 = x_3 = x_4 = x_5$, while at least one of the points x_i', x_i'', x_j', x_j'' lies at a different position. In the application, this cannot occur if $P_i \neq P_j$, hence $P_i' \neq P_j'$, as easily checked from (2.112) and (2.102).

(ii) The following trivial result of essential support theory holds: if $A_1 = A_2 + A_3$, then

$$\mathrm{ES}_P(A_1)\,(= \mathrm{ES}_P\,(A_2 + A_3)) \subset \mathrm{ES}_P(A_2) \cup \mathrm{ES}_P(A_3). \quad (2.119)$$

Hence, if one knows initially that $\mathrm{ES}_P(A_i) \subset S_i$, $i = 1, 2, 3$, then

$$\mathrm{ES}_P(A_1) \subset S_1 \cap (S_2 \cup S_3). \qquad (2.120)$$

The information obtained at that stage from (i) and (ii) and (2.112) and (2.102) on ⟨+⟩ is shown in Figure 2.8a at any non M_0 physical point P ($P_i \neq P_j$, $i, j = 3, 4, 5$, $i \neq j$).

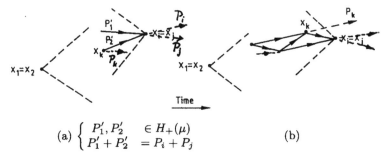

$$(a) \begin{cases} P_1', P_2' & \in H_+(\mu) \\ P_1' + P_2' & = P_i + P_j \end{cases} \qquad (b)$$

Figure 2.8 Possible configurations $(x_1, \ldots, x_5)$ in $\mathrm{ES}_P(T)$ in the first step (Fig. a) if $P_i \neq P_j$, $i, j = 3, 4, 5$, $i \neq j$ and in the third step (Fig. b) of the iteration procedure.

(iii) The final result (2.108) follows from two more iterations of the procedure, by using at each stage the information previously obtained on ⟨+⟩ and on all terms ⟨+⟩$_k$. To carry out the iteration procedure, results in successive steps have to be established under somewhat more general conditions than those of Theorem 2.19 (removal of some of the conditions on P). Theorem 2.19 is the outcome of the third step. More precisely, apart from another *a priori* possible class which can also be eliminated and is omitted here, the final result asserts that the only possible (relative) configurations in the essential support of T_5 are of the form shown in Figure 2.8b, such that there exists a corresponding classical multiple scattering diagram with successive $2 \to 2$ scatterings (and real on-shell external and intermediate particles). However, such a classical diagram is not possible with equal mass particles, below the 4-particle threshold, in view of the mass-shell and energy-momentum conservation constraints, unless all energy-momenta are parallel (i.e. equal in the equal mass case). This condition cannot be satisfied if $P_i \neq P_j$, unless all $2 \to 2$ vertices coincide. QED

5.4. The 6-point function in the 3-particle region: Landau singularities and structure equation

In this subsection, we consider a $3 \to 3$ process $(1, 2, 3) \to (4, 5, 6)$. The chronological function $T(p_1, \ldots, p_6)$, considered in a real neighborhood of the physical region $M_{3,3}$, is known from the results of Sect. 4.1, to be the boundary value of the analytic (amputated) 6-point function from the directions satisfying the conditions

$$\text{Im}(p_i + p_j) \in V_-, \ \text{Im}(p_\ell + p_m) \in V_+, \ \text{Im}(p_\ell + p_m + p_i) \in V_+ \quad (2.121)$$

for all possible choices of indices $i, j, i \neq j$ among 1, 2, 3, and of indices $\ell, m, \ell \neq m$ among (4, 5, 6). The results of Sect. 3.1 directly yield, moreover, the following essential support property at any physical point P:

$$\text{ES}_P(T) \subset \bigcup_{\substack{i=1,2,3 \\ \ell=4,5,6}} C_{i,\ell}, \quad (2.122)$$

where (see Figure 2.9a)

$$C_{3,4} = \{x = (x_1, \ldots, x_6); x_1 = x_2, x_5 = x_6, x_3 - x_1 \in \bar{V}_+,$$
$$x_5 - x_3 \in \bar{V}_+, \ x_4 - x_1 \in \bar{V}_+, \ x_5 - x_4 \in \bar{V}_+\}. \quad (2.123)$$

Other cones $C_{i,\ell}$ are defined similarly.

As a matter of fact, results are better at many points P. In particular, if we consider the part $s = (p_1 + p_2 + p_3)^2 < (4\mu)^2$ of the physical region, the cone $C_{i,\ell}$ is replaced in the right-hand side of (2.122) by a smaller cone $C'_{i,\ell}$ if P lies outside the Landau surfaces $(p_j + p_k + p_\ell)^2 = \mu^2$ (e.g. $(p_1 + p_2 + p_4)^2 = (p_5 + p_6 + p_3)^2 = \mu^2$ if $i = 3$, $\ell = 4$), where

$$C'_{3,4} = C_{3,4} \cap \{x; \ x_4 - x_3 \in \bar{V}_+\} \quad (2.124)$$

as shown in Figure 2.9b. (Other cones $C'_{i,\ell}$ are defined similarly.) In fact, if $x = (x_1, \ldots, x_6)$ is, for example, a point in $C_{3,4}$ but not in $C'_{3,4}$, the set $x(I)$, where $I = (3, 5, 6)$, contains no point in its closed causal future, but $P_I = P_5 + P_6 + P_3$ cannot lie in $\bar{V}_+(2\mu)$ since $s < (4\mu)^2$ (so that $(P_5 + P_6)^2 < (3\mu)^2$ and $P_I^2 < (2\mu)^2$), and it cannot belong to $H_+(\mu)$ if $(P_5 + P_6 + P_3)^2 \neq \mu^2$. Hence, x cannot lie in $\mathrm{ES}_P(T)$.

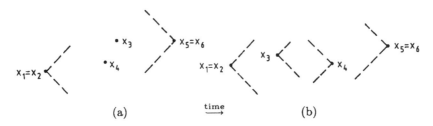

(a) $\overset{\text{time}}{\longrightarrow}$ (b)

Figure 2.9 The cones $C_{3,4}$(Fig. a) and $C'_{3,4}$ (Fig. b).

The aim of this subsection is to improve these results in the region $(3\mu)^2 < s < (4\mu)^2$ (and away from M_0 points). In contrast to the situation of Sect. 5.3, $+\alpha$-Landau singularities of graphs with one internal line and of triangle graphs (see Sect. 4.3 of Chapter I) are encountered in that region, so that either $T_{(6)}$ or its mass-shell restriction $T_{3,3}$ are no longer expected to be analytic, or to be boundary values of analytic functions from the directions $\mathrm{Im}\, s > 0$, everywhere in the region considered. Such results are expected instead from the remainder of Eq. (2.3).

The analysis will proceed as follows. In the direct line of Remark 4 of Sect. 5.1, it is easy (and will be important for our purposes) to first extract one-particle pole singularities at $p_I^2 = \mu^2$ for sets I of the form (ℓ, m, i).

More precisely, let $\begin{smallmatrix}1\\2\\3\end{smallmatrix}\!\!\overline{\underset{\oplus}{\oplus}}\!\!\begin{smallmatrix}4\\5\\6\end{smallmatrix}$ be defined as:

$$= \; \boxed{\oplus}\; (p_1, p_2, p_4, k) \times \boxed{\oplus}\; (-k, p_3, p_5, p_6)\, F_+(k), \qquad (2.125)$$

where $k = -(p_1 + p_2 + p_4)$, bubbles refer to 4-point functions T_4, and F_+ is a Feynman-type propagator. For later purposes (definition of triangle terms), it is convenient to introduce a regularization factor in the definition of F_+, that is, to put

$$F_+(k) = \left[k^2 - \mu^2 + i\varepsilon\right]^{-1} \alpha(k), \qquad (2.126)$$

where $\alpha(k)$ is analytic in a sufficiently large region, will therefore not modify the local properties of F_+ (a pole at $k^2 = \mu^2$ with residue one) and has a sufficient decrease at large k. The remainder of Eq. (2.3) will depend on the choice of α but its essential support or analyticity properties will not depend on it (in the region considered).

Terms are defined similarly. Let $= T^{(1)}(p_1, \ldots, p_6)$ $\times \delta(p_1 + \cdots + p_6)$ be defined as

$$\text{} = \text{} - \sum_{\substack{i=1,2,3 \\ \ell=4,5,6}} \text{}. \qquad (2.127)$$

We then state (in the region $s < (4\mu)^2$):

Lemma 2.4. The distribution $T^{(1)}$ is, in a real neighborhood of the physical region, the boundary value of an analytic function from the directions satisfying the conditions

$$\text{Im}\,(p_i + p_j) \in V_- , \quad \text{Im}\,(p_\ell + p_m) \in V_+. \qquad (2.128)$$

for all possible choices of i, j, $i \neq j$, among 1, 2, 3 and ℓ, m, $\ell \neq m$ among (4, 5, 6). Moreover,

$$\text{ES}_P(T^{(1)}) \subset \bigcup_{\substack{i=1,2,3 \\ \ell=4,5,6}} C'_{i,\ell} \qquad (2.129)$$

at any physical region point P of the region considered.

Remarks. The improvement in the properties of $T^{(1)}$ with respect to those of T is the removal of the conditions $\text{Im}\,(p_\ell + p_m + p_i) \in V_+$ in (2.128), which corresponds to an enlargement of the analyticity domain, and the introduction of the condition $x_4 - x_3 \in \bar{V}_+$ in (2.129) in $C'_{3,4}$ or similar ones in other cones $C'_{i,\ell}$. Such results cannot be expected for T itself, precisely because they do not hold for the terms along their $+\alpha-$Landau singularities $p_I^2 = \mu^2$ $(I = (\ell, m, i))$. For example, configurations $(x_1, \ldots, x_6)$ of the form $(x_1 = x_2 = x_4,\ x_3 = x_5 = x_6,\ x_3 - x_4 = \lambda(P_5 + P_6 + P_3),\ \lambda > 0$, do belong to $\text{ES}_P\left(\text{}\right)$ if $(P_5 + P_6 + P_3)^2 = \mu^2$.

The function $T^{(1)}$ is called "1-particle irreducible" with respect to the nine crossed channels (I_1, I_2), $I_1 = (\ell, m, i)$, $I_2 = (j, k, n)$. Further irreducible kernels will be introduced in Sect. 2.6 but they will be defined in general via integral equations, whereas 1-particle irreducible functions are defined via a mere subtraction of well-defined terms.

Proof of Lemma 2.4. The first part of Lemma 2.4 is directly linked with previous remarks on 1-particle singularities (Remark 4 in Sect. 5.1). In this particular case, the discontinuity of ⊸(+)⊸ across (I, J) where $I = (5, 6, 3)$, $J = (1, 2, 4)$, is equal to ⊸⊕⊸ $(p_1, p_2, p_4, k) \times$ ⊸⊕⊸ $(-k, p_3, p_5, p_6) \times \delta(k^2 - \mu^2)$ in the neighborhood of $p_I^2 = \mu^2$. This discontinuity is the difference between boundary values obtained from the directions (2.121) and from the directions satisfying the same conditions except that $\text{Im}(p_5 + p_6 + p_3) \in V_+$ is replaced by $\text{Im}(p_5 + p_6 + p_3) = -\text{Im}(p_1 + p_2 + p_4) \in V_-$. From direct inspection, the term ⊸⊕⊕⊸ has the same discontinuity, so that the discontinuity of ⊸(+)⊸ − ⊸⊕⊕⊸ vanishes locally, that is, the boundary values obtained with the conditions $\text{Im}(p_5 + p_6 + p_3)$ in V_+ and in V_- respectively, coincide. The edge-of-the-wedge theorem then allows one to remove these conditions. The first part of Lemma 2.4 follows from a refinement of this analysis.

The second part of Lemma 2.4 is established by methods similar to those in Sect. 3.1. The improvement with respect to previous results on T is due to improvements of results derived from the spectral condition (replacement of conditions of the form $p_I \notin H_+(\mu) \cup \bar{V}_+(2\mu)$ by $p_I \notin \bar{V}_+(2\mu)$). QED

Besides the terms ⊸⊕⊕⊸ , it will be useful in later analysis to define "triangle" terms ⊸⊕⊕⊕⊸ , $(i, j, k) = (1, 2, 3)$, $(m, n, \ell) = 4, 5, 6$. They are defined as Feynman-type convolution integrals with propagators F_+ on each internal line (as defined in (2.126)). For example,

$$\text{⊸⊕⊕⊕⊸}(p_1, \ldots, p_6) = \int T_{(4)}(p_1, p_2, k_1, k_2) T_{(4)}(-k_2, p_3, k_3, p_4)$$
$$\times T_{(4)}(-k_1, -k_3, p_5, p_6)\, \delta(p_1 + p_2 + k_1 + k_2)$$
$$\delta(-k_2 + p_3 + k_3 + p_4)\delta(-k_1 - k_2 + p_5 + p_6)$$
$$\prod_{i=1}^{3} F_+(k_i)\mathrm{d}k_i. \tag{2.130}$$

It can be checked from results of essential support theory on products and integrals (and some further considerations), that the integral (2.130) is well defined and that if $P = (P_1, \ldots, P_6)$ is a physical region point such that $P_1 \neq P_2$, $P_5 \neq P_6$, $(3\mu)^2 < (P_1 + P_2 + P_3)^2 < (4\mu)^2$,

$$\mathrm{ES}_P(\,_3\!\!\!\!\xrightarrow{}{}^{\,4}) \subset C''_{3,4}, \qquad\qquad (2.131)$$

$$\mathrm{ES}_P(\,_3\!\!\!\!\xrightarrow{}{}_{\,4}) \subset C'''_{3,4}, \qquad\qquad (2.132)$$

where $C''_{3,4}$ and $C'''_{3,4}$ are the sets of points $(x_1, \ldots, x_6)$ shown in Figure 2.10 allowing corresponding classical multiple scattering diagrams (with on-shell intermediate particles). Similar results hold for other pole or triangle terms.

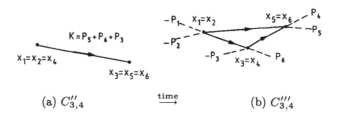

$$(a)\ C''_{3,4} \qquad\qquad \xrightarrow{\text{time}} \qquad\qquad (b)\ C'''_{3,4}$$

Figure 2.10 Points $(x_1, \ldots, x_6)$ in the sets $C''_{3,4}$ and $C'''_{3,4}$.

Correspondingly, the terms $\xrightarrow{}$ and $\xrightarrow{}$ (after factorization of the δ-function) are analytic in the real region considered outside the $+\alpha$-Landau surfaces of respective graphs with one internal line and triangle graphs obtained by replacing bubbles with vertices. The result is true off-shell or on-shell. These terms satisfy also plus $i\varepsilon$ rules along these surfaces and it can be checked that they admit discontinuities of the form $\xrightarrow{}$ or $\xrightarrow{}$ around the latter.

These results are due in particular to the analyticity of individual 4-point functions $T_{(4)}$, involved at each bubble, at points with squared center-of-mass energy between $4\mu^2$ and $9\mu^2$.

Further global results, in particular on triangle terms, follow from the methods of J. Bros briefly recalled in Sect. 6 on general "$\mathcal{G}$-convolution" integrals. Triangle terms then appear to be suitable boundary values of functions satisfying the same type of analytic and algebraic properties as in Sect. 4.1, including analyticity in the primitive domain. The more refined local results mentioned above are also derived in this approach from a more detailed analysis for the terms under consideration.

We then state the following result (which is close to that stated in Sect. 1: see Eq. (2.3) and below):

Theorem 2.20.

$$
\boxed{-\!\!\bigcirc\!\!-} = \sum_{\substack{i=1,2,3\\l=4,5,6}} {}_i\overline{}^{\,l} + \sum_{\substack{i=1,2,3\\l=4,5,6}} {}_i\overline{}_{\,l} \; + R \qquad (2.133)
$$

with

$$
\mathrm{ES}_P(R) \subset \left\{ x; x_1 = x_2 = x_3, x_4 = x_5 = x_6, x_4 - x_1 \in \bar{V}^+ \right\} \qquad (2.134)
$$

at any physical region point P such that $(3\mu)^2 < s < (4\mu)^2$ and $(P_i + P_j - P_k)^2 \neq \mu^2$ where (i,j,k) is any permutation of $(1, 2, 3)$ or of $(4, 5, 6)$.

Remark. By Lorentz invariance, the condition $x_4 - x_1 \in \bar{V}^+$ in (2.134) can, as previously, be replaced by $x_4 - x_1 = \lambda\,(P_4 + P_5 + P_6)$, $\lambda \geq 0$, and the remainder R (after factorization of its δ−function) is again the boundary value (off-shell or on-shell) of an analytic function from the directions Im $s > 0$. Theorem 2.20 and these improvements follow as in Sect. 5.3 from the exploitation of 2-particle asymptotic completeness equations. The use of 3-particle asymptotic completeness equations (and further regularity assumptions) would be needed to show actual analyticity at points under consideration. Submanifolds $(P_i + P_j - P_k)^2 = \mu^2$ are again parasitic (apart from M_0 points).

Proof of Theorem 2.20 (outline). The analysis makes now recourse to the equations:

$$
-\!\!\bigcirc\!\!- \; - \; -\!\!\bigcirc\!\!-_{_4} \; = \; \overline{-\!\!\bigcirc\!\!-\bigcirc\!\!-}_{_4}, \qquad (P_5 \neq P_6)\,, \qquad (2.135a)
$$

$$
-\!\!\bigcirc\!\!- \; - \; {}_3\overline{-\!\!\bigcirc\!\!-} \; = \; {}_3\overline{-\!\!\bigcirc\!\!-\bigcirc\!\!-} \qquad (P_1 \neq P_2) \qquad (2.135b)
$$

and similar equations with 4 replaced by 5 or 6, or 3 replaced by 1 or 2. Information on the essential support of the bubbles $-\!\!\bigcirc\!\!-$ or $-\!\!\bigcirc\!\!-$ is again obtained by the methods of the linear program. Eqs. (2.135) yield, on the other hand, related relations on $-\!\!\bigcirc\!\!-$ of the form:

$$
-\!\!\bigcirc\!\!- \; - \; -\!\!\bigcirc\!\!-_{_4} \; = \; \overline{-\!\!\bigcirc\!\!-\bigcirc\!\!-}_{_4} \; - \; \sum_k {}_k\overline{-\!\!\bigcirc\!\!-\bigcirc\!\!-}^{\,4}, \qquad (2.136)
$$

where ⟨⟩ is the relevant discontinuity of ⟨⟩ (pole terms in (2.133) where $\ell \neq 4$ have no discontinuity in the channel considered). Eq. (2.136) yields in turn:

$$
\text{[diagram]} \cdot \text{[diagram]} = \text{[diagram]} - \sum_k \text{[diagram]}, \qquad (2.137)
$$

where the usual expression of the discontinuity of ⟨⟩ has been used to cancel some terms.

A procedure analogous to that used in Sect. 5.3 is now applied to R (defined as ⟨⟩ minus the triangle terms) and provides announced results. The presence of triangle terms does not block the analysis in view of results on the essential support of differences such as ⟨⟩ − ⟨⟩ . (See details in [EGI]).

§6. The nonlinear program based on irreducible kernels

6.1. Some general preliminary theorems

This section presents methods and results of the nonlinear program relying on the introduction of Bethe-Salpeter type irreducible kernels. These kernels are different from the irreducible kernel U introduced in Sect. 5.2 for the analysis of the 2-particle threshold (or of the $m-$particle threshold in the simplified $m \to m$ theory) and seem potentially best adapted for many-particle structure analysis in more general situations. The link between the kernel U and the 2-particle irreducible $2 \to 2$ Bethe-Salpeter type kernel G or G_α to be introduced in Sect. 6.2 will be explained there.

Bethe-Salpeter irreducible kernels are introduced in perturbation theory (see Chapter III) as (formal) sums of Feynman integrals associated to graphs with given irreducibility properties in a graphical sense. Relations between N-point functions and irreducible kernels then follow in simple cases from graphical analysis, in the form of integral equations. These equations, or regularized versions of the latter (intended to cover more general theories), are used in the axiomatic framework to *define* irreducible kernels from the N-point functions. Specific analyticity properties of these kernels in complex energy-momentum space, expressing irreducibility, are then expected to follow from asymptotic completeness and regularity conditions and to give in turn information on the structure of N-point functions. Cases actually studied so far in this approach are treated in Sect. 6.2 (4-point function in the low-energy region and related results on the 2-particle structure of N-point functions) and in Sect. 6.3 (6-point function in the low-energy 3-particle region). The analysis of Sect. 6.3 is restricted to even theories; otherwise, new problems occur. The analysis of Sect. 6.2 also applies to a large extent to noneven theories, but simplifications still occur for even theories which we shall mainly consider, for simplicity.

Results of the axiomatic approach are based on theorems established by J. Bros and collaborators which we outline below. For detailed statements and proofs, see [B3,B4] and references therein. Diagrams denoting $\mathcal{G}$-convolutions, and related ones, do not include a global emc δ-function. (See Sect. 3.2 of the general Introduction.)

(i) The general preservation of primitive analytic and algebraic properties by "$\mathcal{G}$-convolution." More precisely, $\mathcal{G}$-convolution products are convolution integrals associated to connected graphs $\mathcal{G}$, with analytic 2-point functions on internal lines and analytic $N(v)$-point connected, amputated functions at each vertex v of $\mathcal{G}$ ($N(v)$ is the number of internal and external lines involved at v). These functions are either the actual N-point functions introduced in Sect. 4 or, for some purposes, other functions satisfying the same primitive analytic and algebraic properties. (Functions satisfying these properties are called "general N-point functions.") These integrals are initially defined in Euclidean energy-momentum space due to the remarkable property of the Euclidean space to be contained in the primitive axiomatic domain of all N-point functions. Integration is made over internal energy-momenta in Euclidean space, more precisely over a set of independent loop variables after the energy-momentum conservations at each vertex have been used to remove some of the integrations. The integrand may include regularization factors $\alpha(k)$ on internal lines, here assumed to be analytic in a sufficiently large domain around Euclidean space, and with sufficient decrease at infinity in *Euclidean* directions, so as to ensure the convergence of the integral. For later purposes, this factor is also assumed to be equal to one on the mass-shell ($k^2 = \mu^2$) (away from Euclidean space).

The $\mathcal{G}$-convolution integrals are then defined away from Euclidean space through *local* distortion of the integration contours in order to remain within the analyticity domains of the functions involved in the integrand, and in particular to avoid the poles of the 2-point functions at the physical mass ($k^2 = \mu^2$). The distorted contours are still Euclidean away from bounded regions but may be Minkowskian locally.

Various boundary values in real Minkowski space will be considered as was already the case for the original N-point functions.

(ii) The derivation of discontinuity formulae (relative to absorptive parts) for these integrals. Results obtained include in particular the following, which applies to $\mathcal{G}$-convolutions of the form ⚬A⚬B⚬ with a corresponding well-defined convolution channel. It is derived under regularity conditions on functions A, B satisfying the primitive properties, for $p_{I_1} = -p_{I_2}$ real in the 2-particle region:

$$4\mu^2 \leq p_{I_1}^2 < 9\mu^2$$

and variables z, z' in tubes associated to cells $\mathcal{S}_1, \mathcal{S}_2$:

$$\Delta^{I_1}\left[I_1\left\{ \text{⚬A⚬B⚬} \right\}I_2\right] = \overset{A\quad B}{\text{⚬①○⚬}} + \overset{A\quad B}{\text{⚬○①⚬}} + \text{⚬A⚬B⚬} ,$$

$$(2.138)$$

where bubbles ⚬①⚬ or ⚬①⚬ refer to corresponding absorptive parts of A and B, and functions A_-, B_+ are defined as in (2.97).

(iii) Fredholm theory with complex parameters and in complex space, namely, with distorted contours of integration.

Besides the cases that will be treated in Sects. 6.2 and 6.3, it is believed that one might define, in successive steps, more general irreducible kernels well adapted to the analysis of the multiparticle structure of N-point functions in higher and higher energy regions. However, various difficulties occur in this purely axiomatic approach (due in particular to subchannel and crossed channel interactions) and the adequate systems of integral equations to start from are not even known so far for the general case. A better understanding [I5] can be obtained from the "Neumann series viewpoint" which leads, in a different "semi-axiomatic" approach, to the consideration of suitable (*a priori* formal) expansions of N-point functions in terms of (possibly regularized) $\mathcal{G}$-convolution or similar integrals with suitable 2-point functions on internal lines and irreducible kernels at each vertex. Relevant expansions in cases studied in Sects. 6.2 and 6.3 will be mentioned there. Those to be considered more generally will be presented in Chapter IV, where they are in fact established (and shown to have convergence properties) in the constructive approach. They involve a general class of irreducible kernels with various irreducibility properties in various channels.

One might try to derive, from these expansions, systems of integral equations from which irreducible kernels might indeed be defined in successive steps in the axiomatic framework and from which suitable analyticity properties might in turn be established. Besides technical difficulties, we note that, even if such a program can be carried out, more and more regularity assumptions will be needed and problems will also arise from the possible occurrence of poles (via Fredholm theory) in irreducible kernels.

The analysis will be reconsidered in Chapter IV from a constructive or semi-axiomatic viewpoint.

6.2. The 2-particle structure of 4-point and N-point functions

The analysis starts with the 4-point function and the Bethe-Salpeter type equation (2.2). The variables k, z, z' of (2.77) through (2.79) are here:

$$k = -(p_1 + p_2) = p_3 + p_4$$
$$z = (p_1 - p_2)/2$$
$$z' = (p_3 - p_4)/2.$$

If F and G_α are expressed in terms of these variables, Eq. (2.2) reads:

$$F(k; z, z') = G_{(\alpha)}(k; z, z') + \int F(k; z, z'')G_{(\alpha)}(k; z'', z')$$
$$\times \omega_{(\alpha)}(k, z'')dz'', \tag{2.139}$$

where F is the connected, amputated 4-point function and $\omega_{(\alpha)}(k, z'')$ is a (regularized) product of 2-point functions. Namely,

$$\omega(k, z'') = H_2(k_1)H_2(k_2), \quad k_1 = k + \frac{z''}{2}, \quad k_2 = k - \frac{z''}{2}$$

and a product $\alpha(k_1)\alpha(k_2)$ of regularization factors is, moreover, included in ω_α.

Leaving the index α implicit, Eq. (2.139) written in the form

$$F = G + F \circ G \tag{2.140}$$

is first replaced by

$$F = G_\lambda + \lambda\, F \circ G_\lambda, \tag{2.141}$$

where λ is here a mathematical Fredholm parameter. Under regularity conditions on F at infinity and with a suitable (related) choice of α, this equation allows one to define G_λ in terms of F in Euclidean space. This is first achieved at small λ, in which case G_λ is the convergent sum of the Neumann series

$$F - \lambda \ F \circ F + \lambda^2 F \circ F \circ F - \cdots .$$

By analytic continuation in λ, G_λ is defined more generally as a ratio $N(k; z, z', \lambda)/D(k; \lambda)$, in view of Fredholm-type analysis; k plays the role of a parameter in this analysis. G is well defined at the physical value $\lambda = 1$, up to possible poles in k (the zeroes of $D(k; 1)$). This result is *a priori* obtained only if $D(k, 1)$ does not vanish identically, but this pathology can be excluded from suitable choices of the regularization factors. From invariance properties, the dependence on k can be replaced by a dependence on k^2.

Moreover, the analyticity properties of F (and H_2) away from Euclidean space allow one also to define G or G_α as a meromorphic function (with possible poles in k^2) in the same primitive domain as F. The proof makes recourse to the techniques mentioned in Sect. 6.1. As a matter of fact, G or G_α is still of the form $N(k, z, z')/D(k)$ where N and D depend analytically on k in the same cut domain as F.

On the other hand, the operation $\circ$ or $\circ_\alpha$, namely, convolution with the measure $\omega_{(\alpha)}(k; z'')$ over the distorted contour $\Gamma(k)$, satisfies, at s real, $4\mu^2 < s < 9\mu^2$, the relation:

$$(\circ_\alpha)_+ - (\circ_\alpha)_- = *_+, \tag{2.142}$$

where $(\circ_\alpha)_+$ and $(\circ_\alpha)_-$ denote integrations over distorted contours $\Gamma_+(k)$ and $\Gamma_-(k)$ obtained as k approaches the cut from above and below respectively, and $*_+$ denotes, as before, on-mass-shell convolution; the right-hand side of (2.142), i.e. $*_+$, is independent of the regularization factor α satisfying properties that have been mentioned in Sect. 6.1 (i) (in particular $\alpha = 1$ at $k^2 = \mu^2$). The relation (2.142) has been established by J. Bros [B1,B4] via a geometrical analysis relying on mathematical works of Leray and F. Pham [Ph2]. It also follows from an improved result of [BI3] presented in Sect. 4 of Chapter IV with some indications on the proof. We only mention here that it can be viewed as a (nontrivial) extension of the well-known formula

$$\frac{1}{2i\pi} \left[\frac{1}{x + i\varepsilon} - \frac{1}{x - i\varepsilon} \right] = \delta(x).$$

If we write the latter in the form $(*) = (+) - (-)$ and if the two 2-point functions on internal lines in $\circ$ are replaced for simplicity by mere Minkowskian propagators, which are more precisely of the form $(k_\ell^2 - \mu^2 + i\varepsilon)^{-1}$ in $\circ_+$ or $(k_\ell^2 - \mu^2 - i\varepsilon)^{-1}$ in $\circ_-$, Eq. (2.142) can be viewed as a consequence of the relation

$$(**) = (++) + (--) - (+-) - (-+)$$

obtained from the formula above, applied to the product of two δ-functions (involving k_1 and k_2 respectively). In fact, the last two terms can be shown to yield vanishing contributions when applied to functions with suitable analyticity properties. The fact that they do not generate singularities is due to the conflicting signs of the two propagators, and the actual vanishing follows from a more refined geometrical analysis based essentially on a contour distortion argument.

The limit relations (with α implicit)

$$F_+ = G_+ + F_+ \circ_+ G_+, \tag{2.143}$$

$$F_- = G_- + F_- \circ_- G_-, \tag{2.144}$$

which follow from (2.139) (and regularity conditions), as also $F_+ \circ_+ G_+ = G_+ \circ_+ F_+$ or $F_- \circ_- G_- = G_- \circ_- F_-$, which follow from Fredholm theory, then entail in view of the asymptotic completeness relation ($F_+ - F_- = F_+ *_+ F_-$) that, away from possible poles,

$$G_+ = G_-. \tag{2.145}$$

This result of [B1], which can also be reobtained from a direct application of Lemma 1.1 in Sect. 6.2. of Chapter I (by writing, for example, $G_+ = F_+ - G_+ \circ_+ F_+$, $F_+ = F_- + F_+ *_+ F_-$, $F_- = G_- + F_- \circ_- G_-$ and using $\circ_+ - \circ_- = *_+$), shows that G, in contrast to F, is analytic or meromorphic in a domain without cut in the region $s < 9\mu^2$.

If, as can be expected under some conditions for an even theory (see Chapter IV), G has no pole in a sufficiently large region around Euclidean space including the 2-particle threshold, Eq. (2.139) used in the converse direction allows one to show that F can in turn be analytically continued as a meromorphic function (with possible poles in s) in an *a priori* multisheeted domain around $s = 4\mu^2$. The singularity of F at $s = 4\mu^2$ is here generated by the pinching of the contour $\Gamma(k)$ between the poles of the two 2-point functions as $s \to 4\mu^2$.

If a subscript r now refers (as in Chapter I) to the rth determination obtained by turning around $s = 4\mu^2$, the relations $*_1 = -*_0$ (hence $*_2 = *_0$) if d is even, or $*_1 = *_0$ (hence $*_r = *_0 + r*_0$) if d is odd (see Chapter I), entail that:

$$\circ_0 - \circ_2 = \circ_0 \qquad \text{(d even)}, \tag{2.146}$$

$$\circ_0 - \circ_r = r *_+ \qquad \text{(d odd)}. \qquad (2.147)$$

An analysis completely analogous to that presented in Chapter I allows one to conclude that $F_2 = F_0$ (hence F has a square root type 2-sheeted singularity at $s = 4\mu^2$) if d is even, while F is infinitely sheeted if d is odd. Results can be extended similarly to a simplified theory of the m-particle threshold.

Irreducible kernels G and U. The irreducible kernels U (introduced in Sect. 5.2) and G_α are linked by the relation

$$U = G_\alpha + U\nabla_\alpha G_\alpha = G_\alpha + G_\alpha \nabla_\alpha U, \qquad (2.148)$$

where

$$\nabla_\alpha = \circ_\alpha - \otimes. \qquad (2.149)$$

Proof. Eq. (2.148) is again easily derived from Lemma 1.1 in Sect. 6.2. of Chapter I: write, for example, $U = F - U \otimes F$, $F = G + F \circ_\alpha G$. QED

In contrast to the operations $\circ_\alpha$ and $\otimes$, the operation ∇_α is "uniform" around $s = 4\mu^2$ (since $\circ$ and $\otimes$ have the same discontinuity $*_+$). This result is consistent with the fact that both G_α and U enjoy the property of 2-particle irreducibility (in the analytic sense: uniformity around $s = 4\mu^2$).

A more detailed analysis of the operation $\nabla_{(\alpha)}$ will be given in Chapter IV.

Series expansions. The Neumann series of F in terms of G reads:

$$F = G + \sum_{n \geq 1} G^{O(n+1)}, \qquad (2.150)$$

where

$$G^{O(n)} = G \circ \ldots \circ G \quad \text{(n factors } G),$$

or in diagrammatical notation,

$$F = G + \boxed{G\ G} + \boxed{G\ G\ G} + \cdots \qquad (2.151)$$

The analogous series obtained by introducing a Fredholm parameter is convergent in a suitable domain for small values of this parameter and, under some conditions on G (satisfied in Chapter IV at small coupling), the expansion (2.150) has itself convergence properties (in contrast to perturbative series): see Chapter IV where more details on the links between the expansions of F in terms of G and of U [Eqs. (2.150) and (2.104)] will also be given.

2-particle structure of N-point functions. We start with the following preliminary result:

Lemma 2.5. Let F be a (connected, amputated) analytic N-point function, and let $F_{(G,g)}$ be defined as:

$$F_{(G,g)} = F - \text{⟩·⟨F⟩·⟨G⟩·⟨}\}_g, \qquad (2.152)$$

where g is a given set of two lines. Then the absorptive part $\Delta^g F_{(G,g)}$ vanishes in the region $s_g \leq 9\mu^2$ (or $16\mu^2$ in an even theory).

Proof. The absorptive part of F is equal to $\text{·⟨F⟩⟨F⟩·}\}_g$ in the region considered (see Sect. 5.1). (The bubble on the left implicitly includes a minus sign with respect to the relevant channel.) Formula (2.138) entails, on the other hand, in view of the analyticity of G:

$$\Delta^g\left(\text{⟩⟨F⟩⟨G⟩}\}_g \right) = \text{⟩⟨\overset{F}{\cdot}⟩⟨G⟩} + \text{⟩⟨F⟩⟨G⟩}$$

$$= \text{⟩⟨F⟩⟨F⟩⟨G⟩} + \text{⟩⟨F⟩⟨G⟩}. \qquad (2.153)$$

The absorptive part of the function $F_{(G,g)}$ introduced in (2.152) thus vanishes since $\text{⟨F⟩ - ⟨G⟩} = \text{⟨F⟩⟨G⟩}$. QED

The 2-particle structure of F in the variable s_g is obtained in turn from the equality

$$F = F_{G,g} + \text{·⟨F_{G,g}⟩⟨F⟩·}\}_g. \qquad (2.154)$$

Eq. (2.154) follows again from the relation $\text{⟨F⟩} = \text{⟨G⟩} + \text{⟨F⟩⟨G⟩}$. The latter allows one to replace the term ·⟨F⟩⟨G⟩ in (2.152) by the second term in the right-hand side of (2.154).

The 2-particle cut in s_g is present in the second term of the right-hand side of (2.154) in the low-energy region $s_g \geq 4\mu^2$, but the analyticity properties of ⟨F⟩ established previously, such as second-sheet meromorphy, yield similar properties for this term, and therefore for F, concerning analytic continuation across this cut.

6.3. The 6-point function in the 3-particle region (even theories)

We consider below a given channel, for example, (1, 2, 3); (4, 5, 6), and restrict our attention to an even theory. Following [B3,B4], the analysis will apply to the (connected, amputated) 1-particle irreducible function in the channel considered:

$$
\begin{array}{c}
\text{(diagram)}
\end{array}
\tag{2.155}
$$

The subtraction of the last term removes the pole at $k^2 = \mu^2$ ($k = (p_1 + p_2 + p_3)$). The following sets are still present in the boundary of the primitive domain of $F_{(1)}$ and will play a role in the region to be studied:

- the main channel cut $\underline{\sigma} : k^2 = (3\mu^2) + \rho,\ \rho \geq 0$
- the subchannel cuts $\underline{\sigma}_{i,j} : (p_i + p_j)^2 = 4\mu^2 + \rho,\ \rho \geq 0,\ i, j \in (1, 2, 3)$
  and $\underline{\sigma}'_{\ell,m} : (p_\ell + p_m)^2 = 4\mu^2 + \rho,\ \rho \geq 0,\ \ell, m \in (4, 5, 6)$
- the crossed channel poles $\omega_{i,j,\ell} = (p_i + p_j + p_\ell)^2 = \mu^2,\ i, j \in (1, 2, 3),\ \ell \in (4, 5, 6)\ (i \neq j)$.

While crossed channel poles can be removed, as in Sect. 5.4, a deeper analysis will be made possible through the definition of kernels with more irreducibility properties. The definition of such kernels is made in two steps.

The undressed 6-point function L. The first step introduces an "undressed" function L for which one shows (see explanations below) the absence of the subchannel cuts and of the crossed channel poles; only the main channel cut remains (in the region considered). L is defined by the relation (which is not here an integral equation):

$$
\begin{array}{c}
\text{(diagram)}
\end{array}
\tag{2.156}
$$

where

$$
\Lambda = 1 - \sum {}_i \overline{\underline{}}{}_i .
\tag{2.157}
$$

In (2.157), 1 denotes the identity for the 3-particle convolution, so that

$$
\begin{array}{c}
\text{(diagram)}
\end{array}
$$

The fact that the subchannel cuts are absent in L (in the low-energy region) is linked with the result on 2-particle structure indicated at the end of Sect. 6.2. As a matter of fact, it can be shown by an extension of the argument given there that the action of Λ, on the right, *or alternatively*, on the left of ─(1)─ , removes the three subchannel cuts $\underline{\sigma}_{\ell,m}$ or $\underline{\sigma}_{i,j}$ respectively. Although this result does *not* extend to the double action of Λ on the right *and* on the left (subchannel cuts are not removed from ─(Λ)─(1)─(Λ)─), it can be shown by a more refined analysis that subchannel cuts, as also crossed channel poles, are indeed removed from L. We do not give here the proof. The origin of the result will be clarified later. We only make the following remark in the sense of formal series, easily checked on a formal level by using systematically the equality ─(F)─ = ─(G)─ + ─(G)─(F)─ :

$$
\sum \overline{\underline{(G)}_{(G)}} - \sum \overline{\underline{(G)}\;(G)}_{(F)}
$$
$$
= \Lambda\left[\sum_i \overline{\underline{(F)}_{(F)}} + \sum_i \overline{\underline{(F)}\;(F)}_{(F)} + \sum_i \overline{\underline{(F)}\;(F)\;(F)}_{} + \cdots \right]\Lambda,
$$

$$(2.158)$$

where the bracket in the right-hand side includes all terms associated with the r-loop "truss-bridge" diagrams ⟨∨∧∨∧⟩ ($\{\cdots\}^{\rvert}$, $r \geq 0$. As a consequence, (2.156) can be written in the form:

$$
L = \Lambda\left[\ \overline{\underline{(F)}}\ -\ \overline{(F)\!-\!(F)}\ -\ \sum \overline{\underline{(F)}_{(F)}} \right.
$$
$$
\left. -\ \sum \overline{\underline{(F)}\;(F)}_{(F)}\ -\cdots \right]\Lambda.
$$

$$(2.159)$$

The removal of poles and subchannel cuts necessitates the subtraction in the bracket not only of pole terms, but also of triangle terms and all other truss-bridge terms. See in this connection, the comment that concludes this section.

The later reconstruction of ─(1)─ from L will be obtained from the relation

$$
─(1)─ = \Lambda^{-1}\ ─(L)─\ \Lambda^{-1} + \left(\Lambda^{-1}\right)_{\text{conn.}}, \tag{2.160}
$$

where Λ^{-1} is the inverse of Λ with respect to the 3-particle convolution $(\Lambda^{-1}\Lambda = \Lambda\Lambda^{-1} = \mathbb{1})$ and

$$\left(\Lambda^{-1}\right)_{\text{conn}} = \Lambda^{-1} - \mathbb{1} - \sum \;\overline{\boxed{F}}\; \qquad (2.161)$$

is equal to the bracket [ ] in the right-hand side of (2.158) in the sense of formal series. In fact, this bracket is the sum of connected terms in the expansion of Λ^{-1}, namely (as easily checked):

$$\Lambda^{-1} = \mathbb{1} + \sum \;\overline{\boxed{F}}\; + \sum \;\overline{\boxed{F}\boxed{F}}\; + \sum \;\overline{\boxed{F}\boxed{F}\boxed{F}}\; + \cdots. \qquad (2.162)$$

We note that (2.156) can thus alternatively be written in the form:

$$L = \Lambda \left[\;\overline{\boxed{1}}\; - \left(\Lambda^{-1}\right)_{\text{conn}} \right] \Lambda. \qquad (2.163)$$

Independent of formal series, Λ^{-1} can be defined in the more complete analysis of [B3] (modulo some regularity assumptions) by recourse to the auxiliary Fredholm-type equation $V_{\text{in}} = T_{\text{in}} + T_{\text{in}} V_{\text{in}}$, or to the alternative one $V_{\text{out}} = T_{\text{out}} + V_{\text{out}} T_{\text{out}}$, where

$$T_{\text{in}} = \sum_i \;\overline{\boxed{G}}^{1}_{\boxed{F}}\; , \qquad (2.164)$$

$$T_{\text{out}} = \sum_i \;\overline{\boxed{F}}^{1}_{\boxed{G}}\; . \qquad (2.165)$$

All terms in $T_{\text{in}}, T_{\text{out}}$ are connected and $T_{\text{in}}, T_{\text{out}}$ are therefore analytic 6-point functions in the sense introduced in Sect. 6.1. The above equations then allow one to define $V_{\text{in}}, V_{\text{out}}$. Λ^{-1} is in turn defined, for example, by the formula

$$\Lambda^{-1} = \left(\mathbb{1} + V_{\text{out}}\right)\left(\mathbb{1} + \sum \;\overline{\boxed{F}}\;\right) \qquad (2.166)$$

or the alternative one involving V_{in}. Eq. (2.166) is easily derived from the relations $\left(1 + \sum \;\overline{\boxed{F}}\;\right)\Lambda = \mathbb{1} - T_{\text{out}}$ and $(\mathbb{1} - T_{\text{out}})^{-1} = \mathbb{1} + V_{\text{out}}$.

The 3-particle irreducible kernel G. In the second step, the "totally 3-particle irreducible" kernel $\;\overline{\boxed{G}}\;$ (with respect to the $3 \rightarrow 3$ channel considered) is defined from L by solving a further Fredholm-type integral equation such as

$$L_{\text{in}} = G_{\text{in}} + L_{\text{in}} G_{\text{in}}, \qquad (2.167)$$

where $L_{\rm in}G_{\rm in}$ denotes a 3-particle convolution $L_{\rm in} \circ G_{\rm in} =$,
and $L_{\rm in} = \Lambda^{-1}L$. Eq. (2.167) allows one to define $G_{\rm in}$, and G is defined
in turn by putting $G = \Lambda\, G_{\rm in}$. From Eq. (2.167), one also has

$$ \text{} \tag{2.168} $$

As a matter of fact, the main channel cut is shown, from asymptotic completeness and regularity assumptions, to be indeed absent for in the low-energy region (subchannel cuts and crossed channel poles also being absent as was already the case for L). The proof makes recourse to the 3-particle asymptotic completeness equation

$$ \Delta\, F_{(1)} = F_{(1)+} *_+ F_{(1)-}, \tag{2.169} $$

where $*_+$ denotes 3-particle on-mass-shell convolution (integration over 3 internal energy-momenta k_1, k_2, k_3 such that $k_1 + k_2 + k_3 = k$). In (2.169), $\Delta\, F_{(1)}$ is the discontinuity relative to the main channel (1, 2, 3); (4, 5, 6). There is no 2-particle on-mass-shell convolution because the theory is even (and no 1-particle term because Eq. (2.168) applies to $F_{(1)}$ and not to F).

In the converse direction, if the analyticity properties of are established or assumed, one goes back to , and in turn to $F_{(1)}$ via Eq. (2.160). The analysis yields, in particular, the following structure equation [B3] of interest in the physical sheet (at $s < (5\mu)^2$):

$$ \text{} \tag{2.170} $$

with suitable analyticity properties established for the remainder term R'. Apart from the first term, explicit terms in the right-hand side of (2.170) are the terms with 0, 1, 2 and 3 loops in the expansion of $(\Lambda^{-1})_{\rm conn}$. These terms have explicit Landau singularities (associated to graphs with, respectively, 0, 1, 2 or 3 loops) occurring in the physical sheet, whereas singularities of other terms in the expansion of $(\Lambda^{-1})_{\rm conn}$ are effective only in other sheets. By considering relevant boundary values, Eq. (2.170) gives a structure equation of the same form as in

Sect. 5.3: the singularity of the 2- and 3-loop terms is not effective on the mass-shell.

Formal series expansions. From the viewpoint of formal series expansions, L can be written, in view of (2.168), in the form:

$$L = G + G\Lambda^{-1}G + G\Lambda^{-1}G\Lambda^{-1}G + \cdots, \qquad (2.171)$$

where G denotes the 6-point kernel defined above, and in turn F can be written in the form:

$$(2.172)$$

In (2.171), (2.172), Λ^{-1} can also be replaced by its expansion (2.162) and $\left(\Lambda^{-1}\right)_{\text{conn}}$ is, as already mentioned, obtained by removing from the latter nonconnected terms. Each 4-point function can also be replaced by its expansion (2.151). A typical term in the expansion (2.172) is then of the form

with 4- and 6-point kernels G and inclusion of contributions to $\left(\Lambda^{-1}\right)_{\text{conn}}$ between kernels .

As discussed again in Chapter IV, this expansion has special simplicity properties: assuming the kernels G satisfy the analyticity properties mentioned previously (and have no spurious poles), possible singularities of irreducible kernels at each vertex are not effective (i.e. do not modify the analytic structure of individual integrals) in the region under consideration. In other words, these kernels can be considered as analytic in the analysis of that region, singularities of F arising from the poles of the 2-point functions on internal lines.

The fact that subchannel cuts and crossed channel poles do not occur for L is linked from that viewpoint to the fact that all terms in the expansion of L start and end with a 6-point kernel G. Poles are reintroduced in $F_{(1)}$ by the terms in $\left(\Lambda^{-1}\right)_{\text{conn}}$ and subchannel cuts are reintroduced both in other terms of $\left(\Lambda^{-1}\right)_{\text{conn}}$ (e.g. the term has cuts with respect to the subchannels (1, 2) and (4, 5)) and in remaining terms in the expansion of F that do not end with 6-point kernels G on both sides.

§7. Macrocausal properties: further results and conjectures

Sect. 7.1 presents some general results on macrocausal factorization of chronological N-point functions derived in causal situations from locality and the spectral condition, completed in most cases by asymptotic completeness. Results represent a weak expression of the idea that energy-momentum transfers over large distances take place via real on-shell intermediate particles. More refined conjectures on physical states and on chronological functions are presented in Sect. 7.2. Conjectures on chronological functions (macrocausality and macrocausal factorization) are off-shell versions of those presented in Chapter I-4 for the S matrix (and reduce to them on-shell). They follow from results of Sects. 5 and 6 in the particular situations that have been considered there, and will be more generally confirmed by the analysis of Chapter IV. They now express the idea that energy-momenta transfers over large distances take place more precisely via real on-shell intermediate (stable) particles in accordance with classical ideas (e.g. propagation of particles in the direction of their energy-momentum). However, in contrast to the results of Sects. 3 and 7.1, in which the basic axioms and conditions used induce definite minimal rates of exponential fall-off, these rates will now depend on further features of the theory considered, for example, the position of possible poles of the 4-point function (linked to possible unstable particles) in the second sheet. Rates of fall-off will not be specified, results being stated in terms of "microlocal" essential support ($=$ microsupport) properties. A deeper analysis of macrocausal factorization in a simple case involving possible sets of intermediate particles "travelling together" is finally given in Sect. 7.3 at the 2-particle threshold (or at the m-particle threshold in a simplied theory).

7.1 Macrocausal factorization: some general results

Asymptotic causality properties have been derived in Sect. 3.1 from locality and the spectral condition in the noncausal situations in which $u(I) \gtrsim u(J)$ and $P_I \notin H_+(\mu) \cup \bar{V}_+(2\mu)$. If $P_I \in H_+(\mu) \cup \bar{V}_+(2\mu)$, $(\mathcal{T} - \mathcal{T}_I)(\{\varphi_{i,\tau}\})$ still decays exponentially in the $\tau \to \infty$ limit, but this is no longer true for $\mathcal{T}_I(\{\varphi_{i,\tau}\})$. However, this term can be analyzed through the inclusion of complete sets of intermediate states of $\mathcal{H}$ between relevant chronological products of field operators involved in the definition of $\mathcal{C}_I$, and use of reduction formulae. Intermediate states will include, according to the spectral condition, 1-particle states and states with energy-momentum in $\bar{V}_+(2\mu)$. If $P_I \in H_+(\mu)$, the latter states will, as in Sect. 3, provide contributions to $T_I(\{\varphi_{i,\tau}\})$ with exponential decay in the $\tau \to \infty$ limit. The only contribution that will not decay exponentially then corresponds to 1-particle intermediate states: see Eq. (2.176) below. If $P_I \in \bar{V}_+(2\mu)$, recourse will be made to asymptotic completeness, by replacing $\mathcal{H}$ with $\mathcal{H}_{\text{in}}$ or $\mathcal{H}_{\text{out}}$. In order to get results involving only, as desirable physically, chronological functions or S matrices, it will in fact be useful in this case to insert the identity operator in the form $\mathbb{1} = \mathbb{1}_{\text{out}} \mathbb{1}_{\text{in}}$ with related natural decompositions of $\mathbb{1}_{\text{out}}$ and $\mathbb{1}_{\text{in}}$ into sums of projectors on intermediate states of $\mathcal{H}_{\text{out}}$ and $\mathcal{H}_{\text{in}}$, respectively. In order to state results that follow from this analysis, we first give some definitions.

Given a physical point P of a given process and any > 0 integer ν, we define the following distribution $\tilde{R}_I^{(\nu)}$ in the neighborhood of P

$$\delta(p_1 + \cdots + p_N)\,\tilde{r}_I^{(\nu)} = \sum_{1 \le n, n' \le \nu} \left\{ \begin{array}{c} \text{(diagram)} \end{array} \right\}_c ,$$

(2.173)

where each term in the right-hand side is a sum of double on-mass-shell convolution integrals, over n and n' internal on-shell energy-momenta respectively, in a sense made clear below. Plus boxes represent *non-connected*, amputated chronological functions T (in energy-momentum space) or their partial mass-shell restrictions in the case of internal lines. The minus box represents the kernel of $S^{-1} = S^\dagger$. I'_{in}, I'_{out} and J'_{in}, J'_{out} are the subsets of I and J, corresponding to initial or final indices respectively. For each box, initial and final variables are drawn on the left and right sides respectively. If, for example, the n internal energy-momenta on the left are denoted $k_1, \ldots, k_n$ with $k_i^2 = \mu^2$, $(k_i)_0 > 0$, the plus box on the left depends on $k_1, \ldots, k_n$ and on external energy-momenta associated to J, while the minus box depends on $-k_1, \ldots, -k_n$ and on energy-momenta associated to the n' internal lines on the right. The notation $\{\ \}_c$ means that only globally connected parts obtained from decompositions of the plus and minus boxes into connected components are kept. By convention, internal lines   are removed in the case of 2-point connected components: for example, the term , where bubbles represent connected functions, is to be replaced by . Each term in the right-hand side of (2.173) thus appears as a sum of on-mass-shell convolution integrals involving only connected, amputated chronological functions with 4 points or more (so that ambiguity in the definition of amputation for 2-point functions and related problems does not occur) or kernels of S^{-1}.

If $\nu = 1$, $n = n' = 1$ and the minus box can be removed:

$$\delta\left(p_1 + \cdots + p_N\right)\tilde{r}_I^{(1)} = \left\{ \text{} \right\}_c. \qquad (2.174)$$

A simple example is that of a $3 \rightarrow 3$ process with $(1, 2, 3)$ initial, $(4, 5, 6)$ final, $J = (1, 2, 4)$, and $I = (3, 5, 6)$, in which case:

$$\delta\left(p_1 + \cdots + p_6\right)\tilde{r}_I^{(1)} = \text{}, \qquad (2.175)$$

where bubbles represent connected, amputated chronological functions.

The following results then follow (with $u = (u_1, \ldots, u_N)$, $u(I) \gtrsim u(J)$):

(a) if $P_I \in H_+(\mu)$

$$T\left(p_1, \ldots, p_N\right) \underset{(P,u)}{\simeq} \tilde{r}_I^{(1)}\left(p_1, \ldots, p_N\right) \qquad (2.176)$$

(b) if $P_I \in \bar{V}_+(2\mu)$, $P_I^2 < [(\nu + 1)\mu]^2$:

$$T\left(p_1, \ldots, p_N\right) \underset{(P,u)}{\simeq} \tilde{r}_I^{(\nu)}\left(p_1, \ldots, p_N\right), \qquad (2.177)$$

where $\simeq$ at (P, u) means that $u \notin \mathrm{ES}_P(T - \tilde{r}_I^{(\cdot)})$. In other words, the action of the connected, amputated chronological function on test functions $\varphi_{i,\tau}$ of the form (2.15) is equal to $R_I^{(\cdot)}(\{\varphi_{i,\tau}\})$ up to a remainder that decays exponentially in the $\tau \rightarrow \infty$ limit.

As already mentioned, the derivation of (2.177) makes recourse to asymptotic completeness.

Remarks.

(1) As in Sect. 3, rates of exponential fall-off can be specified. They depend on P, u, γ and the mass μ.

(2) Corresponding results follow, by restriction of external energy-momenta to the mass-shell for the connected S matrix, with related results on rates of exponential fall-off.

(3) The occurrence in general of the minus box in the term $\tilde{R}_I$ can be physically interpreted as needed to compensate possible interactions of the intermediate particles, already taken into account in each one of the plus terms.

(4) The analysis above can be extended in a similar way to situations involving more than two subsets I, J. Let us consider, for example, three subsets I_1, I_2, I_3, with $I_1 \cup I_2 \cup I_3 = (1, \ldots, N)$, $I_i \cap I_j$ is empty if $i \neq j$, and a configuration u such that $u(I_1) \gtrsim u(I_2) \gtrsim u(I_3)$. If P is such that $P_{I_1} \notin H_+(\mu) \cup \bar{V}_+(2\mu)$ or $P_{I_3} \notin H_-(\mu) \cup \bar{V}_-(2\mu)$, results of Sect. 3 can be applied. Otherwise, a result analogous to that above applies, with again specified rates of exponential fall-off: $\tilde{R}_I^{(\nu)}$ is replaced by

$$\delta(p_1 + \cdots + p_N) R_{I_1, I_2, I_3}^{(\nu, \nu')}$$

$$= \sum_{\substack{1 \leq n_1, n_1' \leq \nu \\ 1 \leq n_2, n_2' \leq \nu'}} \left\{ \vphantom{\sum} \right\}_c,$$

(2.178)

where ν, ν' are any > 0 integers such that $(P_{I_3})^2 < [(\nu + 1)\mu]^2$, $(P_{I_1})^2 < [(\nu' + 1)\mu]^2$, and where the right-hand side is defined in a way similar to that above.

Example. An example that will be discussed again in Sect. 7.2 is that of $3 \to 3$ process, 1, 2, 3 initial, 4, 5, 6 final with $I_1 = (1, 2)$, $I_2 = (3, 4)$, $I_3 = (5, 6)$ and for simplicity $(2\mu)^2 < (p_1 + p_2)^2 < (3\mu)^2$, $(2\mu)^2 < (p_5 + p_6)^2 < (3\mu)^2$. The configuration u is, for example, such that $u_1 = u_2$, $u_3 = u_4$, $u_5 = u_6$, with u_1 in the past of u_3, and u_3 in the past of u_5.
Then,

$$T \underset{(P,u)}{\simeq} \tilde{r}_{I_1, I_2, I_3},$$

(2.179)

where $\tilde{r}_{I_1, I_2, I_3}$ is equal in the neighborhood of the physical point P to

This result can be compared with the more refined one stated in Sect. 7.2 for this situation.

7.2 General conjectures

(a) Physical states

The following conjecture will give some information in the $\tau \to \infty$ limit for states of the form

$$\Phi_\tau = \prod_{i=1}^{m} A(f_{i,\tau})\Omega$$

with functions $f_{i,\tau}$ of the form (2.15).

As in Sect. 2.2, $\Psi_\tau^{(\text{in})}$ and $\Psi_\tau^{(\text{out})}$ will denote the states of $\mathcal{H}$ asymptotically tangent, before and after interactions respectively, to the free-particle state composed of the m 1-particle states with wave functions $\hat{f}_{1,\tau}, \ldots, \hat{f}_{m,\tau}$. The conjecture will assert that Φ_τ is close to $\Psi_\tau^{(\text{in})}$, or to $\Psi_\tau^{(\text{out})}$ respectively, up to exponential fall-off with τ, under corresponding conditions on the set of trajectories (P_k, u_k). (A similar conjecture can be stated for functions $f_{i,\tau}$ of the form (2.14) with trajectories replaced by velocity cones and exponential fall-off replaced by rapid fall-off. In the case of functions of the form (2.16), all velocity cones intersect at the origin, so that the conjecture reduces to a result already given in Sect. 2: see Remark 2 in Sect. 2.3.)

In view of the expressions of free-particle states with which a given vector coincides before and after interactions (see the end of Sect. 2.3), this conjecture is related to Conjecture 2.2 stated below on chronological functions.

Conjecture 2.1. $\|\Phi_\tau - \Psi_\tau^{(\text{in})}\|$ (respectively $\|\Phi_\tau - \Psi_\tau^{(\text{out})}\|$) decays exponentially with τ (with a rate of fall-off at least proportional to γ for small γ), if there exists no connected or nonconnected classical diagram $\mathcal{D}_+$ with initial (respectively final) trajectories (P_k, u_k) and one or more u_k in the closed future cone (respectively in the closed past cone) of the interaction vertex of $\mathcal{D}_+$, at which k is incoming (respectively outgoing).

Example 1. We first illustrate this conjecture in Figure 2.11 in the simplest case ($m = 2$).

Example 2 ($m = 3$). If there is no intersection of any pair of the three trajectories, then $\Phi_\tau \approx \Psi_\tau^{(\text{in})} \approx \Psi_\tau^{(\text{out})}$ as in Figure 2.11a. In Figure 2.12, we consider a case in which (P_1, u_1) and (P_2, u_2) intersect at a point v_1 and in which (P_3, u_3) might intersect a possible trajectory issued from v_1 (according to classical laws) at a point v_2 in the future of v_1.

If, as shown in Figure 2.12, u_1, u_2 are in the past of v_1 and u_3 is in the past of v_2, then $\Phi_\tau \approx \Psi_\tau^{(\text{in})}$ (different in general from $\Psi_\tau^{(\text{out})}$). (In

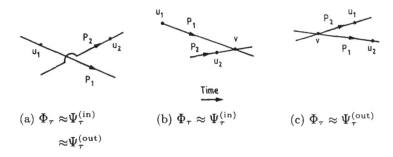

(a) $\Phi_\tau \approx \Psi_\tau^{(in)}$ (b) $\Phi_\tau \approx \Psi_\tau^{(in)}$ (c) $\Phi_\tau \approx \Psi_\tau^{(out)}$

$\approx \Psi_\tau^{(out)}$

Figure 2.11 Trajectories (P_1, u_1) and (P_2, u_2) do not intersect in case a. They intersect at a point v in the future of u_1 and u_2 (case b) or in the past of u_1, u_2 (case c).

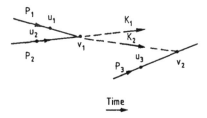

Figure 2.12 A situation, at $m = 3$, in which $\Phi_\tau \approx \Psi_\tau^{(in)}$.

Figure 2.12, K_1, K_2 are on-shell energy-momenta such that $K_1 + K_2 = P_1 + P_2$; for simplicity, we may consider the region $(P_1 + P_2)^2 < 9\mu^2$, in which case, at most, two trajectories can be issued from v_1).

   (b) Macrocausal properties of chronological functions
      As in Sect. 3, T denotes the connected, amputated chronological function.

Conjecture 2.2 (Macrocausality). Given any point $P = (P_1, \ldots, P_N)$ of a $m \to n$ physical region $(m + n = N)$:

$$\mathrm{ES}_P(T) \subset \big\{ u = (u_1, \ldots, u_N); \; \exists \text{ connected } \mathcal{D}_+ \text{ with external}$$
$$\text{trajectories } (P_k, u_k) \text{ such that } u_k \text{ lies at the}$$
$$\text{interaction vertex of } \mathcal{D}_+ \text{ involving line } k \big\}. \qquad (2.180)$$

Corollary. The connected, nonamputated chronological function satisfies the same property except that u_k is only required to lie in the past of the interaction vertex involving k if k is initial, or in the future of this vertex if k is final.

Remark. Similar, more refined conjectures can be made on irreducible kernels, with corresponding restrictions on the diagrams $\mathcal{D}_+$.

Local analyticity properties of T. Conjecture 2.2 yields equivalent local analyticity properties of T (and in turn of scattering functions) in the neighborhood of physical region points. It is here useful to define $+\alpha$-Landau surfaces in the space of (real) points $p = (p_1, \ldots, p_N)$ satisfying energy-momentum conservation but not necessarily the mass-shell constraints. ("Off-shell context": see Sect. 3 of Chapter I).

Conjecture 2.2 then entails in particular that T_N is always locally, along $+\alpha$-Landau surfaces, the boundary value of an analytic function (the analytic N-point function) from plus $i\epsilon$ directions; moreover, it is analytic outside the $+\alpha$-Landau surfaces (of connected graphs). Cones of analyticity obtained at $+\alpha$-Landau points (which are the nonempty intersection of those associated to the various $+\alpha$-Landau surfaces, at points that lie on several such surfaces) are in general locally much larger than those determined previously in Sects. 3 or 4. Refined decomposition theorems are also established at points that lie on several $+\alpha$-Landau surfaces.

Results on scattering functions described in Sect. 4.1 of Chapter I are reobtained by restriction to the mass-shell away from M_0 points.

Macrocausal factorization. Conjectures on macrocausal factorization, which reduce on-shell to those of Chapter I, can also be stated for the amputated chronological functions. They take the same form as in Sect. 4.2 of Chapter I, and involve similar convolution integrals over on-mass-shell internal energy-momenta, except that *external* energy-momenta are allowed to vary off-shell. (Kernels at each vertex involving external lines are amputated chronological functions, restricted to the mass-shell with respect to internal energy-momenta in integration domains.)

Such results follow directly from those established in Sects. 5.3 and 6.3 in the situations considered there, which involve only (particularly in Sect. 5.3) graphs without sets of multiple lines. They can be more generally reobtained (at least formally) from expansions in terms of irreducible kernels: see Chapter IV where the origin, from that viewpoint, of the factor S^{-1} or S_ν^{-1} on sets of multiple lines will be explained on some examples.

Remark. If we come back to the example given at the end of Sect. 7.1, the various conjectures presented in this section are, as a matter of fact, consequences (in the low-energy region) of the results of Sects. 5 or 6. The latter entail that

(i) besides (2.179),

$$T \simeq 0 \qquad \text{at } (P, u) \qquad (2.181a)$$

if the set (P, u) does not allow the existence of a classical multiple scattering diagram $\mathcal{D}_+$ (namely, in this case a triangle diagram in the low-energy region); and

(ii) in the causal case when there exists such a classical triangle diagram,

$$T \simeq d \qquad \text{at } (P, u) \qquad (2.181b)$$

where $\delta \left(p_1 + \cdots + p_6 \right) d = \ _3\overbrace{}^{}\underbrace{}_{}\ _4 \, .$

This result is consistent with (2.179), as can be checked from the decomposition of the G-point bubble in the right-hand side of (2.179) as a sum of pole and "triangle" contributions.

7.3 Macrocausal analysis of the 2-particle threshold

In this subsection, we give a deeper analysis of macrocausal factorization in the simple case of a point P of the 2-particle threshold in a $2 \rightarrow 2$ process. (The same analysis applies also to the m-particle threshold in the simplified $m \rightarrow m$ theory.) In this case, macrocausal factorization has a nonempty content in the "off-shell" context. The causal direction $\hat{u}_+(P)$ is that determined by the configuration:

$$u = (u_1, \ldots, u_4) \, ; u_1 = u_2, u_3 = u_4, u_3 - u_1 = \lambda \left(P_3 + P_4 \right), \lambda > 0.$$
$$(2.182)$$

Macrocausal factorization, as given either in Sects. 7.1 or 7.2, takes the form:

$$\begin{smallmatrix}1\\2\end{smallmatrix}\!-\!\boxed{+}\!-\!\begin{smallmatrix}3\\4\end{smallmatrix} \underset{(P,\tilde{u}_+)}{\sim} \begin{smallmatrix}1\\2\end{smallmatrix}\!-\!\boxed{+}\!-\!\boxed{-}\!-\!\boxed{+}\!-\!\begin{smallmatrix}3\\4\end{smallmatrix} \, . \qquad (2.183)$$

The result is, as a matter of fact, trivial in this case since the right-hand side is equal to $-\!\boxed{+}\!- \ - \ -\!\boxed{-}\!-$ (as easily checked from the unitarity equation) and since, as a minus $i\epsilon$ boundary value, $-\!\boxed{-}\!- \ \simeq 0$ at $(P, \hat{u}_+)$. Our purpose below is to give a deeper interpretation, in terms of multiple scattering, of the expansion of the 4-point function in terms of the irreducible kernel U or G. This interpretation is simple, but not quite trivial.

We assume below that U or G is not only uniform around, but is analytic at, the 2-particle threshold. For definiteness, the argument is given for the expansion in terms of U, but it is presented in a way which also applies (at least at a formal level) in terms of G by replacing $\otimes$ and U with $\circ$ and G respectively. We first fix some notations. We write, in a full real neighborhood $\mathcal{N}(P)$ of P (and not only at $s > 4\mu^2$):

$$F_+ = \sum_{n \geq 1} F_+^{(n)}, \tag{2.184}$$

$$F_+^{(n)} = (U \otimes U \ldots \otimes U)_+ (n \text{ factors } U), \tag{2.185}$$

where F_+, $F_+^{(n)}$ are the plus $i\epsilon$ boundary values of the analytic functions F and $F^{(n)} = U \otimes \ldots \otimes U$. The actual physical on-mass-shell convolution is equal in $\mathcal{N}(P)$ to $\theta(\sigma)*_+$ where, as previously $\sigma = s - 4\mu^2$, $\theta(\sigma) = 1$ if $\sigma > 0$, $\theta(\sigma) = 0$ if $\sigma < 0$; here, $*_+$ is also the plus $i\epsilon$ boundary value of $*$ in $\mathcal{N}(P)$ (and not only at $s > 4\mu^2$). If, for example, $\beta > 0$, $U *_+ U$ is a continuous function which vanishes at $\sigma = 0$ so that multiplication by θ is unambiguous. Multiple on-mass-shell convolutions of n factors U are similarly equal in $\mathcal{N}(P)$ to $\theta(\sigma)(U * \ldots * U)_+ = \theta(\sigma)U *_+ U \ldots *_+ U$.

For $n > 1$, $\hat{u}_+$ does belong in most cases to, and is the unique direction in, $\mathrm{ES}_P(F_+^{(n)})$, since $F_+^{(n)}$ is a plus $i\epsilon$ boundary value locally and is not analytic at P. (It contains several factors $\sigma^{1/2}$ or $\ell n \, \sigma$.)

As we now explain, each term $F_+^{(n)}$ can be interpreted as the contribution to F_+ corresponding to the possibility of n successive $2 \rightarrow 2$ scatterings.

(i) The factor U corresponds to a "pure" $2 \rightarrow 2$ scattering: no possibility of further $2 \rightarrow 2$ scatterings. Correspondingly, $\mathrm{ES}_P(U)$ is empty (local analyticity of U).

(ii) At $n = 2$, the following relation is satisfied:

$$(U \otimes U)_+ \underset{(P, \hat{u}_+)}{\simeq} \theta(\sigma)U *_+ U, \tag{2.186}$$

where $f \simeq g$ at $(P, \hat{u}_+)$ means again that relevant actions on test functions are equal, modulo a term that falls off exponentially in the direction $\hat{u}_+$, that is, when $x_3 = x_4$ is taken further and further away from $x_1 = x_2$ (with $x_3 - x_1 = \lambda(P_3 + P_4)$, $\lambda > 0$).

This result is a particularly simple analogue of those described in Sect. 6.2 of Chapter I (The surface $\sigma = 0$ is a smooth codimension one submanifold in the present "off-shell" context). It follows from the relation

$$(U \otimes U)_+ - (U \otimes U)_- = \theta(\sigma)U *_+ U, \tag{2.187}$$

itself a consequence of $\otimes_+ - \otimes_- = *_+$ at $\sigma > 0$, and from the fact that $(U \otimes U)_-$, being a minus $i\epsilon$ boundary value, satisfies

$$(U \otimes U)_- \simeq 0 \quad \text{at} \quad (P, \hat{u}_+). \tag{2.188}$$

The result (2.186) is in full agreement with the ideas on macrocausal factorization of Chapter I. As $u_3 = u_4$ is taken away from $u_1 = u_2$ in the causal direction, the contribution of order 2 to F_+, integrated with functions $f_{i,\tau}$ of the form (2.15), is equal (up to exponential fall-off) to the integral of the right-hand side of (2.186) (with the same functions $f_{i,\tau}$), that corresponds to two $2 \to 2$ successive scatterings with interactions U at each vertex and on-shell intermediate particles.

(iii) The interpretation is more delicate for $n > 2$, in which case the direct analogue of (2.186), i.e. $F_+^{(n)} \simeq \theta(\sigma) U *_+ U *_+ \cdots *_+ U$ at $(P, \hat{u}_+)$, no longer holds. Instead, for example, at $n = 3$

$$(U \otimes U \otimes U)_+ \underset{(P,\hat{u}_+)}{\simeq}$$
$$[\theta(\sigma)U *_+ F_+^{(2)}] + [\theta(\sigma)F_+^{(2)} *_+ U] - [\theta(\sigma)U *_+ U *_+ U]. \tag{2.189}$$

Eq. (2.189) follows from the discontinuity formula

$$(U \otimes U \otimes U)_+ - (U \otimes U \otimes U)_- =$$
$$\left\{ [(U \otimes U)_+ *_+ U] + [U *_+ (U \otimes U)_+] - [U *_+ U *_+ U] \right\}\theta(\sigma) \tag{2.190}$$

in $\mathcal{N}(P)$ (easily derived by induction from $\otimes_+ - \otimes_- = *_+$ at $\sigma > 0$), and from the fact that $(U \otimes U \otimes U)_-$ is a minus $i\epsilon$ boundary value.

The first and second terms in the right-hand side of (2.189) correspond to two successive interactions with kernels $F_+^{(2)}$ and U, or U and $F_+^{(2)}$ at each vertex. Since $F_+^{(2)}$ itself contains the possibility of two successive $2 \to 2$ scatterings, the term with three successive scatterings is implicity contained in each one of these two terms. The last term in the right-hand side of (2.189) has a minus sign in order to compensate for one of them.

The interpretation extends similarly to other values of n.

Remark. The expansion $F_+ = \Sigma F_+^{(n)}$ can be conversely derived from ideas of macrocausal factorization. The latter give

(i) $F_+^{(1)} = A_1$, A_1 analytic at P

(ii) $F_+^{(2)} \underset{(P,\hat{u}_+)}{\simeq} \theta(\sigma) A_1 *_+ A_1.$ (2.191)

Eq. (2.191) yields in turn

$$F_+^{(2)} = (A_1 \otimes A_1)_+ + A_2 , \quad A_2 \text{ analytic at } P \qquad (2.192)$$

(iii) $F_+^{(3)} \simeq \theta(\sigma) \left\{ \left(F_+^{(2)} *_+ A_1 \right) + \left(A_1 *_+ F_+^{(2)} \right) - A_1 *_+ A_1 *_+ A_1 \right\},$

(2.193)

at $(P, \hat{u}_+)$, and in turn

$$\begin{aligned} F_+^{(3)} = (A_1 \otimes A_1 \otimes A_1)_+ + (A_1 \otimes A_2)_+ + (A_2 \otimes A_1)_+ \\ + A_3 , \quad A_3 \text{ analytic at } P. \end{aligned} \qquad (2.194)$$

Similar expressions follow for each $F_+^{(n)}$ with locally analytic remainders A_n. The expansion of F_+ in terms of U is reobtained (formally) by redefining the pure interaction

$$U = \sum_{n=1}^{\infty} A_n. \qquad (2.195)$$

CHAPTER III

Euclidean Constructive Field Theory

§1. Introduction

As mentioned in Sect. 3 of the general Introduction, notations are different throughout this part from those of Chapter II: x or p will be taken real in Euclidean space, with $x^2 = x_0^2 + \vec{x}^2$, $p^2 = p_0^2 + \vec{p}^2$.

1.1 Historical survey and description of contents

The purpose of this chapter is to present the Euclidean definition of models, and related results, in nonperturbative field theory. We start in Sects. 1.2 and 1.3 with some preliminaries. Sect. 1.2 presents the standard Euclidean axioms [OS,GJ] that are believed to be satisfied by any probability measure $d\mu(\varphi)$ (on the space of distributions φ in Euclidean space-time) characterizing a field theory in the class we wish to consider, or by the corresponding set of Euclidean N-point functions, and which allow one, following the original analysis of Osterwalder and Schrader, to define a Wightman theory by analytic continuation to Minkowski space-time. As described in more detail in later sections, it has been possible, so far in space-time dimensions 2 and 3, to rigorously define nontrivial models (i.e. theories with interactions) satisfying the Euclidean axioms. A heuristic introduction to models is given in Sect. 1.3 on the example of φ_d^4 (interaction $\varphi(x)^4$ in space-time dimension d): up to the modifications to be mentioned later, $d\mu(\varphi)$ is then the measure $[Z(\lambda)]^{-1}e^{-\lambda \int \varphi^4(x)dx}d\nu(\varphi)$, where $d\nu(\varphi)$ is a "free" measure of covariance C, C is the Fourier transform of an energy-momentum propagator $\tilde{C}(p) = 1/(p^2 + m^2)$, and Z is a normalization factor; λ is the coupling and m is a "bare" mass (different from physical masses to appear in Chapter IV). Euclidean N-point functions $S(x_1,\ldots,x_N)$, equal to $\int \varphi(x_1)\ldots\varphi(x_N)d\mu(\varphi)$, are corresponding ratios $N(x_1,\ldots,x_N;\lambda)/Z(\lambda)$.

The perturbative approach, which introduces formal expansions in the (possibly renormalized) coupling and is a useful guide from several viewpoints, is outlined in Sect. 2, where the classification of theories as super-renormalizable, renormalizable or nonrenormalizable is given, according to the *a priori* possible divergences of individual terms in the perturbative series before renormalization. These divergences are treated by "perturbative renormalization." Although other methods are available, we shall present methods of phase-space analysis, developed in the 1980s, which are an introduction to those used later in constructive theory. For definiteness and for our later purposes, we shall mainly consider theories with *a priori* ultraviolet, but not infrared, divergences. An ultraviolet cut-off depending on a parameter ρ, to be ultimately removed ($\rho \to \infty$ "ultraviolet" limit) is then first introduced in Euclidean energy-momentum space. Even though individual terms are well defined after renormalization in the $\rho \to \infty$ limit, the perturbative expansions remain in general divergent, however small the (possibly renormalized) coupling is. A number of models can be reconstructed by the Borel (or related) procedures from the perturbative series which then appear as asymptotic series: this is the case, for example, for φ_2^4, as will be explained in Sect. 3, and also for the Gross-Neveu model presented in Sect. 4. However, this result has not been proved directly so far and is a byproduct of the methods of constructive field theory that we now introduce.

The first important success of constructive theory is the definition, achieved in the first part of the 1970s, of the massive, weakly coupled super-renormalizable $P(\varphi)_2$ models. In the latter, and also in further models to be introduced below, the measure $d\mu(\varphi)$ will be defined as the $\Lambda \to \infty$, $\rho \to \infty$ limit of regularized measures $d\mu_{\Lambda,\rho}$ where Λ is a bounded region in Euclidean space-time and ρ is, as above, an ultraviolet cut-off in energy-momentum space. N-point functions $S_{\Lambda,\rho}$ are corresponding ratios $N_{\Lambda,\rho}(x_1, \ldots, x_N)/Z_{\Lambda,\rho}$. The infinite-volume ($\Lambda \to \infty$) limit is treated via "cluster expansions" originally introduced by Glimm, Jaffe, Spencer [GJS]. Simple cluster expansions, which we shall use below, were later introduced, in particular in [Br,BaF]. The treatment of the ultraviolet limit for $P(\varphi)_2$ models, already needed in a finite box Λ, is based, on the other hand, on original arguments of Nelson [N] in the 1960s. Although it is already nontrivial, this treatment appears today as a largely simplified version of the more detailed "phase-space analysis" needed in further models following original methods of Glimm and Jaffe [GJ] in the 1970s, close to Wilson renormalization group ideas and first applied to (massive) φ_3^4. While φ_3^4 is still super-renormalizable, these methods were developed from a more general viewpoint in the 1980s, starting with works by Magnen, Seneor [MS] and by Gallavotti and collaborators (see [VW2] and references therein), and have led to the rigorous definition [FMRS1,GK1] of new, nonsuper-renormalizable models such as the massive, weakly coupled Gross-Neveu [GN] model in dimension 2 in the middle of the 1980s. To be more precise, the model treated in these references and in our Sect. 4 is that initially introduced in [MW]. (It includes from the outset a strictly positive mass, in contrast to the actual "Gross-Neveu" model considered in [GN]: see comments in Sect. 4.1.) This model, which involves spin and colour indices, is analogous from the viewpoint of renormalization to (massive) φ_4^4 (2 and 4-point relevant or marginal operators) but does exist as a nontrivial theory, whereas this is doubtful for φ_4^4. This is linked with the fact that it is "asymptotically free" (on the ultraviolet side, relevant here), whereas this is not the case for φ_4^4. Another related result of the 1980s [FMRS2,GK2] is the construction of "massless infrared φ_4^4 with fixed ultraviolet cut-off," or infrared φ_4^4 for short, a Euclidean model in which problems now arise on the infrared side, which does *not* satisfy all axioms and which is not of direct interest for our later purposes, but provides a useful illustration of problems and of methods used to treat them. (This model is renormalizable but not super-renormalizable and is asymptotically free, from the infrared viewpoint.)

To conclude this historical survey, and before giving more details on contents of this chapter, we mention further results that will not be treated here. Along lines similar to those above, but with more complications, important partial results on gauge theories in a finite box in dimension $d \leq 4$ have been obtained, in particular by Balaban (see [Bal]) and more recently by Federbush [Fe] and the Ecole Polytechnique group [MRS]. The "construction" of the Gross-Neveu model in dimension 3 at large colour number has been achieved [dCdVMS]. (The latter is nonrenormalizable, though it reduces to the renormalizable case after a suitable treatment.) For other related results, see the treatment of the Gross-Neveu model in $2 + \varepsilon$ dimensions in [GK3] and of "planar $\varphi_{4+\varepsilon}^4$" with a + sign in the interaction (i.e. a term in $\exp +\lambda \int \varphi_{4+\varepsilon}^4$) in [FG]. (This model will exist in spite of this wrong sign because of the "planar" restriction.) In most cases that have been mentioned, results are obtained at weak (renormalized) coupling or for sufficiently small values of other parameters: the smallness of these parameters ensures convergence of relevant expansions. Among other rigorous results of the 1970s and 1980s let us also mention:

- the definition of $P(\varphi)_2$ models in the two-phase region (which also depends on the smallness of some parameters and involves further technical tools beside those already mentioned).
- results based on correlation inequalities for a class of models (in which case the constraint on the smallness of the coupling is to some extent removed).

- special models in space-time dimension 2 (completely integrable models) which can be defined in a more explicit way, such as the Sine-Gordon and related massive Thirring model, or the "Federbush model." (The latter, as GN_2, is renormalizable but not super-renormalizable and does not reduce to the super-renormalizable case for practical purposes, in contrast to the Thirring model. It is the first nonsuper-renormalizable model for which Wightman axioms, and also asymptotic completeness, have been established: see [Ruij] and references therein.) However, the analysis depends on special properties of these models.

This chapter aims to treat the massive, weakly coupled $P(\varphi)_2$, φ_3^4 and GN_2 models in a way which will exhibit the main methods, mentioned above, developed in constructive theory. We first treat, in Sect. 3, the $\Lambda \to \infty$ limit in $P(\varphi)_2$ models. The treatment of the ultraviolet limit in those models is left to Sect. 5.1. The general ideas of phase-space analysis to be applied in either φ_3^4 or GN_2 are similar but each of them presents specific features. φ_3^4 is simpler from the viewpoint of renormalization. However, as explained in more detail below, important technical problems, present in bosonic models, disappear in fermionic models like GN_2, which we shall therefore first consider in Sect. 4, following the somewhat simplified version of [FMRS1] given in [IM3]. We then outline in Sect. 5.2 the analysis of bosonic models. Though a special analysis can still be given for φ_3^4 via the introduction of specific "counterterms" in the interaction from the outset (linked with the super-renormalizable character of this model), we shall here outline a treatment which can be adapted to nonsuper-renormalizable cases such as "infrared φ_4^4." We shall mainly follow [FMRS2] but will also mention the more recent treatment [AIM,IM5] of an important technical problem.

More details are now given in the remainder of this subsection. As mentioned above, Sect. 3 is mainly devoted to $P(\varphi)_2$ models, which are introduced in Sect. 3.1. Cluster expansions, presented in Sect. 3.2, are made with respect to a paving of Λ by boxes (= squares in dimension 2) of, for example, unit size. In contrast to perturbative expansions, they provide at most one explicit propagator between any pair of squares. The simplest one ("pairwise cluster expansion"), to be described first, cannot be used for bosonic models in view of technical problems, but can be used later for fermionic models like the Gross-Neveu model. It is obtained by introducing an auxiliary dependence of the propagator, and in turn of $N_{\Lambda,\rho}$ and $Z_{\Lambda,\rho}$ on couplings $s_{\Delta,\Delta'}$ between pairs (Δ, Δ') of squares, with $s_{\Delta,\Delta'} = 1$ in the actual theory, while the value $s_{\Delta,\Delta'} = 0$ uncouples Δ and Δ' in the interaction. The cluster expansion then follows from a first order Taylor expansion with respect to all variables $s_{\Delta,\Delta'}$ around the origin. A related, more refined procedure ("inductive cluster expansion"), which can be applied to bosonic models (like $P(\varphi)_2$), will then be presented.

Expansions of N-point functions $S_{\Lambda,\rho}$ (including "à la Mayer" links between some pairs of squares) follow and allow one, in view of the decay of the propagator, to establish uniform bounds (independent of Λ) and the existence of the $\Lambda \to \infty$ limit at sufficiently small coupling, as explained in Sect. 3.3. The treatment of the ultraviolet ($\rho \to \infty$) limit is previously needed to obtain uniform bounds, independent of ρ, already in a unit square and on related, more general quantities involved in the expansion of $S_{\Lambda,\rho}$. It will be outlined in Sect. 5.1. Sect. 3.3 also outlines the proof of the Euclidean axioms for $P(\varphi)_2$ models and will present the derivation [EMS], for φ_2^4, of the Borel summability of the N-point functions with respect to the coupling λ: the (divergent) perturbative series then appear as asymptotic series from which actual N-point functions can be recovered by the Borel procedure. Similar results, involving a generalization of the usual Borel procedure, should apply to $P(\varphi)_2$ models with degree of P larger than 4, but more (nontrivial) work would be needed.

A more detailed phase-space analysis is needed in general to treat the ultraviolet limit. To treat models involving ultraviolet but not infrared divergences, it is convenient to divide energy-momentum space into successive slices corresponding roughly to regions $M^{i-1} \lesssim |p| \lesssim M^i$, $i = 1, \ldots, \rho$, $M > 1$, to consider a corresponding decomposition of the propagator as a sum of slice propagators $C^{(i)}$ and to make cluster expansions, for each i, with respect to boxes of dual size M^{-i} in Λ. (In the case of infrared problems, one considers slices going from $M^{-(i-1)}$ to M^{-i} and boxes of size M^i.) A further expansion with respect to auxiliary couplings between boxes of different slices is also needed for bosonic models, but can be avoided in fermionic models like GN_2. The choice of the size M^{-i} of boxes in slice i is linked to renormalization group ideas and cannot probably be avoided in general (in contrast to the case of $P(\varphi)_2$ models where the size of the boxes does not change with i). Two sources of possible divergences in the $\rho \to \infty$ limit are encountered in this analysis. The first one, related to the renormalization problems already present in the perturbative approach, is treated by an adaptation of methods presented in Sect. 2. Following here [FMRS1,2], we shall outline an inductive method from slice ρ to slice 1, in which only "useful renormalization" is made at each step. The second problem, encountered for bosonic models, is related to the divergence of the perturbative series. Its origin can be understood as follows. Consider for simplicity the first step of the induction, in which a cluster expansion is made in slice ρ. Even though the number of explicit propagators attached to a given box in slice ρ is under control, this number becomes arbitrarily large in slice 1 because each box in slice 1 contains $M^{d\rho}$ boxes of slice ρ. This problem, which does not occur in the (massive) $P(\varphi)_2$ models of Sect. 3, already occurs for φ_3^4. The same problem also occurs, on the infrared side, for massless φ_4^4 with fixed ultraviolet cut-off. One of the methods used to treat this problem is a "domination procedure" described in [FMRS2] and in the book by V. Rivasseau (see bibliography). The alternate method of [AIM,IM5], applied in these works to the Edwards model in polymer theory (which can be considered as infared φ_4^4 "at zero components"), has a somewhat different character.

As can be understood heuristically from the Pauli principle, this problem will *not* occur in fermionic models like GN$_2$. An introduction to the model and preliminary results are given in Sect. 4.1. The Euclidean definition of the model is given in Sect. 4.2. Sect. 4.3 presents its large momentum and short-distance properties, including the derivation of Wilson short-distance expansion, following [IM4]. The latter results are based on a refined renormalization procedure in the constructive framework. Besides its own basic interest in field theory, this topic will also be partly relevant in some aspects of Chapter IV.

1.2 The Euclidean axioms

An Euclidean field theory can be defined, following, for example, [GJ] by a probability measure $d\mu(\varphi)$ on the space of (tempered) distributions φ in Euclidean space-time. The characteristic function $\mathcal{C}$ of $d\mu$ is defined, for any test function f (in the Schwartz space $\mathcal{S}(\mathbb{R}^d)$), by the formula:

$$\mathcal{C}(f) = \int e^{i \, \varphi(f)} d\mu(\varphi). \tag{3.1}$$

In view of a general theorem by Minlos, there is a 1-1 correspondence between probability measures $d\mu$ and normalized ($\mathcal{C}(0) = 1$), continuous functions $\mathcal{C}$ satisfying the positivity condition

$$\sum_{i,j=1}^{n} \bar{z}_i \, \mathcal{C} \left(f_i - f_j \right) z_j \geq 0 \tag{3.2}$$

for any n and arbitrary sets of test functions $f_1, \ldots, f_n$ and complex numbers $z_1, \ldots, z_n$.

The Euclidean Schwinger functions are the Nth order moments of $\mathrm{d}\mu$

$$S_N(x_1, \ldots, x_N) = \int \varphi(x_1) \cdots \varphi(x_N) \, \mathrm{d}\mu(\varphi). \qquad (3.3)$$

More precisely, they are, *a priori*, distributions defined by the relation

$$S_N(f_1, \ldots, f_N) \left(\equiv \text{``} \int S_N(x_1, \ldots, x_N) f(x_1) \cdots f(x_N) \mathrm{d}x_1 \cdots \mathrm{d}x_N \text{''} \right)$$

$$= \int \varphi(f_1) \cdots \varphi(f_N) \, \mathrm{d}\mu(\varphi). \qquad (3.4)$$

They can be generated by derivation of $\mathcal{C}(\alpha f)$ or $\mathcal{C}(\alpha_1 f_1 + \cdots + \alpha_N f_N)$ with respect to α, or $\alpha_1, \ldots, \alpha_N$, and are well-defined distributions if $\mathrm{d}\mu$, or $\mathcal{C}$, satisfies relevant regularity properties.

In a scalar theory, Euclidean axioms on $\mathrm{d}\mu$ or $\mathcal{C}$ include, in the version of [Fro] (which is slightly more general than that of [GJ]):

(i) Euclidean invariance: $\mathcal{C}(Ef) = \mathcal{C}(f)$ for any Euclidean symmetry E of $\mathbb{R}^d$ (translations, rotations and reflections), where $Ef(x) = f(Ex)$.

(ii) Osterwalder-Schrader reflection positivity:

$$\sum_{i,j=1}^{n} \bar{z}_i \mathcal{C}(\theta f_i - f_j) z_j \geq 0, \qquad (3.5)$$

where θ is time reflection ($\theta f(\vec{x}, t) = f(\vec{x}, -t)$), for any set $f_1, \ldots, f_n$ of functions f_i in $\mathcal{S}_+$ ($f_i(\vec{x}, t) = 0$ if $t < 0$).

(iii) Cluster property (related to ergodicity[GJ]):

$$\lim_{\tau \to \infty} \mathcal{C}(f + g_{\tau u}) - \mathcal{C}(f)\mathcal{C}(g) = 0 \qquad (3.6)$$

for any point $u \neq 0$ in $\mathbb{R}^d$, where $g_{\tau u}(x) = g(x - \tau u)$.

A further regularity property is also required, for example:

(iv) $\mathcal{C}(\alpha f)$ is C^∞ in a real neighborhood of $\alpha = 0$ and there exists a Schwarz space norm $|\cdots|$ and finite positive numbers a, b, c such that:

$$|\partial_\alpha^n \mathcal{C}(\alpha f)|_{\alpha=0} \leq ab^n(n!)^c |f|^n, \quad \forall n. \qquad (3.7)$$

Previous axioms on dμ yield the original Osterwalder-Schrader (OS) axioms on Schwinger functions:

(i)′ Euclidean invariance: $S_N(Ex_1, \ldots, Ex_N) = S_N(x_1, \ldots, x_N)$ for the subgroup of translations and rotations (and suitable test functions).

(ii)′ OS positivity: $\sum_{i,j=0}^{n} S_{i+j}(\theta f_i^* \times f_j) \geq 0$ for any set of functions $f_0, \ldots, f_n$, $f_0 \in \mathbf{C}$, $f_1 \in S_+$, $f_i \in S_{i,+}$ where $S_{i,+}$ is the set of functions f in $S((\mathbb{R}^d)^i)$ such that $f(x_1, \ldots, x_i) = 0$ unless $0 < t_1 < \cdots < t_i$ (and which vanish, together with all derivatives, if $x_\alpha = x_\beta$, $1 \leq \alpha \neq \beta \leq i$); $f^*(x_1, \ldots, x_i) = f(x_i, \ldots, x_1)$ and $f_i \times f_j(x_1, \ldots, x_{i+j}) = f_i(x_1, \ldots, x_i) f_j(x_{i+1}, \ldots, x_{i+j})$.

(iii)′ Cluster property:

$$\lim_{\tau \to \infty} S_{n+m}(f \times g_{\tau u}) - S_n(f) S_m(g) = 0$$

for all $u \neq 0$ and suitable test functions f, g in $S(\mathbb{R}^{dn})$ and $S(\mathbb{R}^{dm})$ respectively.

(iv)′ $S_0 = 1$, $S_n \in S'((\mathbb{R}^d)^n)$ and there is a Schwartz space norm and finite positive a, b, c such that (for test functions $f_1, \ldots, f_n$ in $S(\mathbb{R}^d)$):

$$|S_n(f_1, \ldots, f_n)| \leq a\, b^n (n!)^c \prod_{i=1}^{n} |f_i|$$

(v)′ Symmetry: $(S_n(x_1, \ldots, x_n) = S_n(\pi(x_1, \ldots, x_n))$ for any permutation π of the set of indices $1, \ldots, n$.

Euclidean invariance analytically continues to Lorentz invariance. Together with reflection positivity, one gets positivity of the norm in the Hilbert space of states and the (general) spectral condition. (Spectrum of the energy-momentum operator in $\bar{V}_+$.) Locality also follows (it is a consequence of the symmetry property in the OS framework). Ergodicity, or the cluster property, yields the uniqueness of the vacuum. These results and their proofs are described in detail in [OS] and [GJ]. They are not directly needed later and are thus omitted here.

1.3 The models

For definiteness, we consider below the models φ_d^4, $d \geq 2$, where d denotes as above the dimension of space-time. (Similar expressions of dμ can be written for other forms of the interaction.) The measure dμ is then, from a heuristic viewpoint and up to modifications needed for reasons explained below, of the form:

$$\mathrm{d}\mu(\varphi) =$$
$$\frac{1}{Z} \exp\left(-\lambda \int \varphi^4(x)\,\mathrm{d}x - \frac{m^2}{2}\int \varphi^2(x)\,\mathrm{d}x - \frac{b^2}{2}\int (\nabla\varphi)^2(x)\,\mathrm{d}x\right)\mathrm{D}\varphi$$
$$(3.8)$$

where $\mathrm{D}\varphi \equiv \text{``}\prod_{x\in\mathbf{R}^d}\mathrm{d}\varphi_x\text{,''}$ φ_x is, for each given point x, a real one-dimensional variable and $\mathrm{d}\varphi_x$ is the corresponding usual Lebesgue measure; λ is a coupling constant for the φ^4 interaction; and m is the bare mass. Finally, Z is a normalization factor (so that $\int \mathrm{d}\mu(\varphi) = 1$):

$$Z = \int \left\{\exp\left(-\lambda\int \varphi^4(x)\,\mathrm{d}x - \cdots\right)\right\}\mathrm{D}\varphi. \qquad (3.9)$$

Schwinger N-point functions are given, according to (3.3), as

$$S(x_1,\ldots,x_N) = \frac{1}{Z}\int \varphi(x_1)\cdots\varphi(x_N)\exp\left(-\lambda\int \varphi^4(x)\,\mathrm{d}x - \cdots\right)\mathrm{D}\varphi. \qquad (3.10)$$

In a lattice approximation in which Euclidean space-time is replaced by the set of all sites of, for example, a cubic lattice of size χ (that will tend to zero) contained in a finite volume Λ (that will tend to all $\mathbf{R}^d$), there is one real variable φ_i at each site i in Λ, $\mathrm{D}\varphi$ is replaced by $\prod_{i\in\Lambda}\mathrm{d}\varphi_i$, and $\int \varphi^4(x)\mathrm{d}x$ is replaced in the exponential by $\sum_{i\in\Lambda}\varphi_i^4$, with similar changes for other terms. For example, $\int (\nabla\varphi)^2(x)\mathrm{d}x$ can be replaced by a sum of the form $\sum_{i,j,i\neq j}\left[(\varphi_i - \varphi_j)/\chi\right]^2$ restricted to nearest neighbors.

Schwinger functions S_N, depending on N sites $i_1,\ldots,i_N$, are then given by the formula:

$$S_N(i_1,\ldots,i_N) = \frac{1}{Z}\int \varphi_{i_1}\ldots\varphi_{i_N}\exp\left(-\lambda\sum_{i\in\Lambda}\varphi_i^4 - \cdots\right)\prod_{i\in\Lambda}\mathrm{d}\varphi_i, \qquad (3.11)$$

where

$$Z = \int \exp\left(-\lambda\sum_{i\in\Lambda}\varphi_i^4 - \cdots\right)\prod_{i\in\Lambda}\mathrm{d}\varphi_i. \qquad (3.12)$$

Integrals (3.11) and (3.12) are well defined for $\chi > 0$ and finite Λ. However, the number of variables φ_i tends to infinity as $\chi \to 0$, $\Lambda \to \infty$. The limit is not, *a priori*, well defined and as a matter of fact, some modifications will be needed in order to get a well-defined, nontrivial measure in the limit. Before pursuing the heuristic discussion, we note that (3.8) can also be written, if, for example, $b = 1$, in the form:

$$d\mu(\varphi) = \frac{1}{Z} \exp\left(-\lambda \int \varphi^4(x)\,dx\right) d\nu(\varphi). \tag{3.13}$$

In (3.13), $d\nu(\varphi)$ is the "free" measure. (It would correspond to a theory of particles of mass m without interaction in Minkowski space.) It has covariance C, with

$$C(x, y) = \int \tilde{C}(p)e^{-ip(x-y)}\,dp, \tag{3.14}$$

$$\tilde{C}(p) = 1/\left(p^2 + m^2\right), \tag{3.15}$$

where $C(x, y)$ is the 2-point function, or propagator, associated to $d\nu$:

$$C(x, y) = \int \varphi(x)\varphi(y)\,d\nu(\varphi). \tag{3.16}$$

The characteristic function $\mathcal{C}$ associated to $d\nu$ is more generally given by the formula:

$$\mathcal{C}(f) \equiv \int e^{i\varphi(f)}\,d\nu(\varphi) = \exp\left(-\frac{1}{2}\int f(x)C(x, y)f(y)\,dx\,dy\right) \tag{3.17}$$

and N-point functions of $d\nu$ are sums of products of 2-point functions. (In other words, connected N-point functions vanish for $N > 2$.)

The size of the lattice mentioned above can be considered to correspond to an ultraviolet cut-off in the theory. Below, we remain in continuous space-time and introduce an ultraviolet cut-off (to be ultimately removed) directly in energy-momentum space, by replacing $d\nu$ with the free measure of covariance C_ρ with

$$\tilde{C}_\rho(p) = \tilde{C}(p) \exp\left[-M^{-2\rho}\left(p^2 + m^2\right)\right] \tag{3.18}$$

($\tilde{C}_\rho \to \tilde{C}$ in the $\rho \to \infty$ limit).

In order to define a satisfactory measure in the $\rho \to \infty$ limit, one has (renormalization) to replace the coupling λ, the bare mass m and possibly other couplings by constants λ_ρ, and m_ρ, that will depend on ρ and may tend to zero or to infinity in the $\rho \to \infty$ limit, actual physical quantities such as the renormalized coupling $\lambda_{\text{ren}}, \ldots$ being finite in that limit. We therefore start from a family of regularized measures $d\mu_{\Lambda,\rho}$ of the form:

$$d\mu_{\Lambda,\rho}(\varphi) = \frac{1}{Z_{\Lambda,\rho}} \int \exp\left(-\lambda_\rho \int_\Lambda \varphi^4(x)\,dx - \mu_\rho \int_\Lambda \varphi^2(x)\,dx\right.$$
$$\left. - \zeta_\rho \int (\nabla\varphi)^2(x)\,dx\right) d\nu_\rho(\varphi),$$

$$(3.19)$$

where coefficients μ_ρ, ζ_ρ (and possibly other couplings) are still introduced in the exponential, even though $D\varphi$ has been replaced by $d\nu_\rho(\varphi)$. The actual measure $d\mu$ will be obtained, when possible, in the $\Lambda \to \infty$, $\rho \to \infty$ limits. In the models φ_d^4, λ_ρ can be chosen independent of ρ at $d = 2$ and $d = 3$. At $d = 2$, it is sufficient to introduce the term $\mu_\rho = -6\lambda C_\rho(0,0)$ (which, as $C_\rho(0,0)$, becomes infinite in the $\rho \to \infty$ limit), for reasons already transparent in the perturbative approach (Sect. 2). Equivalently, $\varphi^4(x)$ in the exponential can be replaced, as easily seen by the Wick product:

$$:\varphi^4(x):_\rho = \varphi(x)^4 - 6C_\rho(0,0)\varphi^2(x) + 3C_\rho(0,0)^2, \qquad (3.20)$$

where the terms $C_\rho(0,0)$ and $C_\rho(0,0)^2$ correspond to contractions ⃝ and ⬯ of some of the fields involved at x ($C_\rho(0,0) = C_\rho(x,x)$).

At $d = 4$, one would have to consider a coupling λ_ρ that would tend to infinity, but there are strong doubts on the existence of φ_4^4 as a nontrivial theory (different from a free theory, e.g. with $\lambda_{\text{ren}} \neq 0$) in the limit. Triviality in the limit has been established at $d > 4$. In the Gross-Neveu model, λ_ρ will tend to zero as $\rho \to \infty$, but the measure $d\mu$ will be well defined and nontrivial in the limit ($\lambda_{\text{ren}} > 0$).

We start in Sect. 2 with the perturbative approach in which Λ can be removed from the outset. We consider there, for definiteness, the models φ_d^4 including $d = 4$: φ_4^4 is nontrivial from the perturbative viewpoint. As already mentioned, models $P(\varphi)_2$ and GN_2 will be treated nonperturbatively in Sects. 3 (and 5.1) and 4 respectively. The treament of φ_3^4 and of (massless, infrared) φ_4^4 with ultraviolet cut-off will be mentioned in Sect. 5.2.

§2. The perturbative approach

2.1 Perturbative series

For definiteness, we again consider models φ_d^4 unless otherwise stated with, so far, a given coupling λ. Starting initially from the expression (3.13) of $d\mu$ and expanding the exponential $e^{-\lambda \int \varphi^4}$ $(= 1 - \lambda \int \varphi^4 + \cdots)$, one obtains the following formal expansion of $N(x_1, \ldots, x_N; \lambda) = \int \varphi(x_1) \cdots \varphi(x_N) e^{-\lambda \int \varphi^4(x)dx} d\nu(\varphi)$:

$$N(x_1, \ldots, x_N; \lambda) = \sum_{n \geq 0} \frac{\lambda^n}{n!} N^{(n)}(x_1, \ldots, x_N) \qquad (3.21)$$

$$N^{(n)}(x_1, \ldots, x_N) =$$
$$\int dy_1 \cdots dy_N \int \varphi(x_1) \cdots \varphi(x_N) \varphi^4(y_1) \cdots \varphi^4(y_n) d\nu(\varphi). \qquad (3.22)$$

A similar expansion, in which the product $\varphi(x_1) \cdots \varphi(x_n)$ in (3.22) is removed, is obtained for $Z(\lambda) = \int e^{-\lambda \int \varphi^4} d\nu(\varphi)$. The integral $\int \varphi(x_1) \cdots \varphi(x_N)[\prod_{j=1}^n \varphi^4(y_j)] d\nu(\varphi)$ can be (formally) evaluated as follows: to each point x_i, or y_j, is associated a vertex $\bullet\!\!-\!\!$, or $\times$, with one or four legs respectively (according to the power of φ). The integral is then the sum, over all "contraction schemes" (i.e. ways of pairing together two legs in order to get internal lines, each of which joins two vertices), of the product of propagators $C(t_\ell, t'_\ell)$ associated to each line ℓ joining vertices t_ℓ, t'_ℓ (which may be points x_i or y_i): see example in Figure 3.1.

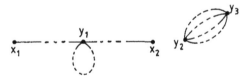

Figure 3.1 A contraction scheme ($N = 2$, $n = 3$).

Each contraction scheme yields a graph made of one or several connected components. The corresponding contribution to $N^{(n)}(x_1, \ldots, x_N)$ is the integral, over $y_1, \ldots, y_n$, of the corresponding product of propagators. It is then easily seen (formally) that the expansion of $S(x_1, \ldots, x_N; \lambda) = N(x_1, \ldots, x_N; \lambda)/Z(\lambda)$ is of the same form as above, except that the only graphs to be kept are those in which each point y_j belongs to a connected component with at least one point x_i. Finally, it is easily seen, in view of the definition of connected functions, that the only relevant graphs for connected N-point functions are connected graphs (one connected component):

$$S_N^c(x_1, \ldots, x_N; \lambda) = \sum_{n \geq 0} \frac{\lambda^n}{n!} \sum_{\substack{\text{connected graphs G} \\ \text{with } n \text{ 4-leg vertices}}} I_G(x_1, \ldots, x_N),$$

(3.23)

where

$$I_G(x_1, \ldots, x_N) = \int dy_1 \ldots dy_n \prod_{\ell \in G_{\text{int}}} C(y_\ell' - y_\ell'') \prod_{\ell \in G_{\text{ext}}} C(x_\ell - y_\ell).$$

(3.24)

The products in the integrand of (3.24) are over internal lines ℓ joining two 4-leg vertices y_ℓ', y_ℓ'' (among $y_1, \ldots, y_n$) and over "external" lines joining a 4-leg vertex y_ℓ to a point x_ℓ. The Fourier transform $\tilde{I}_G(p_1, \ldots, p_N)$ of $I_G(x_1, \ldots, x_N)$ is of the form:

$$\tilde{I}_G(p_1, \ldots, p_N) = \left[\prod_{i=1}^N \tilde{C}(p_i) \right] \int \prod_{\ell \in G_{\text{int}}} \tilde{C}(k_\ell) \prod_v \delta_v(\text{e.m.c.}) \prod_\ell dk_\ell$$

(3.25)

with one energy-momentum variable k_ℓ for each internal line ℓ and one energy-momentum conservation (emc) δ-function for each (4-leg) vertex v (between internal and external energy-momenta involved at v). An overall emc δ-function $\delta(p_1 + \cdots + p_N)$ can be extracted. We thus also write

$$\tilde{I}_G(p_1, \ldots, p_N) = \left[\Pi \, \tilde{C}(p_i) \right] \delta(\Sigma p_i) \tilde{F}_G(p_1, \ldots, p_N)$$

(3.26)

and in turn,

$$\tilde{F}_N(p_1, \ldots, p_N; \lambda) = \sum_{n \geq 0} \frac{\lambda^n}{n!} \sum_{\substack{\text{connected graphs G} \\ \text{with } n \text{ 4-leg vertices}}} \tilde{F}_G(p_1, \ldots, p_N), \quad (3.27)$$

where, at $N > 2$, $\tilde{F}_N$ is obtained from $\tilde{S}_N^c(p_1, \ldots, p_N; \lambda)$ by factorizing out $[\Pi \, \tilde{C}(p_i)]\delta(\Sigma p_i)$.

Some of the integrals $\tilde{F}_G$ are not well-defined functions (or distributions) in view of ultraviolet divergences as some of the integration variables (obtained after elimination of remaining δ-functions) tend to infinity. They are due equivalently in x-space to local singularities of the propagators. (Integration at infinity in x-space can be made without problem, if $m > 0$, due to the exponential fall-off, in $\mathrm{e}^{-m|x|}$, of $C(x)$ as $|x| \to \infty$ and to the fact that graphs under consideration are connected, so that all points y_j are linked to the external points x_i.) The simplest case of a divergent diagram, at $d \geq 2$, is due to the occurrence of factors

$$C(y, y) \equiv C(0, 0) = \int \tilde{C}(p)\mathrm{d}p = \bigcirc \,,$$

for example, in the connected part containing x_1, x_2 in Figure 3.1. Other divergent diagrams, of the form $-\!\!\ominus\!\!-$, occur for the 2-point function at $d = 3$ and further ones, such as $\rangle\!\bigcirc\!\langle$ for the 4-point function at $d = 4$.

Given a graph G, its "superficial degree of divergence" $\omega(G)$ is

$$\omega(G) = (d - 4)n + \frac{2 - d}{2}N + d \tag{3.28}$$

which is in fact equal to

$$\omega(G) = dL(G) - 2\ell(G), \tag{3.29}$$

where $\ell(G)$ is the number of internal lines and $L(G)$ the number of independent closed loops. The definition is similarly extended (with relevant changes) to more general models (involving, for example, spin) and leads to the following classification:

- super-renormalizable theories: only a finite number of graphs with superficial divergences ($\omega(G) \geq 0$), for example, φ_2^4, φ_3^4.
- renormalizable theories: the number of graphs with superficial divergences is infinite but the number N of external lines of these graphs is bounded, e.g. φ_4^4 (with N bounded by four).
- nonrenormalizable theories: an infinite number of graphs with superficial divergences with an arbitrary large number of external lines.

It can be shown (and a proof is given in Sect. 2.2) that $\tilde{F}_G$ is well defined (and is analytic in Euclidean p-space for $m > 0$) if $\omega(G) < 0$, and also $\omega(G') < 0$ for any subgraph G' of G. This is no longer the case if $\omega(G) \geq 0$ (or $\omega(G') \geq 0$ for a subgraph G'). Standard perturbative renormalization is the formal redefinition of the measure $\mathrm{d}\mu$, as the limit of measures $d\mu_\rho$, in a way such that the expansion with respect to a renormalized coupling λ_{ren} will involve only convergent "renormalized" integrals. In φ^4 at $d = 2$, the inclusion of the term $6\lambda C_\rho(0,0) \int \varphi^2(x) dx$ in the exponential amounts to remove formally all Feynman diagrams involving loops $\bigcirc$ (which are exactly cancelled by new diagrams with 2-leg vertices and coupling $\mu_\rho = 6\lambda C_\rho(0,0)$). As a consequence, all integrals are now well defined, even in the $\rho \to \infty$ limit. At $d = 4$, the usual perturbative expansion is made with respect to a renormalized coupling $\lambda_{\text{ren},\rho}$ defined by a relation of the form:

$$\lambda_{\text{ren},\rho} = (\tilde{F}_4)_\rho (p_1, \ldots, p_4) \qquad (3.30)$$

at a given point $p_1, \ldots, p_4$, for example, $p_1 = \cdots = p_4 = 0$. As $(\tilde{F}_4)_\rho(0, \ldots, 0)$, $(\lambda_{\text{ren}})_\rho$ admits a formal perturbative expansion in the bare coupling λ. By introducing similarly a renormalized mass $m_{\text{ren},\rho}$ and a renormalized constant $\zeta_{\text{ren},\rho}$ linked to the 2-point function and its second derivative, for example,

$$1/m^2_{\text{ren},\rho} = \left[\tilde{S}_2^{(\rho)}(p_1, p_2) / \delta(p_1 + p_2) \right]_{|p_1 = -p_2 = 0}, \qquad (3.31)$$

one gets formally perturbative expansions of connected N-point functions in $\lambda_{\text{ren},\rho}$ analogous to the above, except that each Feynman integral is replaced by a "renormalized" integral. For each connected graph G, the latter is a sum of terms associated with all possible "forests"; a forest of G is a collection of connected superficially divergent ($\omega(G') \geq 0$) subgraphs of G, including possibly G itself, that do not overlap. Given two subgraphs G', G'' of the forest, they are either disjoint (no common vertex) or one of them is strictly contained in the second one. Hence, a forest is characterized by a first level of disjoint subgraphs G' contained in no other subgraph of the forest, in each of them zero, one or more disjoint subgraphs, and so forth. The empty forest (no subgraph) is included and gives the usual Feynman integral. The contribution corresponding, for example, to a nonempty forest whose subgraphs all have 4 outgoing legs ($\omega = 0$) is the Feynman integral of the graph in which all subgraphs of the first level are replaced by points, multiplied by a product of constants $c(G')$ for each subgraph G' of the forest. $c(G')$ is minus the value, at zero external energy-momenta, of the Feynman integral of the subgraph in which all further subgraphs (included in it) of the next level have been reduced to points. This contribution corresponds to a suitable first order Taylor expansion of the integrand. More generally, higher order Taylor expansions are needed (of order two when subgraphs G' with 2 outgoing legs, $\omega(G') > 0$, are involved).

For each G and each given ρ, all contributions above are well defined. Although many of the terms involved (including the constants) tend to infinity as $\rho \to \infty$, it can be shown that the *sum* of all contributions is well defined (i.e. finite) in the $\rho \to \infty$ limit.

In section 2.2 we outline more recent approaches to perturbative renormalization, namely, expansions in terms of "effective couplings" rather than "renormalized" couplings. This approach has some advantages, as discussed in [Ri], and is closer to that used later in the constructive approach.

2.2 Perturbative renormalization: phase-space analysis and effective expansions

We consider, for definiteness, φ_4^4 (simpler treatments are available for φ_2^4 and φ_3^4). The perturbative series of N-point functions are considered in Euclidean space-time, and one starts from a decomposition of each propagator C or C_ρ obtained by Fourier transformation from a decomposition of $\tilde{C}$ or $\tilde{C}_\rho$ in energy-momentum space. Although a continuous version can be given, we use a "discrete" version in view of later purposes in constructive theory. A usual decomposition is, for example:

$$C_\rho(x) = \sum_{i=1}^{\rho} C^{(i)}(x), \qquad (3.32)$$

where, for some $M > 1$

$$\tilde{C}^{(1)}(p) = \tilde{C}(p)\, \mathrm{e}^{-V_1(p)} \qquad (3.33)$$

$$\tilde{C}^{(i)}(p) = \tilde{C}(p)\left[\mathrm{e}^{-V_i(p)} - \mathrm{e}^{-V_{i-1}(p)}\right] \qquad \text{for } i > 1 \qquad (3.34)$$

with

$$V_i(p) = M^{-2i}\left(p^2 + m^2\right). \qquad (3.35)$$

The propagator $\tilde{C}^{(i)}$ is (roughly) localized in the slice $M^{i-1} \lesssim |p| \lesssim M^i$, due to the exponential factor $\exp -V^{(i)}(p)$ at large values of p ($|p| > M^i$), or to the difference between the two exponentials, which produces a factor of the order of $(|p|M^{-i})^2$, in the region $|p| < M^{i-1}$. On the other hand, $C^{(i)}$ satisfies the bound

$$\left|C^{(i)}(x)\right| < K\, M^{2i}\mathrm{e}^{-\delta M^i|x|}, \qquad (3.36)$$

where $K > 1$ and $\delta > 0$ are fixed constants. The bound (3.36) (and in fact a better one with an exponential factor $\exp -\delta M^{2i}|x|^2$) is established by direct inspection. (We note that, in contrast to $\tilde{C}$, or $\tilde{C}_\rho$, which has a pole at $p^2 = -m^2$ in complex space away from Euclidean space, $\tilde{C}^{(i)}$ has no pole, the pole of $\tilde{C}$ being cancelled by a factor $p^2 + m^2$ coming from the difference $e^{-V_i} - e^{-V_{i-1}}$. The fact that $C^{(i)}$ decays faster than any exponential is then a consequence of the ordinary Fourier-Laplace transform theorem.) Both the behaviour at small x (bound in M^{2i}, which tends to infinity as $i \to \infty$ as a souvenir of the divergence of $C(x)$ at $x = 0$) and the fall-off at large x are relevant for later bounds. Exponential fall-off is not crucial, but will give some technical simplifications. A fall-off in $(1 + M^i|x|)^{-\alpha}$, $\alpha > 4$, ensuring integrability of $|C^{(i)}(x)|$ and the bound

$$\int |C^{(i)}(x)| \, dx < \text{cst } M^{-2i} \tag{3.37}$$

would be sufficient for present purposes.

Other decompositions of C with more or less accurate localization of $C^{(i)}$ and $\tilde{C}^{(i)}$ in space-time (around the origin) and in energy-momentum space (around $|p| \sim M^i$) respectively, can be considered according to their purposes. One may, for example, consider functions $\tilde{C}^{(i)}$ of the form $\tilde{C}(p)\chi_i(p)$, each χ_i being a sufficiently smooth function with compact support around the region $M^{(i-1)} \leq |p| \leq M^i$ and equal to one locally. $\tilde{C}^{(i)}$ then has a good localization in energy-momentum space which may be useful for "energy-momentum conservation" arguments (whereas the previous choice gives a rather rough localization at small $|p|$), but $C^{(i)}(x)$ decays at best rapidly as $|x| \to \infty$ (if χ_i is C^∞). An alternative choice providing, in some nontrivial sense, localization up to exponential fall-off on both sides is also possible (see Sect. 5 of the Appendix). For present purposes we may stick to the choice (3.32)–(3.35).

We first present a proof of the convergence, in the $\rho \to \infty$ limit, of ordinary Feynman integrals for graphs without superficial divergences. The proof is explained for φ_4^4, in which case graphs G considered here do not include subgraphs with only 2 or 4 outgoing lines, but it can be adapted similarly to more general theories. Given G, we treat for definiteness the integral

$$A_G \equiv \int dy_2 \cdots dy_n \prod_{\ell \in G_{\text{int}}} C\left(y'_\ell - y''_\ell\right) \tag{3.38}$$

also equal to $\tilde{F}_G(0, \ldots, 0)$. The passage to integrals in which external energy-momenta are nonzero is similar.

Theorem 3.1. There exists a constant K (independent of n) such that for any connected φ_4^4 graph G without superficial degrees of divergences ($\omega(G') < 0$, $\forall G' \subseteq G$), the Feynman amplitude (3.38) is absolutely convergent and the following bound holds:

$$|A_G| \leq K^n, \tag{3.39}$$

where $n = n(G)$ is the number of vertices of G.

Proof. We follow [FMRS3] with some slight modifications. (For earlier proofs by different methods, see [dCR].) We consider below a theory with ultraviolet cut-off (i.e. C is replaced by C_ρ) and establish the bounds (3.39) with a constant K uniform with respect to ρ. The existence of the $\rho \to \infty$ limit (and the bound (3.39) in the limit) are established similarly.

Using decompositions (3.32) for each C_ρ involved, one gets

$$A_G = \sum_\mu A_{G,\mu}, \tag{3.40}$$

$$A_{G,\mu} = \int A_{G,\mu}(y_1, \ldots, y_n)\, dy_2 \cdots dy_n, \tag{3.41}$$

where the sum (3.40) is over all possible values, called "momentum assignments," of $\mu = \{i_\ell\}$, with one i_ℓ, $1 \leq i_\ell \leq \rho$, for each internal line ℓ of G and

$$A_{G,\mu}(y_1, \ldots, y_n) = \prod_{\ell \in G_{\text{int}}} C^{(i_\ell)}(y'_\ell - y''_\ell). \tag{3.42}$$

A contribution $A_{G,\mu}(y_1, \ldots, y_n)$ associated in general to a graph, such as the graph G of Figure 3.2a, is represented diagrammatically in Figure 3.2b.

A bound on each $A_{G,\mu}$ is obtained as follows. Given any slice i, the part of the graph in slices $\geq i$ is made up of one or several connected parts $G_{i,\alpha}$, $\alpha = 1, \ldots, \alpha_0(i)$. (The latter are connected together, if $\alpha_0(i) > 1$, by parts of the graph below i.) The number of legs issued from $G_{i,\alpha}$ below i will be denoted $e_{i,\alpha}$. It will be convenient to use the formula (if $M \geq 2$)

$$\exp\left(-\delta M^i |x|\right) \leq \exp\left(-\frac{\delta}{2} \sum_{j \leq i} M^j |x|\right), \tag{3.43}$$

which will allow one to "distribute" each factor $\exp\left(-\delta M^i |y'_\ell - y''_\ell|\right)$ in the bounds (3.36) on each $C^{(i)}$, of slice i, into related factors in all slices

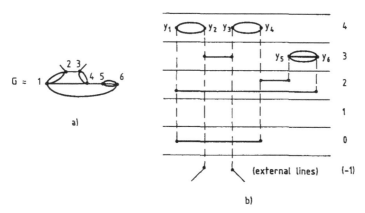

Figure 3.2 (a) A graph G. (b) A term $A_{G,\mu}(y_1,\ldots,y_6)$. Each vertex of G has 4 legs, each of which belongs to a given slice, joined by a dotted vertical line.

$\leq i$. The procedure below is inductive, from slice ρ to slice 1. First, one "fixed" vertex occurring in slice ρ is chosen in each $G_{\rho,\alpha}$. There will later be integration over the positions of all other vertices encountered in $G_{\rho,\alpha}$ with the help of the exponential fall-off factors associated to each propagator of $G_{\rho,\alpha}$. We next consider, in each $G_{\rho-1,\alpha}$, the set of all new vertices encountered in slice $\rho - 1$, including, possibly, previous "fixed" vertices of slice ρ. Among this set, one fixed vertex will again be chosen, and there will be integration over the positions of all others, and so forth. We note that, if we denote $z_1,\ldots,z_r$, $r \equiv r(i,\alpha)$, the set of vertices thus considered in each given $G_{i,\alpha}$ (with, for example, z_1 "fixed"), the exponential fall-off factors arising, as explained above, from each propagator (before integrations) yield at least an exponential fall-off factor $\exp -\delta' M^i \ell(z_1,\ldots,z_r)$ for each $G_{i,\alpha}$, where $\ell(z_1,\ldots,z_r)$ is the minimal length of all trees joining $z_1,\ldots,z_r$. To see this, one may use the inequality (see, e.g. [DIS])

$$(\ell(z_1,\ldots,z_r) \geq)\ \ L(z_1,\ldots,z_r) \geq \ell(z_1,\ldots,z_r)/2, \qquad (3.44)$$

where L is the minimal length of all connected graphs joining $z_1,\ldots,z_r$ and possibly other points.

Integration over $z_2,\ldots,z_r$ then yields a factor cst $M^{-4i(r-1)}$ in the bounds. At that stage, a bound of the following form is then easily established:

$$|A_{G,\mu}| \leq \mathrm{cst}^n \prod_{\ell \in G_{\mathrm{int}}} M^{2i(\ell)} \prod_{(i,\alpha)} M^{-4i[r(i,\alpha)-1]}. \qquad (3.45)$$

A line ℓ in slice $i(\ell)$ belongs to $i(\ell)$ parts $G_{i,\alpha}$ such that $i \leq i(\ell)$. The factor $M^{2i(\ell)}$ can then be written as a product of factors M^2 for each $G_{i,\alpha}$ that contains ℓ. If, on the other hand, $M^{-4i(r(i,\alpha)-1)}$ is written as a product of factors $M^{-4(r(i,\alpha)-1)}$ attributed to all $G_{i',\alpha}$, $i' \leq i$, containing the $r(i,\alpha)$ vertices of $G_{i,\alpha}$, one obtains

$$|A_{G,\mu}| \leq \text{cst}^n \prod_{(i,\alpha)} \left[(M^2)^{\text{total number of lines in } G_{i,\alpha}} \right.$$
$$\left. \times M^{-4(\text{total number of vertices in } G_{i,\alpha}-1)} \right] \quad (3.46)$$

and in turn,

$$|A_{G,\mu}| \leq \text{cst}^n \prod_{(i,\alpha)} M^{-(e_{i,\alpha}-4)}. \quad (3.47)$$

We note that $e_{i,\alpha} - 4 \geq 2$ for the graphs under consideration, so that, for example, $M^{-(e_{i,\alpha}-4)} \leq M^{-e_{i,\alpha}/3}$. It is then easy to see that

$$\prod_{(i,\alpha)} M^{-e_{i,\alpha}/3} \leq \prod_v M^{-h(v)/3}, \quad (3.48)$$

where the product in the right-hand side runs over all vertices of G and $h(v)$ is the difference between the highest and lowest slices in which v occurs. It can finally be checked that the "vertical" exponential fall-off factors $M^{-h(v)/3}$ between slices occuring in (3.48) allow one to sum over μ and to obtain the bounds (3.39) (external lines of G being by convention fixed in slice 0, or -1, as shown in figure 3.2). QED

Renormalization. The previous proof no longer applies if there are subgraphs with 2 or 4 outgoing lines (since $e_{i,\alpha}$ may be equal in some cases to 2 or 4, so that $e_{i,\alpha} - 4 \leq 0$). For simplicity, we restrict our attention to subgraphs with possibly 4 outgoing lines ($\omega = 0$). Following [FMRS3], the "usefully renormalized" term $A_{G,\mu}^{\text{UR}}$, which will be finite in the $\rho \to \infty$ limit, is defined for each given ρ inductively, starting from slice ρ. Four-point functions $G^{(\rho)}(z_1, \ldots, z_4)$ of slice ρ, associated to connected parts $G_{\rho,\alpha}$ with $e_{\rho,\alpha} = 4$ (and equal to the corresponding product of propagators of $G_{\rho,\alpha}$ integrated over all vertices, except the possibly four vertices $z_1, \ldots, z_4$ at which outgoing lines below slice ρ are attached) are "regularized," namely $G^{(\rho)}$ is replaced by

$$G_{\text{reg.}}^{(\rho)}(z_1, \ldots, z_4) = G^{(\rho)}(z_1, \ldots, z_4) - \left\{ \delta(z_1 - z_2)\delta(z_1 - z_3)\delta(z_1 - z_4) \right.$$
$$\left. \times \int G^{(\rho)}(z_1, z_2', z_3', z_4') dz_2' \, dz_3' \, dz_4' \right\}. \quad (3.49)$$

We note that $\int G^{(\rho)}(z_1, \ldots, z_4)\, dz_2 dz_3 dz_4$ is equal to $(\tilde{F}_G)^{(\rho)}(0, \ldots, 0)$, where $(\tilde{F}_G)^{(\rho)}$ is the Fourier transform of $G^{(\rho)}$ after factorization of $\delta(p_1 + \cdots + p_4)$.

The integral of the product $G^{(\rho)}_{\text{reg}}(z_1, \ldots, z_4)$ with the propagators $C^{(i_1)}(z_1, \cdot)$, $C^{(i_2)}(z_2, \cdot)$, $C^{(i_3)}(z_3, \cdot)$, $C^{(i_4)}(z_4, \cdot)$ in slices $< \rho$ can be written in the form:

$$
\int dz_1 \cdots dz_4 G^{(\rho)}(z_1, \ldots, z_4)
$$
$$
\times \Big\{ C^{(i_1)}(z_1, \cdot)\big[C^{(i_2)}(z_2, \cdot) - C^{(i_2)}(z_1, \cdot)\big].C^{(i_3)}(z_3, \cdot)C^{(i_4)}(z_4, \cdot)
$$
$$
+ C^{(i_1)}(z_1, \cdot)C^{(i_2)}(z_1, \cdot)\big[C^{(i_3)}(z_3, \cdot) - C^{(i_3)}(z_1, \cdot)\big]C^{(i_4)}(z_4, \cdot)
$$
$$
+ C^{(i_1)}(z_1, \cdot)C^{(i_2)}(z_1, \cdot)C^{(i_3)}(z_1, \cdot)\big[C^{(i_4)}(z_4, \cdot) - C^{(i_4)}(z_1, \cdot)\big] \Big\}.
$$
$$
(3.50)
$$

Each difference $C^{(i_\ell)}(z_k, \cdot) - C^{(i_\ell)}(z_1, \cdot)$ will then be written in the form:

$$
C^{(i_\ell)}(z_k, \cdot) - C^{(i_\ell)}(z_1, \cdot) = (z_k - z_1) \cdot \nabla C^{(i_\ell)}\big(z_1 + \theta.(z_k - z_1)\big). \quad (3.51)
$$

The factor $z_k - z_1$ will be attributed to slice ρ and will generate (in view of bounds of the form $\int |x\, C^{(i)}(x)|\, dx < \text{cst }M^{-3i}$) a factor $M^{-\rho}$ in the bounds. It is called below a regularization factor. The gradient, acting in slice i_k, with generate a factor M^{i_k}.

The previous procedure is pursued inductively from slice ρ to slice 1 with, at each stage, regularization of 4-point functions.

The term $A^{\text{UR}}_{G,\mu}$ finally obtained can be bounded in modulus by methods similar to those already described in the proof of the bounds (3.39). Apart from some technical unessential problems, the main fact is that each $e_{i,\alpha}$ is replaced, in view of the new factors $M^{-\rho}$, M^{i_k} mentioned above, by $e'_{i,\alpha}$ where

$$
e'_{i,\alpha} = e_{i,\alpha} + [\text{number of internal regularizations}-
$$
$$
\text{number of internal gradients}] \quad (3.52)
$$

and where internal regularizations or gradients in (3.52) are those involved in $G_{i,\alpha}$. If $e_{i,\alpha} = 4$, the bracket is at least equal to 1 (namely the number of gradients on *outgoing* lines below i). As a consequence, $A^{\text{UR}}_{G,\mu}$ can be treated as $A_{G,\mu}$ in the previous theorem. Uniform bounds as $\rho \to \infty$ and the existence of the $\rho \to \infty$ limit follow.

The proof above is not sufficient to obtain results on the renormalized amplitude $(A_G)_{\text{ren}}$ introduced in Sect. 2.1, which is *not* equal to

$\sum_\mu A^{\mathrm{UR}}_{G,\mu}$. The latter sum has to be completed in this approach in a nontrivial way, by introducing "useless" renormalizations, in the terminology of [FMRS3], in order to reobtain $(A_G)_{\mathrm{ren}}$. This is not the purpose of this text, and we therefore omit this part, but describe instead the following perturbative expansion in terms of "effective" couplings λ_ρ, $\lambda_{\rho-1}, \ldots, \lambda_1$ attached to each momentum index i. The latter are defined inductively as follows. First,

$$\lambda_{\rho-1} = \lambda_\rho + \delta\lambda_\rho, \qquad (3.53)$$

where $\delta\lambda_\rho$ is the sum, over all possible connected graphs of slice ρ with 4 outgoing lines in slices $< \rho$, of the functions $(\tilde{F}_G)^{(\rho)}(0)$ introduced for each such graph. Other quantities $\delta\lambda_{\rho-1}, \delta\lambda_{\rho-2}, \ldots$, and in turn $\lambda_{\rho-2}$, $\lambda_{\rho-3}, \ldots$, are defined inductively by the value at zero energy-momenta of the 4-point functions of slice $\rho-1$, $\rho-2$, $\ldots$, (themselves defined inductively according to the procedure previously described. At stage $i < \rho$, $\delta\lambda_i$ is a sum over contributions similar to above, in which there is at least one propagator in slice i).

The "biped free" connected functions (in which subgraphs with 2 outgoing lines are still disregarded) can then be written in the form:

$$S_{bf} = \sum_{G,\mu} \left[\prod_{v \in G} -\lambda_{i(v,\mu)} \right] A^{\mathrm{UR}}_{G,\mu}, \qquad (3.54)$$

where, for each vertex v of G, $i(v, \mu)$ is the largest momentum slice in which a propagator line is attached to the vertex v.

Formula (3.54) is checked (formally) by inspection for each given ρ, starting from the original expression in $\lambda = \lambda_\rho$, in view of cancellations between various terms involved when $\lambda_{\rho-1}$, $\lambda_{\rho-2}, \ldots$, are reexpressed in terms of λ_ρ.

For similar expansions of the complete connected functions, see [Ri].

§3. The $P(\varphi)_2$ models

3.1 Preliminaries

For definiteness, we restrict our attention to φ_2^4. Other $P(\varphi)_2$ models are treated similarly. We shall mainly consider 4-point functions. The analysis of N-point functions is made with the same methods, with results given at the end of Sect. 3.3. As mentioned in Sect. 1, we first consider a theory with cut-off Λ, ρ, namely N-point functions

$S_{\Lambda,\rho}(x_1,\ldots,x_N) = N_{\Lambda,\rho}(x_1,\ldots,x_N)/Z_{\Lambda,\rho}$. $N_{\Lambda,\rho}$ and $Z_{\Lambda,\rho}$ are given by the formulae:

$$N_{\Lambda,\rho}(x_1,\ldots,x_N) = \int \varphi(x_1)\cdots\varphi(x_N)$$
$$\times \left\{ \exp\left(-\lambda\int_\Lambda {:}\varphi^4(x){:}_\rho\,\mathrm{d}x\right)\right\}\mathrm{d}\nu_\rho(\varphi), \quad (3.55)$$

$$Z_{\Lambda,\rho} = \int \left\{ \exp\left(-\lambda\int_\Lambda {:}\varphi^4(x){:}_\rho\,\mathrm{d}x\right)\right\}\mathrm{d}\nu_\rho(\varphi), \quad (3.56)$$

where $\mathrm{d}\nu_\rho$ is the free measure of covariance C_ρ, with C_ρ given, for example, in (3.18), and ${:}\varphi^4(x){:}_\rho$ is the Wick product defined in (3.20).

The parameter ρ will be left implicit and N is taken for definiteness equal to 4, unless otherwise stated. Note that the variable N introduced below in (3.64) has a different meaning. It will be convenient for later purposes in Chapter IV to "extract" the external propagators $C(x_i,\cdot)$. This is made nonperturbatively via an "integration by parts" of the fields $\varphi(x_i)$ that yields

$$N_\Lambda(x_1,\ldots,x_4) = \int_{\substack{z_i\subset\Lambda\\i=1,\ldots,4}} \mathrm{d}z_1\cdots\mathrm{d}z_4 \prod_{i=1}^{4} C(x_i-z_i)$$
$$\times N'_\Lambda(z_1,\ldots,z_4) + R_\Lambda(x_1,\ldots,x_4), \quad (3.57)$$

where R_Λ will not contribute to the connected 4-point function and N'_Λ is a sum of terms containing 0, 1, 2 or 3 δ-functions between $z_1,\ldots,z_4$. The term with three δ-functions, once divided by Z_Λ, gives $\lambda\,\delta(z_1-z_2)\delta(z_1-z_3)\delta(z_1-z_4)$ and corresponds to the trivial graph with only one vertex. The term without δ-functions, to which we restrict our attention for definiteness, is

$$I_\Lambda(z_1,\ldots,z_4) = \int\left[\prod_{i=1}^{4} -4\lambda\,{:}\varphi^3(z_i){:}\right]\exp\left(-\lambda\int_\Lambda {:}\varphi^4(x){:}\,\mathrm{d}x\right)\mathrm{d}\nu(\varphi),$$
$$(3.58)$$

where the Wick product ${:}\varphi^3(x){:}$ is equal to $\varphi(x)^3 - 3C_\rho(0,0)\varphi(x)$.

We then define

$$H_\Lambda(x_1,\ldots,x_4) = I_\Lambda(x_1,\ldots,x_4)/Z_\Lambda, \quad (3.59)$$
$$H^c_\Lambda(x_1,\ldots,x_4) = \text{connected part of } H_\Lambda. \quad (3.60)$$

In Sect. 3.2, we first present cluster expansions of I_Λ. They will provide expansions of the form:

$$I_\Lambda\left(x_1,\ldots,x_4\right) = \sum_{q\geq 1}\sum_{X_1,\ldots,X_q}\prod_{i=1}^{q}I\left(X_i;\{x_r\}_i\right), \qquad (3.61)$$

where the sum runs over partitions of the set of unit squares of Λ into subsets $X_1,\ldots,X_q$, $\{x_r\}_i$ is the set of points $x_1,\ldots,x_4$ that belong to a square of X_i, and (leaving the index i implicit)

$$I\left(X;\{x_r\}\right) = \Sigma_G I\left(X;\{x_r\};G\right). \qquad (3.62)$$

In (3.62), the sum runs over a class of *connected* graphs G linking squares of X with at most one line between any pair of squares (lines of G join here squares rather than points; G is connected if the graph obtained by associating a vertex to each square, that is, by replacing squares with points, is connected). An explicit propagator C is associated to each line. $I(X;\cdot)$ and $I(X;\cdot;G)$ depend on X but not on Λ. If X is composed of only one square Δ containing no point x_r, $I(X)$ is equal to a constant $a \neq 0$, independent of Δ, so that (3.61) can be rewritten in the form:

$$I_\Lambda\left(x_1,\ldots,x_4\right) = a^{|\Lambda|}\sum_{q\geq 1}{\sum_{X_1,\ldots,X_q}}'\prod_{i=1}^{q}\left(I\left(X_i;\{x_r\}_i\right)a^{-|X_i|}\right), \qquad (3.63)$$

where the second sum runs now over (nonoverlapping) subsets $X_1,\ldots,$ X_q of squares of Λ, whose union is not in general Λ, such that $|X_i| \geq 2$, or $|X_i| = 1$ and $x_r \in X_i$ for at least one index r. ($|\Lambda|$ is the number of unit squares in Λ.)

Cluster expansions of Z_Λ are similar (with no point x_r). It will then be shown in turn, if, for example, $x_1,\ldots,x_4$ belong to different squares $\Delta_{0,1},\ldots,\Delta_{0,4}$ that H_Λ^c admits an expansion of the form

$$H_\Lambda^c\left(x_1,\ldots,x_4\right) = \sum_{N\leq|\Lambda|-4}\frac{1}{N!}\sum_{\substack{\Delta_1,\ldots,\Delta_N \\ \Delta_i\subset\Lambda,i=1,\ldots,N}}\sum_{\mathbf{G}}\mathbf{I}\left(X;\{x_r\};\mathbf{G}\right),$$

$$(3.64)$$

where squares $\Delta_1,\ldots,\Delta_N$ now vary independently in Λ, X is the set of squares $\Delta_{0,1},\ldots,\Delta_{0,4}$ and $\Delta_1,\ldots,\Delta_N$, and the last sum runs over graphs $\mathbf{G}$. These graphs are composed of propagator lines and new "Mayer" lines joining pairs of the squares $\Delta_{0,1},\ldots,\Delta_{0,4},\Delta_1,\ldots,\Delta_N$, and are connected when both types of lines are taken into account. The factor **I** is a product of factors $a^{-|X|}I$ associated with subgraphs already connected by propagator lines and of factors χ associated to Mayer lines. Given a Mayer line joining Δ_α, Δ_β, $\chi = 0$ if $\Delta_\alpha \neq \Delta_\beta$ and $\chi = -1$ if Δ_α and Δ_β coincide in Λ.

From bounds on the terms $|I(X; \cdot; G)|$ (and partial regroupings of terms in (3.64)), uniform bounds on H_Λ^c (independent of Λ) and the existence of the $\Lambda \to \infty$ limit of H_Λ^c will then be established in Sect. 3.3 for any given ρ and sufficiently small values of λ. In order to treat the $\rho \to \infty$ limit, bounds on the previous terms a and I that are uniform in ρ are first established. This is briefly mentioned in Sect. 3.2 and is described in more detail in Sect. 5.1.

3.2 Cluster expansions and related expansions of N-point functions

Pairwise cluster expansion. A dependence of I_Λ on auxiliary variables $s_{\Delta,\Delta'}$ is introduced by replacing the propagator $C(x,y)$, which defines the measure $d\nu(\varphi)$, by $C(x,y;s)$, $s = \{s_{\Delta,\Delta'}\}$, where

$$
\begin{aligned}
C(x,y;s) \;\; &= s_{\Delta,\Delta'}C(x,y) \text{ if } x \in \Delta, \; y \in \Delta' \\
&\qquad \text{or } x \in \Delta', \; y \in \Delta \\
C(x,y;s) \;\; &= C(x,y) \text{ if } x,y \text{ lie in the same square.}
\end{aligned} \tag{3.65}
$$

This procedure is only formal for bosonic models like $P(\varphi)_2$ because $C(x,y;s)$ is not in general of a positive type, so that the associated measure $d\nu(\varphi;s)$ is no longer a well-defined probability measure, but it already gives the main ideas. If $s_{\Delta,\Delta'} = 0$, $C(x,y;s) = 0$ for $x \in \Delta$, $y \in \Delta'$ or vice versa, and Δ, Δ' are decoupled in the measure. A first order Taylor expansion around the origin in all variables $s_{\Delta,\Delta'}$ yields:

$$
I_\Lambda(x_1,\ldots,x_4) \equiv I_\Lambda(x_1,\ldots,x_4;s)|_{s=1} = \sum_G I(\Lambda;\{x_r\};G), \tag{3.66}
$$

where $s = \{s_{\Delta,\Delta'}\}$, $s = 1$ means $s_{\Delta,\Delta'} = 1$, $\forall(\Delta,\Delta')$, the sum in the right-hand side runs over all connected or nonconnected graphs G composed of lines joining squares of Λ, with at most one line between any pair of squares, and

$$
I(\Lambda;\{x_r\};G)
$$

$$
= \left[\int_0^1 \cdots \int_0^1 \prod_{\ell \in G} ds_\ell \left(\prod_{\ell \in G} \frac{d}{ds_\ell} \right) I_\Lambda(\{x_r\};s) \Bigg|_{\substack{s_{\Delta,\Delta'}=0 \\ \text{if}(\Delta,\Delta')\notin G}} \right]. \tag{3.67}
$$

In (3.67), $s_\ell \equiv s_{\Delta_\ell, \Delta'_\ell}$ if $\Delta_\ell, \Delta'_\ell$ are the squares joined by line ℓ, and $(\Delta, \Delta') \notin G$ means that there is no line of G joining Δ, Δ'. As an example, a graph G is shown in Figure 3.3.

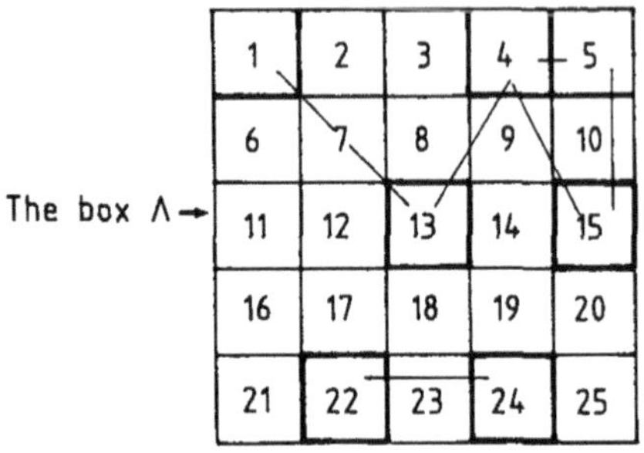

The box $\Lambda \rightarrow$

1	2	3	4	5
6	7	8	9	10
11	12	13	14	15
16	17	18	19	20
21	22	23	24	25

Figure 3.3 A graph G in the cluster expansion; G divides Λ into 20 connected subsets X_i, where $X_1 = \{1, 13, 4, 5, 15\}$, and $X_2 = \{22, 24\}$. Remaining subsets are composed of one square.

By direct calculation, one gets from (3.67) a more explicit expression of $I(\Lambda; \cdot; G)$. In view of later purposes, we give it for an arbitrary set X of squares of Λ, whose union is not necessarily Λ, and a graph G with lines joining squares of X; $\{x_r\}$ then denotes the set of points $x_1, \ldots, x_4$ that belong to X:

$$I(X; .; G) =$$

$$\left\{ \prod_{\ell \in G} \int_0^1 ds_\ell \int_{(u_\ell, v_\ell) \in \Delta_\ell \times \Delta'_\ell} du_\ell\, dv_\ell\, C(u_\ell, v_\ell) \frac{\delta}{\delta\varphi(u_\ell)} \frac{\delta}{\delta\varphi(v_\ell)} \right\}$$

$$\left(\prod_{r; x_r \in X} [-4\lambda : \varphi^3(x_r):] \right) e^{-\lambda \int_\Lambda :\varphi^4(x): dx} d\nu(\varphi; s) \Bigg|_{\substack{s_{\Delta, \Delta'}=0 \\ \text{if } (\Delta, \Delta') \notin G}} .$$

$$(3.68)$$

The graph G divides Λ into subsets X_i of squares joined by connected subgraphs G_i (or isolated squares): see example in Figure 3.3. All variables $s_{\Delta, \Delta'}$ associated to pairs (Δ, Δ') such that Δ, Δ' do not belong to a common subset are fixed at zero, so that there is factorization of the measure and in turn,

$$I_\Lambda(x_1, \ldots, x_4; G) = \prod_{i=1}^q I(X_i; \{x_r\}_i; G_i). \qquad (3.69)$$

Remark. Formula (3.69) can be understood as follows from a perturbative viewpoint. $I_\Lambda(s)$ is a (formal) sum of diagrams of the type described in Sect. 2 with propagators $C(s)$. All diagrams involving propagators joining points belonging to different subsets X_i, X_j thus vanish and (3.69) follows. Each $I(X_i; \cdot; G_i)$ is a sum of diagrams including explicit propagators $C(u_\ell, v_\ell)$ associated to lines of G_i, and also an arbitrary number of nonexplicit propagators $C(\cdot; s)$ joining vertices belonging to squares joined by a line of G and propagators $C(\cdot)$ joining vertices inside each square. These diagrams may here be connected or nonconnected, considered as diagrams between interaction vertices; the total number of vertices, over which there is integration in respective squares, is arbitrarily high.

If X_i contains only one square Δ, G_i is empty. If, moreover, there is no point x_r in Δ, $I(X_i; \cdot)$ is equal to constant a independent of Δ (which can be shown to be nonzero):

$$a = \int e^{-\lambda \int_\Delta :\varphi^4(x): \mathrm{d}x} \mathrm{d}\nu_\Delta(\varphi), \qquad (3.70)$$

where $\mathrm{d}\nu_\Delta$ has covariance $C(x, y)$ if $x, y \in \Delta$ and zero otherwise.

Formulae (3.61)–(3.63) follow from (3.66), (3.69) and (3.70).

Inductive cluster expansion. The inductive cluster expansion is a modified version of the pairwise expansion based on similar ideas and in which all $C(x, y; s)$ will be of a positive-type as linear combinations, with ≥ 0 coefficients, of positive-type terms. First, an (arbitrary) square Δ_1 is chosen. $C(x, y; s_1)$ is then defined as

$$C(x, y; s_1) = C(x, y) \text{ if } x, y \text{ both belong to } \Delta_1 \text{ or to } \Lambda \backslash \Delta_1, \quad (3.71)$$

$$C(x, y; s_1) = s_1 C(x, y) \text{ otherwise.} \qquad (3.72)$$

Fixing s_1 at zero will now uncouple Δ_1 and $\Lambda \backslash \Delta_1$. A Taylor formula entails:

$$I_\Lambda(x_1, \ldots, x_4) \equiv I_\Lambda(x_1, \ldots, x_4; s_1)|_{s_1=1}$$

$$= I_\Lambda(x_1, \ldots, x_4; 0) + \int_0^1 \mathrm{d}s_1 \frac{\mathrm{d}}{\mathrm{d}s_1} I_\Lambda(\cdot; s_1), \quad (3.73)$$

where

$$\frac{\mathrm{d}}{\mathrm{d}s_1} I_\Lambda = \sum_{\Delta_2 \in \Lambda \backslash \Delta_1} I_{\Delta_1, \Delta_2}(\cdot; s_1), \qquad (3.74)$$

$$I_{\Delta_1,\Delta_2}\left(\cdot\,;s_1\right) = \int_{(u,v)\in\Delta_1\times\Delta_2} \mathrm{d}u\,\mathrm{d}v\,C(u,v)\frac{\delta}{\delta\varphi(u)}\frac{\delta}{\delta\varphi(v)}$$
$$\left(\prod_{r;x_r\in\Delta_1\cup\Delta_2}\left[-4\lambda{:}\varphi^3\left(x_r\right){:}\right]\right)\mathrm{e}^{-\lambda\int_\Lambda{:}\varphi^4(x){:}\,\mathrm{d}x}\mathrm{d}\nu\left(\varphi;s_1\right). \tag{3.75}$$

The term $I_\Lambda(\cdot\,;0)$ in (3.73) is equal to the product $I_{\Delta_1}\,I_{\Lambda\setminus\Delta_1}$ (where each term includes respective points x_r in Δ_1 or $\Lambda\setminus\Delta_1$). This factorization, and also formulae (3.74) and (3.75), are due to the same reasons as those described previously in the pairwise expansion. The same procedure as that presently described will be applied to $I_{\Lambda\setminus\Delta_1}$. Coming back to $I_{\Delta_1,\Delta_2}(\cdot\,;s_1)$, it is treated by introducing a new variable s_2 and the propagator

$$C\left(x,y;s_1,s_2\right) = C\left(x,y;s_1\right)\text{ if }x,y\text{ both belong to}$$
$$\Delta_1\cup\Delta_2\text{ or to }\Lambda\setminus\left(\Delta_1\cup\Delta_2\right), \tag{3.76}$$
$$C\left(x,y;s_1,s_2\right) = s_2 C\left(x,y;s_1\right)\text{ otherwise.} \tag{3.77}$$

Here, $\Delta_1\cup\Delta_2$ is thus decoupled at $s_2 = 0$ from its complement $\Lambda\setminus\left(\Delta_1\cup\Delta_2\right)$. One then writes (leaving variables x_r implicit)

$$I_{\Delta_1,\Delta_2}\left(s_1\right) \equiv I_{\Delta_1,\Delta_2}\left(s_1,s_2=1\right)$$
$$= I_{\Delta_1,\Delta_2}\left(s_1,s_2=0\right) + \int_0^1\frac{\mathrm{d}}{\mathrm{d}s_2}I_{\Delta_1,\Delta_2}\left(s_1,s_2\right)\mathrm{d}s_2 \tag{3.78}$$

with

$$\frac{\mathrm{d}}{\mathrm{d}s_2}I_{\Delta_1,\Delta_2}\left(s_1,s_2\right) = \sum_{\substack{\Delta_3\neq\Delta_1,\Delta_2\\\Delta_3\subset\Lambda\setminus(\Delta_1\cup\Delta_2)}} I_{\Delta_1,\Delta_2,\Delta_3}\left(s_1,s_2\right) \tag{3.79}$$

and so forth.

The final result is again an expansion of I_Λ of the form (3.61)–(3.63), the expressions of $I(X;\cdot\,;G)$ being, however, somewhat different from those introduced in the pairwise expansion. The sum Σ in (3.62) now runs over *trees* joining squares of X and the contribution obtained for each tree is a sum over various procedures P that give rise to the same tree:

$$I(X;G) = \sum_P I(X;G,P), \tag{3.80}$$

where the set $\{x_r\}$ has been left implicit. Each procedure P is characterized by the order in which successive lines of the tree appear. The expression of $I(X; G, P)$ is analogous to (3.68) except that $\mathrm{d}\nu(\varphi; s)$ has the covariance $C(x, y; s)$ that has been described here, and the integrand now includes a term $M_{G,P}(s)$ which is a product of some of the variables $s_\ell(\ell \in G)$, depending on the procedure P. The following equality can be checked for each tree:

$$\sum_{\substack{\text{procedures } P \\ \text{giving rise to } G}} \int M_{G,P}(s) \prod_{\ell \in G} \mathrm{d}s_\ell = 1. \tag{3.81}$$

An example of tree G and a procedure P giving rise to G is shown in Figure 3.4.

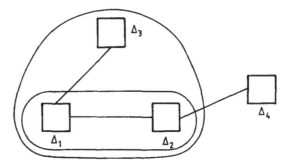

Figure 3.4 A procedure P generating a tree G between 4 squares.

Bounds on $I(X; \cdot; G)$. The following basic result can be established. Some indications on the proof are given below, the independence of the bounds on ρ being explained in a more detailed way in Sect. 5.1.

Lemma 3.1. If there is no point x_r in X, there exist, for any $\varepsilon > 0$, constants $C_1(\varepsilon)$, C_2, independent of X, G (and Λ), as also of ρ, such that

$$|I(X; G)| < \left[\prod_{\ell \in G} |\lambda|^{1/2} C_1(\varepsilon) e^{-(m-\varepsilon)\mathrm{d}(\ell)}\right] C_2^{|X|}, \tag{3.82}$$

where $\mathrm{d}(\ell)$ is the distance of the centers of the squares of X joined by line ℓ.

If there are points x_r in X, a similar bound applies, with possibly a loss of a fixed power of $|\lambda|$ (depending on the number of points x_r in X) and a further multiplicative factor $(1 + |\ln \ \mathrm{Inf}|x_r - x_{r'}||)^6$, where the Inf is over pairs (r, r'), $x_r, x_{r'} \in X$, $r \neq r'$, for small values of $|x_r - x_{r'}|$.

Proof (outline). Lemma 3.1 applies in particular to the integral (3.70) in a unit square Δ (with no point x_r). The uniformity with respect to ρ as $\rho \to \infty$ is nontrivial, already in this case, because the product $\varphi^4(x)$ in the exponential has been replaced by $:\varphi^4(x):$ which, in contrast to $\varphi^4(x)$, is no longer always positive so that $\exp -\lambda \int :\varphi^4:$ is no longer bounded by one. (This replacement is needed: see remark below this proof.) A bound in $\exp(\text{cst } \rho^2)$ can be established easily by using the fact that $C_\rho(0,0) \approx \text{cst } \rho$ at large ρ, the inequality

$$\alpha^4 - 6\rho\alpha^2 + 3\rho^2 \geq -6\rho^2 \tag{3.83}$$

for any real α, and $\int d\nu_\rho(\varphi) = 1$. However, it is insufficient to control the $\rho \to \infty$ limit. A bound by a constant independent of ρ can be established from the more refined procedure explained in Sect. 5.1. (It is also shown that $a_\rho \to a \neq 0$ in that limit.) In order to get more general bounds on $|I(X;G)|$, we first note that, in view of (3.81), it is sufficient to bound $|I(X;G,P)|$. The exponential fall-off factors $e^{-(m-\varepsilon)d(\ell)}$ in (3.82) arise from the decrease of the propagators $C(u_\ell, v_\ell)$ occurring in the integrand in the expression of $I(X;G,P)$ (analogous to (3.68)). These propagators decay as do $e^{-m|u_\ell - v_\ell|}$ at large $|u_\ell - v_\ell|$. Factors $e^{-\varepsilon|u_\ell - v_\ell|}$ are kept to control the sum over the terms generated by the derivatives $\delta/\delta\varphi$ and, for each of these terms, the sum over contraction schemes of the fields. The independence of the bound on P and s arises from the inequality

$$0 \leq C(x,y;s) \leq C(x,y), \tag{3.84}$$

which gives directly the inequality $\int \pi\varphi \, d\mu(\varphi;s) \leq \int \pi\varphi \, d\mu(\varphi)$ and can also be shown to yield the inequality $\int \exp -\lambda \int_X :\varphi^4: d\nu(\varphi;s) \leq \text{cst}^{|X|}$, with a constant independent of s. The factor $\lambda^{1/2}$ in (3.82) for each line arises from the fact that there are four fields by vertex and that propagators are obtained by contraction of two fields. The power of λ is different if there are points x_r in X because some propagators can be obtained by contraction of fields $\varphi(x_r)$. On the other hand, the factor $(1+\ln)^6$ accounts for the fact that a propagator obtained by contraction of $\varphi(x_r)$, $\varphi(x_{r'})$ behaves at short distances like $\ln|x_r - x_{r'}|$.

Remark. The replacement of $\varphi^4(x)$ by $:\varphi^4(x):$ in the exponential was previously justified by perturbative arguments (in order to avoid divergences of Feynman integrals). The nonperturbative analysis would yield N-point functions that would all be identically zero otherwise in the $\rho \to \infty$ limit.

Expansion of 4-point functions. While the procedure below might be applied in a simpler form for the purposes of this section (introduction of nonoverlap factors $1 + \chi$ between subsets X_i rather than between squares), we present it in a form that is also convenient in Chapter IV for the study of irreducible kernels. Let $\Delta_{0,1}, \ldots, \Delta_{0,\sigma}$ denote the different squares of Λ containing one or more points x_r. Equation (3.63) can be rewritten, via elementary manipulations, in the form:

$$
\begin{aligned}
I_\Lambda\left(x_1, \ldots, x_4\right) = a^{|\Lambda|} \sum_{N \leq |\Lambda| - \sigma} & \frac{a^{-(N+\sigma)}}{N!} \\
\times \sum_{\substack{\Delta_1, \ldots, \Delta_N \\ \Delta_i \subset \Lambda, i=1, \ldots, N}} & \prod_{\substack{(\alpha,\beta), \alpha < \beta \\ \alpha,\beta=(0,1),\ldots,(0,\sigma),1,\ldots,N}} \left(1 + \chi\left(\Delta_\alpha, \Delta_\beta\right)\right) \\
\times \sum_{\substack{\text{partitions } \omega_1, \ldots, \omega_q \\ \text{of } (0,1),\ldots,(0,\sigma),1,\ldots,N; \\ q \geq 1}}' & \sum_{G_1, \ldots, G_q} \prod_{j=1}^q I\left(X_j; \{x_r\}_j; G_j\right),
\end{aligned}
\tag{3.85}
$$

where $\chi\left(\Delta_\alpha, \Delta_\beta\right) = -1$ (i.e. $1 + \chi = 0$) if $\Delta_\alpha = \Delta_\beta$, $\chi = 0$ (i.e. $1 + \chi = 1$) if $\Delta_\alpha \neq \Delta_\beta$. The factors $1 + \chi$ account for the fact that $\Delta_1, \ldots, \Delta_N$ have *a priori* to be different from each other and from $\Delta_{0,1}, \ldots, \Delta_{0,\sigma}$, and the sum $\sum'$ runs over partitions including no subset ω_j of one element among $1, \ldots, N$. The factor $(N!)^{-1}$ accounts for the fact that each configuration of given subsets X_j is obtained $N!$ times when $\Delta_1, \ldots, \Delta_N$ vary in Λ. The graphs G_j are connected graphs between the indices of ω_j with the same properties as above and $X_j = \{\Delta_\gamma\}_{\gamma \in \omega_j}$. Factors $1 + \chi\left(\Delta_\alpha, \Delta_\beta\right)$ for $\alpha, \beta = (0,1), \ldots, (0,\sigma)$ are equal to one and can thus be removed. The introduction of factors $1 + \chi$ is analogous to a related procedure used in the Mayer expansion in statistical mechanics.

Through expansion of the product of factors $1 + \chi$, one obtains an expression of the form:

$$
\begin{aligned}
I_\Lambda\left(x_1, \ldots, x_4\right) = a^{|\Lambda|} \sum_{N \leq |\Lambda| - \sigma} & \frac{1}{N!} \sum_{\substack{\Delta_1, \ldots, \Delta_N \\ \Delta_i \subset \Lambda, i=1, \ldots, N}} \sum_{\substack{\omega_1, \ldots, \omega_q \\ q \geq 1}} \\
& \sum_{\mathbf{G}_1, \ldots, \mathbf{G}_q} \prod_{j=1}^q \mathbf{I}\left(X_j; \{x_r\}_j; \mathbf{G}_j\right),
\end{aligned}
\tag{3.86}
$$

where the graphs $\mathbf{G}_j$ include lines associated with propagators as before and "Mayer" lines associated with factors χ. They must be connected when both types of lines are taken into account. The factors **I** are defined as indicated below (3.64). In particular, **I** vanishes if squares $\Delta_\alpha, \Delta_\beta$ joined by a Mayer line do not coincide in Λ. The value of $I(X; G)$ when some squares of X coincide is arbitrary. It can be taken equal to zero in this section, but a different choice is useful in Chapter IV.

The separation of those ω_j that contain one or more of the indices $(0, 1), \ldots, (0, \sigma)$ and of remaining ones leads to a factorization of I_Λ into the term Z_Λ (no point x_r) and a term which is equal to H_Λ in view of (3.59). Finally, the definition of connected functions yields in turn the expression (3.64), if $x_1, \ldots, x_4$ belong to different squares. Otherwise, H_Λ^c is expressed as a sum of similar terms associated either with the partition π of $(1, \ldots, 4)$ into subsets such that points x_r in each subset belong to a common square, or with (nontrivial) subpartitions of the latter.

Remark. From a perturbative viewpoint, H_Λ^c is a sum of Feynman integrals of *connected* graphs (with vertices integrated in Λ). This perturbative content does *not* directly appear on the expression (3.64) in which the graphs **G** that are involved are connected only with respect to squares. From the perturbative viewpoint, each contribution to H_Λ^c in (3.64), corresponding to given squares $\Delta_1, \ldots, \Delta_N$ and a given graph **G**, appears as a sum of graphs involving explicit propagator lines and other lines (either inside each square or between squares joined by an explicit line). These propagators join vertices that belong to one of the squares involved, but the graphs are *not* connected in general. They become connected only if all squares are replaced by points, with all squares joined by a Mayer line being replaced by the same point. The absence of such nonconnected graphs in the sum comes from cancellations between contributions corresponding to different sets of Mayer lines. For example, the perturbative graph

can be obtained in several ways: for instance, y_1 may lie in $\Delta_{0,1}$ *or* in a new square Δ_i occupying the same position as $\Delta_{0,1}$ and joined to $\Delta_{0,1}$ by a Mayer line that gives a factor -1.

3.3 Infinite-volume limit, Euclidean axioms, Borel summability

Theorem 3.2. H_Λ^c has a well-defined limit H^c when $\Lambda \to \infty$. Moreover, $\forall \varepsilon > 0$, $\exists \lambda_\varepsilon > 0$ and a constant C_ε (independent of λ) such that for any (finite or infinite) Λ and any $\lambda, |\lambda| \leq \lambda_\varepsilon$

$$|H^c_\Lambda (x_1, \ldots, x_4)| < |\lambda|^4 C_\varepsilon \mathrm{e}^{-(m-\varepsilon)L(x_1,\ldots,x_4)} (1 + |\ln \ \mathrm{Inf}_{i\neq j}|x_i - x_j||)^6, \tag{3.87}$$

where $L(x_1, \ldots, x_N)$ is (as in Sect. 2) the minimal length of all connected graphs (or trees) joining $x_1, \ldots, x_N$ and possibly intermediate points.

Remark. The last factor $(1 + |\ln \ \mathrm{Inf}|x_i - x_j||)^6$ is included to take into account the short distance behaviour of H^c_Λ. The behaviour at large distances in $\exp\{-\mathrm{cst}\ L(x_1, \ldots, x_4)\}$, or more generally $\exp\{-\mathrm{cst}\ L(x_1, \ldots, x_N)\}$ for N-point functions (see below), has been established in [EMS]. It is analogous to related results first given in statistical mechanics in [DIS]. Results including the existence of the $\Lambda \to \infty$ limit with a somewhat weaker type of exponential decay have been obtained previously (see [GJ] and references therein to works of these authors). In view of (3.44), the decay in $\exp\{-\mathrm{cst}\ L(x_1, \ldots x_N)\}$ is equivalent, as indicated in [DIS], to $\exp -\{\mathrm{cst}'\ \ell(x_1, \ldots, x_N)\}$ where ℓ is the minimal length of all *trees* joining $x_1, \ldots, x_N$ *without* intermediate points. This fact is useful for, among other things, the later derivation of Borel summability properties for φ^4_2, again as in the related problems of statistical mechanics (in a simpler framework) treated in [DIS].

Proof of Theorem 3.2. We follow here [IM1]. The proof of the uniform bounds (3.87) for any finite Λ is obtained from the expression (3.64) of H^c_Λ and Lemma 3.1 of Sect. 3.2, but partial regroupings of terms are needed; there are otherwise too many terms (due to Mayer lines). If we first consider the sum (3.64) restricted to graphs that are already connected by propagator lines, it is convenient in this section to regroup together all terms associated with graphs that differ only by Mayer lines. Equivalently, we consider only graphs without Mayer lines and the summation over $\Delta_1, \ldots, \Delta_N$ is restricted as originally by the nonoverlap conditions between squares. A common uniform factor $\mathrm{e}^{-(m-\varepsilon')L(x_1,\ldots,x_4)}$ is first extracted for all terms by extracting a factor $\mathrm{e}^{-(m-\varepsilon')\mathrm{d}(l)}$ from each factor $\mathrm{e}^{-(m-\varepsilon)\mathrm{d}(l)}$ of the bounds (3.82) of Lemma 3.1. If $\varepsilon' > \varepsilon$, we still have at our disposal a factor $\mathrm{e}^{-(\varepsilon'-\varepsilon)\mathrm{d}(l)}$ for each line. The result then follows easily from inequalities of the form:

$$\sum_N \frac{C^N}{N!} \sum_{\substack{\Delta_1,\ldots,\Delta_N \\ \Delta_i \cap \Delta_j = \phi \ \mathrm{if} \ i\neq j}} \sum_{T(0,1,\ldots,N)} \prod_{l \in T} \mathrm{e}^{-\eta\mathrm{d}(l)} \leq C', \tag{3.88}$$

if the constant C is sufficiently small, a condition fulfilled in the application for sufficiently small λ. In (3.88), the last sum runs over trees T

between vertices $0, 1, \ldots, N$. To prove (3.88), one may note that the left-hand side is bounded, for each tree $T(0, \ldots, N)$, by $[C \sum_{\Delta \neq \Delta_0} \mathrm{e}^{-\eta \mathrm{d}(l)}]^N$, and (3.88) follows from the fact that the number of trees is bounded by $\mathrm{cst}^N N!$.

The treatment of the actual sum (3.64) is slightly more complicated. Given a connected graph $\mathbf{G}$, let $G_1, G_2, \ldots$ be its subgraphs already connected by propagator lines. We again regroup together all graphs that differ only by Mayer lines *inside* each G_i, so that it is sufficient to consider graphs without Mayer lines (but with nonoverlap conditions between squares) inside each G_i. Further regroupings of terms are still needed. A skeleton graph $\hat{\mathbf{G}}$ is associated to each graph $\mathbf{G}$ as follows. Starting from the subgraph, denoted G_1, that contains the index $(0, 1)$, we consider all subgraphs G_i, $i \neq 1$, of a "first shell," namely those linked to G_1 by at least one Mayer line. If there are several Mayer lines between G_1 and G_i, all are removed, except that line attached to the vertex of lowest index in G_1. The order of vertices of G_1 is their natural order in the sequence $(0, 1), \ldots, (0, \sigma), 1, \ldots, N$. All Mayer lines between the subgraphs G_i of the first shell are also removed. Next, one considers all subgraphs of a "second shell," namely those attached to subgraphs of the first shell by Mayer lines. The same procedure as above is applied (using, for example, the natural order of all vertices of all subgraphs of the first shell), and so forth. The summation over all graphs $\mathbf{G}$ that have the same skeleton $\hat{\mathbf{G}}$ amounts to the reintroduction of nonoverlap conditions between squares that belong to different subgraphs of the same shell, and also various further nonoverlap conditions between some of the squares that belong to different shells. The uniform bounds of Theorem 3.2 will follow from uniform bounds in modulus on all contributions corresponding to different skeleton graphs, the nonoverlap functions just mentioned, and also the factors -1 of each Mayer line, being bounded in modulus by 1. A common uniform factor $\mathrm{e}^{-(m-\varepsilon')L(x_1, \ldots, x_4)}$, $\varepsilon' > \varepsilon$, is again extracted for all terms.

For any given N, skeleton graphs that give nonzero contributions (taking into account nonoverlap conditions) are of the form shown in Figure 3.5. After some combinatorics, one is then led to a summation over all possible graphs G_1 with an arbitrary number of vertices and zero or one subgraphs of a next shell attached to each vertex by a Mayer line, and so forth.

The bound (3.88), or more precisely, its analogue in which the constant C is replaced by $2C$, is then applied starting from subsets in the most remote shell up to the first one. (The fact that subsets may be attached to each vertex of a subgraph leads to add a fixed multiplicative constant,

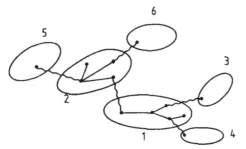

Figure 3.5 A skeleton graph giving a nonzero contribution; ⌁⌁⌁ denotes here a Mayer line.

for example, 2 for each vertex in the bounds.) A bound independent of the number of shells is then obtained.

The fact that H_Λ^c converges to a well-defined function H^c in the $\Lambda \to \infty$ limit can be established with a precise control of how fast the limit is obtained by showing in a way similar to above that $|H_\Lambda^c - H_{\Lambda'}^c|$ is uniformly bounded by $C(\Lambda)$, $C(\Lambda) \to 0$ as $\Lambda \to \infty$, $\forall \Lambda' \supset \Lambda$. In fact, exponential fall-off factors with the distance between the set of points $x_1, \ldots, x_4$ and the boundary of Λ arise since $H_\Lambda^c - H_{\Lambda'}^c$ is a sum over connected graphs and over squares $\Delta_1, \ldots, \Delta_N$, one of which at least belongs to $\Lambda' \backslash \Lambda$. The independence of H^c on the way the limit is obtained (i.e. on the sequence of boxes Λ) can also be similarly established. QED

N-point functions. The analysis of N-point functions is similar. The existence of the $\Lambda \to \infty$ limit is established and decay in $e^{-(m-\varepsilon')L(x_1,\ldots,x_N)}$ is obtained for connected functions when the "distance" $L(x_1, \ldots, x_N)$ between the points $x_1, \ldots, x_N$ tends to infinity. In view of their expression in terms of connected functions, one directly obtains, in turn, the factorization of nonconnected functions $S(x_1, \ldots, x_N)$ into the product $S(x(I))S(x(J))$ for great separation of two subgroups $x(I), x(J)$ of points, with a remainder that falls exponentially like $e^{-(m-\varepsilon')d(x(I),x(J))}$, where d is the distance between the two subgroups.

The following more precise bounds on connected functions can be established. Let us consider the N-point function $S_{(\alpha)}(x_1, \ldots, x_N) = N_{(\alpha)}(x_1, \ldots, x_N)/Z$, where $\alpha = \{\alpha_i\}$ and

$$N_{(\alpha)}(x_1, \ldots, x_N) = \int \left(\prod_{i=1}^N :\varphi^{\alpha_i}(x_i): \right) e^{-\lambda \int :\varphi^4(x): \, dx} d\nu(\varphi). \quad (3.89)$$

Then, it can be shown by an adaptation of previous methods that the connected function $S_{(\alpha)}^c(x_1, \ldots, x_N)$ satisfies, after integration of each

x_i in a square Δ_i with possibly an L^2 test function f_i, bounds of the form, $\forall \varepsilon > 0, |\lambda| < \lambda_\varepsilon, \operatorname{Re} \lambda > 0$:

$$\left| \int S^c_{(\alpha)}\left(x_1, \ldots, x_N\right) \left[\prod f_i(x_i) \right] \mathrm{d}x_1 \ldots \mathrm{d}x_N \right| \leq C_1 \left(C_2^{\sum \alpha_i} \right) \prod_{i=1}^{\alpha} \left\| f_i \right\|$$

$$\cdot \left[\left[\prod_{j=1}^{k} \left(\sum_{i; \Delta_i = \Delta'_j} \alpha_i \right)!^{1/2} \right] N! e^{-(m-\varepsilon) L(\Delta_1, \ldots, \Delta_N)} \right], \qquad (3.90)$$

where $\Delta'_1, \ldots, \Delta'_k$ are the distinct squares among $\Delta_1, \ldots, \Delta_N$. (The constants C_1, C_2 may depend on ε. On the other hand, $L(\Delta_1, \ldots, \Delta_N) \equiv L(\Delta'_1, \ldots, \Delta'_k)$ is the minimal length of connected graphs joining the squares involved and possibly intermediate points.)

Euclidean axioms. We briefly outline the derivation of the Euclidean axioms (i)' to (v)' on N-point functions given in Sect. 1.2. (For the adaptation to the axioms (i) to (iv) on the measure $\mathrm{d}\mu$, see [Fro].)

(i)' Euclidean invariance: for 4-point functions or similarly for N-point functions, it is obtained in the same way as the independence of the $\Lambda \to \infty$ limit with respect to sequences of boxes Λ (see end of the proof of Theorem 3.2): if τ is a given Euclidean transformation, $S^c_\Lambda(\tau(x_1, \ldots, x_N))$ is clearly equal to $S^c_{\tau^{-1}\Lambda}(x_1, \ldots, x_N)$. On the other hand, the limits of S_Λ and $S_{\tau^{-1}\Lambda}$ coincide: the differences can again be bounded by uniform factors times exponential fall-off factors in the distance between the given set of points $x_1, \ldots, x_N$ and the union of the boundaries of Λ and $\tau^{-1}\Lambda$.

(ii)' O.S. positivity: a possible proof can be obtained by first considering a lattice approximation of $\mathbb{R}^2$, in which case, O.S. positivity can be established by explicit inspection, whatever the size of the lattice is. Hence, the result is still valid in the limit when this size tends to zero. (It is shown, by arguments similar to above, that this limit does coincide with that previously obtained.)

(iii)' Cluster property: it follows from the factorization property already mentioned of N-point functions for great separation of subgroups of points.

(iv)' Particular aspect of the bounds of the form (3.90) (which provide in particular more precise values of the constant c).

(v)' Symmetry holds in a straightforward way.

Borel summability. The perturbative series $\sum_n a^{(c)}_{N,n}(x_1, \ldots, x_N) \lambda^n$ of each N-point (or connected N-point) function S (or S^c) is in general divergent, however small λ is. This result, which can be proved in many cases, is expected from results on individual Feynman integrals and on the number of these integrals. For example, individual integrals of φ_2^4 are bounded by (and are for large n of the order of) cst^n and the number of graphs for each n is of the order of $\mathrm{cst}^n n!$, so that $a_{N,n}$ behaves like $\mathrm{cst}^n n!$ at large n. It is easily shown that the Borel transformed series $\sum_n a_{N,n}^{(c)} \frac{t^n}{n!}$ is absolutely convergent in a circle around the origin in complex t-space, where it defines an analytic function $B_N^{(c)}(.\,;t)$ of t. As a nontrivial byproduct of previous methods and results, it can moreover be shown, as outlined below, that this function can be analytically continued in a strip around the real axis (see Figure 3.6b), where it satisfies bounds including exponential increase at most in $\mathrm{e}^{\mathrm{cst}\ t}$ as $t \to \infty$, and that $S^{(c)}(\cdot\,;\lambda)$ can in turn be recovered from $B^{(c)}(\cdot\,;t)$ through the Borel formula

$$S^{(c)}(\cdot\,;\lambda) = \int_0^\infty \mathrm{e}^{-t/\lambda} B(\cdot\,;t)\mathrm{d}t. \qquad (3.91)$$

Moreover, $S^{(c)}(.\,;\lambda)$, considered as a function of λ in the region $\lambda \geq 0$, is infinitely differentiable and its nth derivatives at $\lambda = 0$ are equal to $n! a_{N,n}^{(c)}$.

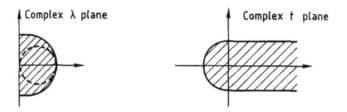

Figure 3.6 Initial domains of analyticity of B in t (Fig. b) and of S in λ (Fig. a)

The results just mentioned are a consequence, via a general mathematical theorem of Nevanlinna on Borel summability, of the following preliminary results (i) and (ii):

(i) Analyticity of $S^{(c)}(.\,;\lambda)$ in λ in a domain of the form shown in Figure 3.6a.

In fact, analyticity holds by direct inspection for $\mathrm{Re}\ \lambda > 0$, $|\lambda|$ sufficiently small, for the function $S_\Lambda^{(c)}$ in a finite box Λ. The uniform bounds obtained at small $|\lambda|$ with respect to Λ and Vitali's theorem then ensure that the result still holds in the $\Lambda \to \infty$ limit.

(ii) Bounds in $\mathrm{cst}^n (n!)^2$ on the derivatives $|\mathrm{d}^n/\mathrm{d}\lambda^n S^{(c)}_{(\alpha)}(\lambda)|$ for λ in the previous region shown in Figure 3.6a.

These bounds are a consequence of the following expression of $\mathrm{d}^n/\mathrm{d}\lambda^n S^{(c)}_\alpha(\lambda)$, which is easily checked inductively (as in related problems in statistical mechanics) by direct inspection:

$$\frac{\mathrm{d}^{n-1}}{\mathrm{d}\lambda^{n-1}} S^c_{(\alpha)}(x_1,\ldots,x_N;\lambda) =$$
$$\int S^c_{(\alpha,\cdot)}(x_1,\ldots,x_N,y_1,\ldots,y_n;\lambda)\,\mathrm{d}y_1\ldots\mathrm{d}y_n$$
$$(3.92)$$

or the corresponding formula on $\mathrm{d}^{n-1}/\mathrm{d}\lambda^{n-1} S_{(\alpha)}(.\,;\lambda)$ which expresses it as the integral of the partially connected function with respect to $y_1,\ldots,y_n$ and the set $(x_1,\ldots,x_N)$. (In (3.92), the new coefficients α_i in the right-hand side are equal to 4.)

The desired bound on $\mathrm{d}^n/\mathrm{d}\lambda^n S^{(c)}_{(\alpha)}(x_1,\ldots,x_N;\lambda)$, with $x_1,\ldots,$ x_N integrated in given squares, then follows from the bounds (3.90) (in which N is now replaced by $N+n$, N fixed). The fact, already mentioned, that $L(\Delta'_1,\ldots,\Delta'_k)$ is larger than one half the minimal length of trees joining $\Delta'_1,\ldots,\Delta'_k$ without intermediate points can be used to that purpose.

Remarks.

(1) Analyticity of $S^{(c)}$ is not expected, and in fact does not hold, at $\lambda = 0$. It has on the other hand been proved in a domain larger than that of Figure 3.6a and is valid in a cut-domain with a cut along the negative axis: see [RE].

The function B is expected (and this can also be proved) to have a pole on the real negative t-axis.

(2) Analyticity in a smaller region (interior of the circle shown in Figure 3.6a) would be sufficient to apply Nevanlinna's theorem.

§4. The massive Gross-Neveu model in dimension two

4.1 Preliminary results

This model [MW,GN] is a fermionic model in two dimensions, with quadratic interaction $\bar{\psi}\psi(x)^2$, leaving implicit spin indices (which take two values) and "colour" indices. The colour number is ≥ 2. (The value 1 corresponds to the massive Thirring model which is equivalent to the

Sine-Gordon model and has simpler renormalization properties. It belongs to the special class of completely integrable systems in dimension 2 which can be treated by more direct and explicit methods, as already mentioned.) In the version of the model that will be treated, the mass is already present, as in [MW], from the outset ($m_0 \neq 0$ in Eq. (3.107)) and the model will be defined at small values of the renormalized coupling λ_{ren}. In an alternative version, which is as a matter of fact that mainly considered in [GN], the initial mass is zero, but an actual (strictly positive) mass is generated by a more complicated mechanism. This version can also probably be defined rigorously by methods of constructive theory, with, however, a further nontrivial work and more complete arguments: see [KMR]. It is believed that it also belongs to the class of completely integrable models, but this is probably not the case for the model we consider for $m_0 \neq 0$.

The 4-point function is initially of the form: $N_{\Lambda,\rho} Z_{\Lambda,\rho}^{-1}$ with (formally and up to some changes) $N_{\Lambda,\rho}$ of the form:

$$N_{\Lambda,\rho}\left(x_1,\ldots,x_4\right) = \int \psi\left(x_1\right)\psi\left(x_2\right)\bar{\psi}\left(x_3\right)\bar{\psi}\left(x_4\right) e^{\int_\Lambda \lambda(\bar{\psi}\psi)^2(y)\,\mathrm{d}y}\mathrm{d}\nu_\rho\left(\bar{\psi}\psi\right)$$

$$(3.93)$$

where λ is the bare (unrenormalized) coupling constant; the measure $\mathrm{d}\nu_\rho$ is associated with the propagator $\tilde{C}_\rho(p) = e^{-(p^2+m^2)M^{-2\rho}}\tilde{C}(p)$, $\tilde{C}(p) = (-\not{p}+m)/(p^2+m^2)$; and $\not{p} = p_0\sigma_0 + p_1\sigma_1 : \sigma_0, \sigma_1$ are 2×2 matrices such that $\sigma_0^2 = \sigma_1^2 = -1$, and $\sigma_0\sigma_1 + \sigma_1\sigma_0 = 0$ (hence $\not{p}^2 = -p^2$, $p^2 = p_0^2 + p_1^2$).

In order to obtain an actual nontrivial theory in the $\rho \to \infty$ limit, one will have to consider, more precisely, sequences of bare couplings λ_ρ and bare masses m_ρ depending on ρ: see Sect. 4.2. In particular, λ_ρ will tend to zero as $\rho \to \infty$ ("asymptotic freedom") in a specified way that will ensure that the renormalized coupling λ_{ren} will be different from zero in that limit. (This will ensure the nontriviality of the theory.) Methods of constructive theory will provide an actual (rigorous) definition of the theory for small enough values of λ_{ren}. In the remainder of this subsection, in which the $\rho \to \infty$ limit is not treated, we consider possible general values of λ.

Whereas this is not appropriate in bosonic models, it is convenient here to expand the exponential $\exp \lambda \int (\bar{\psi}\psi)^2$ in λ. By some algebraic calculations, one gets the following form of $N_{\Lambda,\rho}$ and an analogous form of $Z_{\Lambda,\rho}$, that will be, as a matter of fact, our starting point:

$$N_{\Lambda,\rho}\left(x_1,\ldots,x_4\right) = \sum_{n\geq0} \frac{\lambda^n}{n!} \int_{\Lambda^n} \mathrm{d}y_1\ldots\mathrm{d}y_n \left\{ \begin{matrix} x_1x_2y_1y_1 & \cdots & y_ny_n \\ x_3x_4y_1y_1 & \cdots & y_ny_n \end{matrix} \right\}_\rho,$$

$$(3.94)$$

where

$$\left\{ \begin{matrix} u_1 & \cdots & u_n \\ v_1 & \cdots & v_n \end{matrix} \right\}_\rho = \det\left(\{C_\rho(u_\alpha, v_\beta)\}\right). \tag{3.95}$$

More precisely, (3.94) applies up to modifications arising from the inclusion of spin and colour indices which have again been left implicit. (Relevant determinants therefore do not vanish. In fact, if ↑ and ↓ denote spins $1/2$ and $-1/2$ respectively, then terms corresponding to $\bar\psi_\uparrow \psi_\uparrow \bar\psi_\uparrow \psi_\uparrow$ in the interaction $(\bar\psi\psi)^2$ will vanish, so that $(\bar\psi\psi)^2$ can be replaced by $2\bar\psi_\uparrow \psi_\uparrow \bar\psi_\downarrow \psi_\downarrow$. Correspondingly, the determinant above can be replaced by a product of two determinants. Further factorizations also follow from the considerations of colour indices.)

Lemma 3.2 below, which takes into account potential cancellations between individual products of terms of the determinant, first shows that, for any given ρ and Λ, the expansion (3.94) is convergent and defines in fact an entire function of λ. However, the result thus obtained does not allow one to treat the $\Lambda \to \infty$, $\rho \to \infty$ limits. Completed by, for example, pairwise cluster expansions of $N_{\Lambda,\rho}$ and $Z_{\Lambda,\rho}$ (i.e. here, replacement of $C_\rho(x,y)$ by $C_\rho(x,y;s) = s_{\Delta,\Delta'} C(x,y)$ if $x \in \Delta$, $y \in \Delta'$ or vice versa, and Taylor expansions of $N_{\Lambda,\rho}$ and $Z_{\Lambda,\rho}$ around the origin in all variables $s_{\Delta,\Delta'}$) and, as in Sect. 3, by an a la Mayer procedure, it will also allow one to treat the $\Lambda \to \infty$ limit for the 4-point (or N-point) functions $N_{\Lambda,\rho}/Z_{\Lambda,\rho}$. Although modifications will be made from the outset in Sect. 4.2, to treat the $\rho \to \infty$ limit also, we first describe this analysis, which already introduces important aspects useful later. If we again consider the 4-point function $H^c_{\Lambda,\rho}$, it is here expressed in the form:

$$H^c_{\Lambda,\rho}(x_1,\ldots,x_4) = \sum_{\mathbf{G}} \left(\prod_{V \in \mathbf{G}} \frac{\lambda^{n_V}}{n_V!} \right) \sideset{}{'}\sum_{\{\Delta_V\}} \frac{1}{N(\mathbf{G})!} \int_{y^{(V)} \subset \Delta_V, \forall V}$$
$$R\left(\left\{y_1^{(V)},\ldots,y_{n_V}^{(V)}\right\}, x_1,\ldots,x_4\right) \prod_{V \in \mathbf{G}} \left(\mathrm{d}y_1^{(V)}\ldots \mathrm{d}y_{n_V}^{(V)}\right), \tag{3.96}$$

where the sum runs over connected graphs $\mathbf{G}$ with an arbitrary number of vertices V to which will be associated (unit) squares Δ_V in a paving D of Λ. Each vertex V, or square Δ_V, is equipped with n_V interaction vertices with 2 legs ψ and 2 legs $\bar\psi$. Given any pair of vertices V, V', or squares $\Delta_V, \Delta_{V'}$, there are zero or one explicit propagator lines, and zero or one Mayer lines, between them. $\mathbf{G}$ must be connected when all lines are taken into account. Explicit propagator lines join here, more precisely, specified interaction vertices with one leg ψ and one leg $\bar\psi$ on each side. The sum $\sum'$ runs over all possible sets $\{\Delta_V\}$ of squares varying independently in Λ, subject to the condition that two squares joined by a Mayer line coincide. Each interaction vertex $x_1, \ldots, x_4$ must be contained in one Δ_V (and has 3 legs apart from that associated to the external propagator that has been factored out). Interaction vertices $y_1^{(V)}, \ldots, y_{n_V}^{(V)}$ are integrated in Δ_V. R is a product of factors (which are explicit propagators and determinants) associated to each part of $\mathbf{G}$ already connected by explicit propagator lines, of factors -1 for each Mayer line and of a symmetry factor (smaller than one). The first factors are those arising from the cluster expansion. For example, the factor associated with the term

with one explicit propagator line is equal, up to a sign (and correct inclusion of spin and colour indices), to

$$C_\rho\left(y_2, y_3\right) \int_0^1 \left\{ \begin{array}{ccccccc} y_1 & y_1 & y_2 & y_3 & y_3 & y_4 & y_4 \\ y_1 & y_1 & y_2 & y_2 & y_3 & y_4 & y_4 \end{array} \right\}_\rho (s)\mathrm{d}s,$$

where s is the variable associated to the pair of squares (Δ_1, Δ_2) in the (pairwise) cluster expansion. In $\{\ldots\}_\rho$, there is one y_α in the first line for each leg ψ attached to y_α, and one y_β in the second line for each $\bar\psi$ attached to y_β, except those ψ or $\bar\psi$ already linked by the explicit propagator line. Propagators involved in the determinant are $C_\rho\left(y_\alpha, y_\beta; s\right)$ if $y_\alpha \in \Delta_1$, $y_\beta \in \Delta_2$ or vice versa (or $C_\rho\left(y_\alpha, y_\beta\right)$ if y_α, y_β belong to the same square).

The $\Lambda \to \infty$ limit is treated here with the help of the following preliminary lemmas which replace considerations mentioned in the proof of Lemma 3.1 of Sect. 3.2. We state them in a form that will also be useful in Sect. 4.2. In the latter, C will be a slice propagator $C^{(i)}$, for any given $i = 1, \ldots, \rho$, and χ will depend on i (and will be equal to M^i), while K will be independent of i. In this subsection, C is the propagator C_ρ, χ is a constant close to the bare mass m (and does not depend on ρ) and K is of the order of cst M^ρ. The bounds (3.97) needed in Lemma 3.2 on $C^{(i)}$ or C_ρ are checked by direct inspection in either case. For purposes of this subsection, the result of Lemma 3.2 is a bound on the determinant in (cst $M^\rho)^n$.

Lemma 3.2. Let C be a propagator which, together with a given number of low-order derivatives $D^{\underline{q}} C$, satisfies bounds of the form:

$$|D^{\underline{q}} C(x, y)| \leq K \chi^{|\underline{q}|+1} \mathrm{e}^{-\chi|x-y|}. \tag{3.97}$$

Let D be a paving of $\mathbb{R}^d$ by boxes of size cst χ^{-1} and let n'_Δ and n''_Δ be the number of points u_α and v_β, respectively, in each box Δ with $\sum_\Delta n'_\Delta = \sum_\Delta n''_\Delta = n$. Then,

$$\left| \det \left(\{ C(u_\alpha, v_\beta) \} \right) \right| \leq (\text{cst } K \chi)^n. \tag{3.98}$$

Proof. This result is due originally to [GK1] (in a stronger version). We briefly outline here a simpler proof given in an Appendix of [IM2] (including the collaboration of J. Feldman). The most naïve bound, obtained by simply expanding the determinants, ignoring potential cancellations and using the bound (3.97) at $\underline{q} = 0$ only, is

$$\left| \det \left(\{ C(u_\alpha, v_\beta) \} \right) \right| \leq \sum_{\substack{\text{permutations } \pi \\ \text{of } (1,\ldots,n)}} \prod_{i=1}^n \left[(K \chi) e^{-\chi |u_i - v_{\pi(i)}|} \right]. \tag{3.99}$$

Using a small amount of the exponential decay allows one to "turn" the global $n!$ arising from the sum over permutations into local $n_\Delta!$'s. One thus obtains, for any $\varepsilon > 0$ ($\varepsilon < 1$),

$$\left| \det\{\ldots\} \right| \leq \text{cst}_\varepsilon (K \chi)^n \prod_\Delta (n'_\Delta!)^{1/2} (n''_\Delta!)^{1/2}$$
$$\times \text{Sup}_\pi \prod_{i=1}^n e^{-\chi(1-\varepsilon)|u_i - v_{\pi(i)}|}. \tag{3.100}$$

The elimination of the factors $(n'_\Delta!)^{1/2}(n''_\Delta!)^{1/2}$ then follows from a more refined division of each box Δ into $n'_\Delta/2\sigma d^p$ (or $n''_\Delta/2\sigma d^p$) boxes Δ_j, with centers $z_{\Delta,j}$, where σ is the number of values of spin indices (here 2) and p is some (small) integer to be chosen conveniently. A Taylor expansion of order p (with integral remainder) is then applied to each $C(u_\alpha, .)$ (or alternatively $C(., v_\beta)$) around the point $z_{\Delta,j}$ if u_α (or v_β) belongs to the box Δ_j of Δ. The desired result follows from the analysis of the new terms obtained, using the bounds (3.97) for the various $\underline{q}$ involved. QED

We next state:

Lemma 3.3. Given any set of squares $\Delta_0, \Delta_1, \ldots, \Delta_N$ of a paving D (of size cst χ^{-1}), with $\Delta_\alpha \neq \Delta_\beta$ if $\alpha \neq \beta$, and in each square Δ_α, $q(\alpha)$ points $y_{1,\alpha}, \ldots, y_{q(\alpha),\alpha}$ (some of which may coincide), the following bound holds if C_1 is a sufficiently small constant, with a constant C_2 independent of $\chi, N, \Delta_0, \ldots, \Delta_N$ and of the points y:

$$\sum_G \prod_{\ell \in G} \left(C_1 e^{-\text{cst } \chi d(\ell)} \right) < (C_2)^q, \tag{3.101}$$

where the sum runs over all connected or nonconnected graphs G composed of lines joining points of different squares, with at most one line between any pair of squares. $d(\ell)$ is the distance between the squares linked by line ℓ and $q = \sum_{\alpha=0}^{N} q(\alpha)$.

Proof. The sum Σ is bounded by the number of ways of drawing, from each point y of each square Δ_α, zero, one or more lines joining y to points of squares $\Delta_\beta \neq \Delta_\alpha$, with at most one line between y and each square Δ_β. For each y, the sum is bounded by a constant independent of N, y and of the number and position of the squares if:

$$C_1 < \left[\sum_{\Delta \in D} e^{-\text{cst } \chi d(\Delta_\alpha, \Delta)} \right]^{-1} \quad (= \text{cst independent of } \chi).$$

The bounds (3.101) follow. QED

We finally recall

Lemma 3.4.

$$\sum_{N \geq 0} \frac{C'^N}{N!} \sum_{\substack{\Delta_1, \ldots, \Delta_N \in D \\ \Delta_\alpha \neq \Delta_\beta \text{ if } \alpha \neq \beta}} e^{-\text{cst } \chi L(\Delta_0, \ldots, \Delta_N)} < C'', \qquad (3.102)$$

where the constant C'' is arbitrarily small if the constant C' is chosen sufficiently small ($L(\Delta_0, \ldots, \Delta_N)$ is the shortest length of all graphs joining $\Delta_0, \ldots, \Delta_N$, with possible intermediate points or squares, and connected with respect to $\Delta_0, \ldots, \Delta_N$).

Proof. This bound can be obtained by noting (as in cases considered previously) that $L(\Delta_0, \ldots, \Delta_N)$ is larger than one half the minimal length of trees joining $\Delta_0, \ldots, \Delta_N$ without intermediate points or squares. The left-hand side of (3.102) is thus bounded by introducing for each N a sum over all trees with factors $\exp[-\text{cst}' \chi |\ell|]$ for each line of the tree. The end of the proof is then analogous to that of (3.88) in Sect. 3. QED

With the help of these lemmas, the following result is established, starting from (3.96):

Theorem 3.3. For any $\rho > 0$ and λ sufficiently small, $H_{\Lambda,\rho}^c$ has a well-defined limit H_ρ^c when $\Lambda \to \infty$. Moreover, $\forall \varepsilon, \varepsilon > 0, \exists \lambda_\varepsilon > 0$ independent of Λ and ρ such that, if $|\lambda| < \lambda_\varepsilon M^{-2\rho}$,

$$\int_{\substack{x_2 \in \Delta_2 \ldots x_4 \in \Delta_4 \\ \Delta_2, \ldots, \Delta_4 \in D}} dx_2 \ldots dx_4 \left| H_{\Lambda,\rho}^c (x_1, \ldots, x_4) \right|$$

$$< \text{cst} \, |\lambda_\varepsilon|^4 \, e^{-(m-\varepsilon)L(x_1, \Delta_2, \ldots, \Delta_4)}. \qquad (3.103)$$

Details of the proof are omitted. The condition $|\lambda| < \lambda_\epsilon M^{-2\rho}$ is *a priori* needed to compensate factors M^ρ arising from bounds on the propagator C_ρ. This result does not allow one to define a nontrivial theory in the $\rho \to \infty$ limit. This will be achieved in Sect. 4.2.

4.2 Phase-space analysis and renormalization

Phase-space expansion before renormalization. Using a decomposition of the form (3.32)–(3.35) of C_ρ (with m replaced by m_ρ and, for some technical reasons, M^{-2i} replaced in (3.35) by $M^{-2(i+1)}$), one obtains expressions of $N_{\Lambda,\rho}$ and $Z_{\Lambda,\rho}$ in which the determinant $\{\ldots\}_\rho$ in (3.94) is replaced by a sum of determinants where each propagator has a momentum index i. Some of the vertices x or y may occur in several slices. Cluster expansions are made independently in each slice for each determinant. The size of squares in each slice is no longer uniform but is chosen to be M^{-i}; this is the optimal way of decomposing integration domains in space-time into cells where $C^{(i)}$ is approximately constant and such that the number of cells is controlled by the exponential fall-off of $C^{(i)}$ uniformly in i. In fact, $C^{(i)}$ satisfies here a bound of the form:

$$\left| C^{(i)}(x) \right| < \text{cst } M^i e^{-\delta\, M^i |x|} \tag{3.104}$$

and

$$\sum_{\Delta \in D_i} e^{-\delta M^i \mathrm{d}(0,\Delta)} < \text{cst independent of } i, \tag{3.105}$$

where D_i is the set of squares of size M^{-i}. Thus, in the bounds, the local and nonlocal aspects (in x-space) will be decoupled. The couplings between squares of the same slice will be controlled by the (scaled) fall-off of the propagators; each vertex can then be localized in a square, and bounds on each contribution follow from power counting, with factors M^i for each propagator $C^{(i)}$ and M^{-2i} for the summation of one vertex in a square of D_i (= the surface of the square). Each paving D_i is obtained from D_{i-1} by replacing each square by 4 new squares if $M = 2$.

More precisely, the 4-point function $H^c_{\Lambda,\rho}$ is then expressed as a sum of the form (3.96) with the following differences (see Figure 3.7). The sum runs over connected graphs $\underline{G}$ defined as follows. In each slice i, $\underline{G}$ has $N_i \geq 0$ vertices $V = 1, \ldots, N_i$ to which will be associated squares Δ_V of D_i. Each vertex V, or square Δ_V, in each slice is equipped with $n_V \geq 0$ "original" interaction vertices and with $n'_V \geq 0$ "derived" interaction vertices, with $n_V + n'_V \geq 1$. Each interaction vertex has 4 legs, 2 "ψ" and 2 "$\bar{\psi}$," which are distributed between 1, 2, 3 or 4 slices; the vertex is said to belong to each of these slices. It is (by definition) an original vertex in the slice of highest index (to which it belongs). It is a derived vertex in the other slices (to which it belongs). Original and derived vertices in different slices corresponding to a common interaction vertex (to which the same point y or x will be associated) will be joined by lines $---$. The factor $1/N!$ is replaced by the product of the factors $1/N_i!$

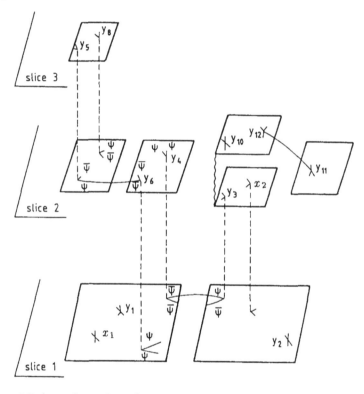

Figure 3.7 A configuration of squares contributing to a connected function $H^c_{\Lambda,\rho}$. Each vertex has two legs ψ and two legs $\bar{\psi}$, only some of which are indicated explicitly. All fields of slice i are to be labelled by i. ⎓⎓ denotes a Mayer line.

In each slice i, vertices V may be joined by propagator and/or Mayer lines in the same conditions as in Sect. 4.1. The graph $\underline{G}$ must be connected when all lines are taken into account (including lines $---$ between slices).

The sum Σ' runs over all possible sets of squares subject, in each slice, to the same Mayer conditions and to the following one: a square in slice i must be contained in a square of lower slice whenever corresponding vertices V are linked by a line $- - -$, since these squares contain a common interaction vertex y or x. (The square Δ_V must also contain the point x_r if the latter is an original or derived interaction vertex of V).

The factor R is again a product of factors associated in each slice to each part of $\underline{G}$ already connected by propagator lines, of factors -1 for each Mayer line and of a symmetry factor (less than one).

An example of a set of squares $\{\Delta\}$ for a given graph $\underline{G}$, with indication of corresponding points y_α and x_r, and of lines arising from those of $\underline{G}$, is shown in Figure 3.7. The factor associated, for example, to the connected part in slice 2 on the left is equal, up to a sign, to

$$C^{(2)}\left(y_5, y_6\right) \int_0^1 \left\{ \begin{matrix} y_8 & y_4 & y_4 \\ y_8 & y_5 & y_6 \end{matrix} \right\}^{(2)} (s)\,\mathrm{d}s.$$

The analysis, by a combination of the methods of Sect. 4.1 and of those already explained in the perturbative approach in the proof of the Theorem 3.1 in Sect. 2.2, yields an improved version of Theorem 3.3 of Sect. 4.1, for $|\lambda| < \lambda_\varepsilon M^{-\rho/2}\rho^3 m_\rho^2$, where the dependence on m_ρ (if the bare mass is varied with ρ) is indicated.

Renormalization. The improved version of Theorem 3.3 that has just been mentioned shows appreciable progress in comparison to the result obtained without phase-space analysis ($\lambda \to 0$ as $M^{-2\rho}$), but (as expected) is still insufficient to control the $\rho \to \infty$ limit, that is, to obtain a nontrivial theory in that limit; a rearrangement of terms of the expansion, called renormalization, is needed. As outlined below, it will yield a nontrivial theory in the $\rho \to \infty$ limit, and also decay properties of connected functions, starting with λ_ρ and m_ρ of the form:

$$\lambda_\rho = \frac{1}{-\beta_2 \ln M^\rho + \frac{\beta_3}{\beta_2}\ln\ln M^\rho + D}, \tag{3.106}$$

$$m_\rho = m_0 \left(\frac{D}{-\beta_2 \ln M^\rho}\right)^\gamma, \tag{3.107}$$

where D is a constant; β_2 and β_3 are standard coefficients of the "β-function" linked to Feynman graphs of lowest order in perturbation theory, and $\gamma > 0$; β_2, β_3 or γ are independent of M, Eqs. (3.106) and (3.107) have been written in a form that exhibits the dependence of λ_ρ and m_ρ on the actual ultraviolet cut-off, that is, M^ρ; $\ln M^\rho$ and $\ln\ln M^\rho$ can alternatively be replaced by $\rho \ln M$ and $\ln\rho$ up to unessential changes, this corresponding to the choice

$$\lambda_\rho = \left[-\beta_2 (\ln M)\rho + \frac{\beta_3}{\beta_2} \ln\rho + D \right]^{-1} \qquad (3.108)$$

and m_ρ can be written in the form

$$m_\rho = m\,\rho^{-\gamma}, \qquad (3.109)$$

where $m = m_0 (D/ - \beta_2 \ln M)^\gamma$.

Although λ_ρ still tends to zero in the $\rho \to \infty$ limit, the constant λ_{ren} will be shown to be nonzero (in fact close to D^{-1} for large D); and the renormalized mass m_{ren} as well as the physical mass will be of the order of m_0.

As in Sect. 2.2, renormalization will be made inductively from slice ρ to 1 and will involve only useful parts. Starting from slice ρ, the scalar $\delta\lambda_\rho$ is defined by the formula:

$$\delta\lambda_\rho = \int S^c_{(\rho)}\,(u_1, u_2, u_3, u_4)\,|_{m_\rho = 0}\;du_2 du_3 du_4, \qquad (3.110)$$

where $S^c_{(\rho)}$ is the connected 4-point function of slice ρ. Namely, it is the sum, i.e. over connected graphs $\underline{G}_\rho$ of slice ρ with 2, respectively 4, external legs hooked at $u_1, \dots, u_4$, of corresponding functions $\underline{G}_{(\rho)}\,(u_1, \dots, u_4)$. ($S^c_{(\rho)}$ is independent of the slices $< \rho$ to which external legs belong. Two or three of these legs may on the other hand be attached to the same vertex u_α. This gives a contribution to $S^c_{(\rho)}$ including δ-functions $\delta\,(u_i - u_j)$.) Related formulae define δm_ρ and $\delta\zeta_\rho$ in terms of the 2-point function $S^c_{(\rho)}\,(u_1, u_2)$. The fact that the functions $S^c_{(\rho)}$ and their integrals are well defined in the $\Lambda \to \infty$ limit is proved by the same methods as in Sect. 4.1. The functions $\underline{G}_{(\rho)}$ and $S^c_{(\rho)}$ are also Euclidean invariant at Λ infinite. Hence, $\delta\lambda_\rho, \delta m_\rho$ and $\delta\zeta_\rho$ are independent of u_1.

We then define the "regularized" functions:

$$\begin{aligned}
S^c_{(\rho)\text{reg}}\,(u_1, \dots, u_4) &= S^c_{(\rho)}\,(u_1, \dots, u_4) \\
&\quad - \delta\lambda_\rho \delta\,(u_1 - u_2)\,\delta\,(u_1 - u_3)\,\delta\,(u_1 - u_4) \\
&= \sum \underline{G}_{(\rho)\text{reg}}\,(u_1, \dots, u_4)\,, \qquad (3.111) \\
S^c_{(\rho)\text{reg}}\,(u_1, u_2) &= S^c_{(\rho)}\,(u_1, u_2) - \delta m_\rho \delta\,(u_1 - u_2) - \delta\zeta_\rho \partial\!\!\!/\delta\,(u_1 - u_2) \\
&= \sum \underline{G}_{(\rho)\text{reg}}\,(u_1, u_2)\,, \qquad (3.112)
\end{aligned}$$

where the functions $\underline{G}_{(\rho)\mathrm{reg}}$ are defined in the same way as $S^c_{(\rho)\mathrm{reg}}$. (The function $S^c_{(\rho)}$ is replaced by functions $\underline{G}_{(\rho)}$, and $\delta\lambda_\rho$, $\delta m\rho$ and $\delta\zeta_\rho$ are replaced by the corresponding quantities.)

Via an analysis similar to that of Sect. 2.2 and up to specific details needed to treat mass renormalization, the procedure again yields "internal regularization factors" and gradients acting on external lines, and rearrangement of terms yields a new phase-space expansion with the following differences:

(i) The coupling constant attached to each interaction vertex is either $\lambda_\rho \equiv \lambda$ if this vertex has a leg of index ρ (i.e. an original vertex of slice ρ) or equal to

$$\lambda_{\rho-1} = \lambda_\rho + \delta\lambda_\rho. \tag{3.113}$$

(ii) A new class of vertices with 2 legs in slices $i, j \leq \rho - 1$, of the type $\delta m_\rho \bar{\psi}_i \psi_j$ or $\delta\zeta_\rho \bar{\psi}_i \partial\!\!\!/ \psi_j$, is introduced.

(iii) The class of graphs $\underline{G}$ is thus replaced by a new class of graphs $\underline{G}^{(\rho)}$ which may include links between interaction vertices in slice ρ that will correspond in particular to internal regularization factors $u_1 - u_2$, and attributions of gradients to some legs in slices $< \rho$.

We next define

$$\delta\lambda_{\rho-1} = \int S^{c(\rho)}_{(\rho-1)}(u_1, \ldots, u_4)\Big|_{m_\rho = 0} \mathrm{d}u_2 \, \mathrm{d}u_3 \, \mathrm{d}u_4, \tag{3.114}$$

where $S^{c(\rho)}_{(\rho-1)}$ is the connected 4-point function of slice $\rho - 1$, that is, the sum, over connected graphs $\underline{G}^{(\rho)}_{\rho-1}$ with 4 external legs and no external gradient in slices $< \rho - 1$, of functions $G^{(\rho)}_{(\rho-1)}(u_1, \ldots, u_4)$. The quantities $\delta m_{\rho-1}$ and $\delta\zeta_{\rho-1}$ are defined similarly. Rearrangements of terms analogous to those above are made. This procedure is pursued up to slice 1. This gives an expansion in which:

(i) The coupling constant attached to an interaction vertex of highest slice i is

$$\lambda_i = \lambda_{i+1} + \delta\lambda_{i+1} = \lambda_\rho + \delta\lambda_\rho + \delta\lambda_{\rho-1} + \cdots + \delta\lambda_{i+1}. \tag{3.115}$$

(ii) The coefficient attached to a 2-leg vertex of highest slice i is $\sum_{j=i+1}^{\rho} \delta m_j (= m_\rho - m_i)$, respectively $\sum_{j=i+1}^{\rho} \delta\zeta_j (= 1 - \zeta_i)$.

(iii) The class of graphs is obtained from the inductive procedure. In particular, all 4-point connected subgraphs, respectively all 2-point subgraphs, obtained in slices $\geq i$ have at least one external gradient attached to one of their external legs, respectively at least two external gradients.

Starting with the expressions (3.106) and (3.107) of λ_ρ and m_ρ (and $\zeta_\rho = 1$), the following result is obtained if the constant D which is chosen in (3.106) or (3.108) is sufficiently large:

Theorem 3.4. H^c is well defined (and nonzero) in the $\rho \to \infty$ limit and the following bounds hold (uniformly in ρ, for any ρ finite or infinite) with $\lambda_1 \neq 0$, for any $\epsilon > 0$:

$$\left| \int_{x_2 \in \Delta_2, \dots, x_4 \in \Delta_4} H^c_\rho (x_1, \dots, x_4) \, dx_2 \dots dx_4 \right| \tag{3.116}$$
$$< \mathrm{cst}(\varepsilon) \, |\lambda_1| \, \mathrm{e}^{-(m-\varepsilon)L(x_1, \Delta_2, \dots, \Delta_4)},$$

where $\Delta_2, \dots, \Delta_4$ are any given squares of the paving D_1.

Remark. We note that, for fixed ρ, convergence and decay are similarly obtained for bare coupling constants whose absolute values are smaller than $|\lambda_\rho|$, where λ_ρ is defined in (3.106), but λ_1 would tend in general toward zero in the $\rho \to \infty$ limit for choices of bare coupling constants significantly different from (3.106).

Proof of Theorem 3.4 (outline). To obtain (3.116), it is proved by induction that

$$\delta\lambda_i = a\lambda_i^2 + b\lambda_i^3 + O\left(\lambda_i^4\right) + \lambda_i^2 \left[O\left(\mathrm{e}^{-(\rho-i)} + O\left(\mathrm{e}^{-i}\right) \right) \right], \tag{3.117}$$

with related results for $\delta m_i, \delta\zeta_i$, where $a = -\beta_2 \ln M$, $b = -\beta_2 \ln M[-\beta_2 \ln M + \beta_3]$. Coefficients of λ_i^2 and λ_i^3 are obtained by explicit computation of first order graphs. Convergence properties of the phase-space expansion, and also the control of remainders at each stage, are ensured if the constant D which is chosen in (3.106) or (3.108) is sufficiently large so that, in particular, all $|\lambda_i|$ will be of sufficiently small size uniformly in i; λ_1, is close to λ_{ren} and is of the order of $1/D$.

In fact, the formulae above entail in turn that:

$$\lambda_i = \left[(-\beta_2 \ln M)i + \beta_3 \ln i + D + f(i) \right]^{-1}, \tag{3.118}$$

with $|f(i)| < 1$ and related results on m_i and ζ_i, for example,

$$m_i = mi^{-\gamma}(1 + g(i)), \quad i \gtrsim \frac{D}{-\beta_2 \ln M}, \quad |g(i)| < \frac{1}{2}, \qquad (3.119)$$

$$m_j \leq \text{cst} m_0, \quad j \lesssim \frac{D}{-\beta_2 \ln M}. \qquad (3.120)$$

4.3 Large-momentum and short-distance properties—Wilson short-distance expansion

This subsection is a summary of [IM4]. The renormalization procedure of Sect. 4.2 can be re-expressed for our present purposes as follows in a Fourier-transformed formulation. Any 4-point function associated with a subdiagram with 4 outgoing lines below j, can be written (in Euclidean energy-momentum space) as:

$$\tilde{D}^{(j)}(k_1, \ldots, k_4) = \left[\prod_v \lambda(v)\right] \tilde{d}^{(j)}(k_1, \ldots, k_4), \qquad (3.121)$$

where $k_1, \ldots, k_4$ are the 4 outgoing energy-momenta ($k_1 + \cdots + k_4 = 0$) and, for each vertex v of the subdiagram, $\lambda(v)$ is the effective coupling (obtained at that stage). One then writes:

$$\tilde{d}^{(j)}(k_1, \ldots, k_4) = \tilde{d}^{(j)}_{\text{reg}}(k_1, \ldots, k_4) + \tilde{d}^{(j)}(0, 0, 0, 0), \qquad (3.122)$$

that is, $\tilde{d}^{(j)}_{\text{reg}} = \tilde{d}^{(j)} - \tilde{d}^{(j)}(0)$. This subtraction yields in Euclidean space-time gradients on outgoing lines and internal regularization factors. From the viewpoint of power counting, this is equivalent to having two more outgoing lines. On the other hand, the factor $\sum_{\mathbf{D}^{(j)}} [\prod \lambda(v)] \tilde{d}^{(j)}(0, \ldots, 0) = \delta \lambda_j$ yields a redefinition of the effective coupling constants of vertices whose lines belong to slices $< j$. Diagrams with 2 outgoing lines below j are also treated.

This renormalization procedure is appropriate for the definition of the model but is not sufficient for the study of the large-momentum behaviour of 4-point or N-point functions in Euclidean energy-momentum space (or their short distance behaviour in space-time). A preliminary difficulty is that some of the outgoing lines of some subdiagrams may correspond to actual external variables. In this case, gradients coming from renormalization will act on the functions $e^{i\Sigma \pm p_k x_k}$ involved in the Fourier transformation and will *a priori* yield unwanted factors $|p_k|$ in bounds on momentum-space functions.

The way which seems most convenient to study large momentum properties of, for example, the 4-point function is to avoid this difficulty by modifying the renormalization procedure as follows.

Given the external energy-momenta p_k, $k = 1, \ldots, 4$, let $i_k \equiv i(p_k)$ be the ≥ 1 integer such that $M^{i_k-1} < |p_k| \leq M^{i_k}$: the external line k will be considered by convention to belong to slice i_k. We assume (without loss of generality) in this section that

$$i_1 \geq i_2 \geq i_3 \geq i_4. \tag{3.123}$$

Similarly, $i_{\alpha,\beta} \equiv i(p_\alpha + p_\beta)$, $i_{\alpha,\beta,\gamma} \equiv i(p_\alpha + p_\beta + p_\gamma)$ and $i_{\alpha,\beta,\gamma} = i_\delta$ ($\alpha, \beta, \gamma, \delta$ are different indices among $1, \ldots, 4$).

For any connected subdiagram $D^{(j)}$ containing $n_{\text{ext}}(D)$ external lines with indices $\geq j$ among its 2 or 4 outgoing lines, the subtraction will now be as follows. The previous rule is unchanged if $n_{\text{ext}} = 0$. On the other hand, no subtraction is needed if $n_{\text{ext}}(D) = 1$, or if $n_{\text{ext}}(D) = 2$ and $j < i_{\alpha,\beta}$, where α, β are the indices of the 2 external lines in slices $\geq j$, or if $n_{\text{ext}}(D) = 3$ and $j < i_{\alpha,\beta,\gamma}$, where α, β, γ are the indices of the 3 external lines in slices $\geq j$. In fact, energy-momentum conservation and the ultraviolet cut-off factors in the definition of the propagators $\tilde{C}^{(l)}$, $l < j$, of the outgoing lines below j, ensure exponential fall-off factors much better than $M^{-(j-l_1)}$, where l_1 is the index of the highest outgoing line below j, and make renormalization useless. In remaining cases:

(a) $n_{\text{ext}}(D) = 2$, $j > i_{\alpha,\beta}$ and 2 outgoing lines below j. From $[\prod_v \lambda_v]$ $\tilde{d}^{(j)}(p_\alpha, p_\beta, k_3, k_4)$, we now subtract:

$$\left[\prod_v \lambda(v) \right] \tilde{d}^{(j)} \left(\frac{p_\alpha - p_\beta}{2}, -\frac{p_\alpha - p_\beta}{2}, 0, 0 \right),$$

where the outgoing energy-momenta k_3, k_4 below j, hence, also $k = k_1 + k_2$, have been fixed at zero.

We note that the effective coupling $\lambda(v)$ and the terms $\tilde{d}^{(j)}$ differ here from those of Sect. 4.2, although we have used the same notation. In particular, the effective couplings of vertices involving external lines will now depend on their energy-momenta.

(b) $n_{\text{ext}}(D) = 3$, 1 outgoing line below j. Subtracted term:

$$\left[\prod_v \lambda(v) \right] \tilde{d}^{(j)} (p_1, p_2, -(p_1 + p_2), 0).$$

The sum over diagrams $D^{(j)}$ defines counterterms δm_j, $\delta \zeta_j$ and $\delta \lambda_j$, which are the usual ones, and new counterterms $\delta \lambda_j(p_\alpha, p_\beta), \delta \lambda_j(p_\alpha, p_\beta, p_\gamma)$ in cases (a) and (b). They also depend in these cases on the channel defined by the external lines. Effective coupling constants of vertices with 4 lines will be the same as in Sect. 4.2 except for (at most 4) vertices involving external lines. The new procedure yields gradients acting only on outgoing lines below j, and still yields internal regularization factors. Exponential factors in $M^{-l_{\mathrm{sup}}}$ are, however, replaced by $M^{-(l_{\mathrm{sup}} - l_{\mathrm{inf}})}$, where l_{sup} is the index of the highest internal line of the diagram and l_{inf} is the lowest index of all internal and external lines (l_{inf} was always equal to 1 in Sect. 4.2, since external lines were by convention in slice 1, and was thus omitted). All summations over slice indices are thus again controlled by those factors as before, except the summation over l_{inf} from slice 1 to i_4 at most. This summation, which would *a priori* give at most a factor $i_4 \sim \ln|p_4|$ in bounds on connected functions, will be made by taking into account the behaviour of effective couplings of vertices involving external lines, so that better results will be obtained.

Effective couplings. Given any vertex, let i denote the highest slice of the 4 lines attached to it. The effective coupling is λ_i (as defined in Sect. 4.2) if the vertex involves no external line, or if at least one of the two highest lines is an internal one. In remaining cases:

(i) If the two highest lines are external lines of indices α, β and if the third one is an internal line of index j, then the effective coupling is:

$$\lambda_j(p_\alpha, p_\beta) \equiv \lambda_{i_\alpha} + \sum_{k=j+1}^{i_\beta} \delta\lambda_k(p_\alpha, p_\beta) \quad \text{if} \quad j \geq i_{\alpha,\beta}, \quad (3.124)$$

or

$$\lambda_{i_{\alpha,\beta}}(p_\alpha, p_\beta) \quad \text{if} \quad j < i_{\alpha,\beta}.$$

(ii) If the three highest lines are external lines of indices $i_\alpha, i_\beta, i_\gamma (i_\alpha \geq i_\beta \geq i_\gamma)$ and the fourth line is of index j, the effective coupling is:

(a) $(\alpha, \beta, \gamma) = (1, 2, 3)$

$$\lambda_j(p_\alpha, p_\beta, p_\gamma) \equiv \lambda_{i_3}(p_1, p_2) + \sum_{k=j+1}^{i_3} \delta\lambda_k(p_1, p_2, p_3) \quad \text{if} \quad j \geq i_4,$$

$$(3.125)$$

or

$$\lambda_{i_4}(p_1, p_2, p_3) \quad \text{if} \quad j < i_4.$$

This case includes the vertex with 4 external lines with effective coupling $\lambda_{i_4}(p_1, p_2, p_3)$.

(b) $(\alpha, \beta, \gamma) \neq (1, 2, 3)$

$$\lambda_{i_\gamma}(p_\alpha, p_\beta) \quad \text{if} \quad i_\gamma \geq i_{\alpha,\beta}, \tag{3.126}$$

or

$$\lambda_{i_{\alpha,\beta}}(p_\alpha, p_\beta) \quad \text{if} \quad i_\gamma < i_{\alpha,\beta}.$$

The term $\delta\lambda_j(p_\alpha, p_\beta)$, and in turn $\lambda_j(p_\alpha, p_\beta)$, is evaluated via expansions of the phase-space diagrams contributing to $\delta\lambda_j(p_\alpha, p_\beta)$ as terms of second order in the effective couplings and remainders. The sum of the latter is bounded by $\text{cst}\lambda_j(p_\alpha, p_\beta)\lambda_j^2$. (Convergence is ensured, as usual, by the smallness of the effective couplings and there are at least 3 couplings, one of which is bounded by $\lambda_j(p_\alpha, p_\beta)$ and the others by λ_j.) Terms with two couplings, that is, two vertices v_1, v_2, have two internal (propagator) lines in slices j, j' respectively, $j' \geq j$, running between v_1, v_2 and two more outgoing lines at each vertex, including the two external lines α and β, which may or not be attached to the same vertex. The result is as follows: $\lambda_j(p_\alpha, p_\beta)$ behaves like $\frac{1}{i_\beta^{1-\sigma}}\frac{1}{j^\sigma}$ at large i_β and j, $j \leq i_\beta$, where $\sigma \leq 1/2$ depends on the channel defined by particles 1, 2 (fields ψ or $\bar{\psi}$, spin and colour indices). Related results apply to $\delta\lambda_j(p_1, p_2, p_3)$.

As a consequence, one gets the following results on the (semi-amputated) connected 4-point function, denoted H^c, obtained after factorizing out external propagators.

(i) Generic situations

Theorem 3.5.

$$H^c(\tau q_1 + r_1, \ldots, \tau q_4 + r_4) \sim \frac{\text{cst}(q_1, \ldots, q_4, r_1, \ldots, r_4)}{\ln\tau} \tag{3.127}$$

as $\tau \to \infty$, whenever the Euclidean energy-momenta $q_1, \ldots, q_4$ satisfy $q_i + q_j \neq 0$, $\forall(i, j), i \neq j$.

(ii) "Exceptional" situations

Let $k = p_1 + p_2$, $z = \frac{p_1 - p_2}{2}$, $z' = \frac{p_3 - p_4}{2}$. We give below results in cases where k is fixed.

Theorem 3.6.

$$H^c(k, z, z') \sim \frac{\text{cst}(k, z')}{\left[\ln|z|\right]^{1-\sigma}} \qquad (3.128)$$

as $|z| \to \infty$, where $\sigma \leq 1/2$ depends on the channel defined by particles $1, 2$.

Theorem 3.7.

$$H^c(k, z, z') \sim \frac{\text{cst}(k)}{(\ln|z|)^{1-\sigma}(\ln|z'|)^{\sigma}} \qquad \text{if} \qquad \sigma < 1/2, \quad (3.129)$$

$$H^c(k, z, z') \sim \frac{\text{cst}(k)\ln\ln|z'|}{(\ln|z|)^{1/2}(\ln|z'|)^{1/2}} \qquad \text{if} \qquad \sigma = 1/2, \quad (3.130)$$

as $|z|, |z'| \to \infty, |z| \geq |z'|$.

For purposes of Chapter IV, we note that these results apply equally to the connected, "fully" amputated 4-point functions F obtained by factorizing out 2-point functions instead of propagators for each $k = 1, \ldots, 4$, in view of the relation

$$\omega_b(k, z)H(k, z, z')\omega_b(k, z') = \omega(k, z)F(k, z, z')\omega(k, z'), \qquad (3.131)$$

where ω and ω_b are, respectively, products of 2-point functions or bare propagators (e.g. $\omega(k, z) = S(p_1)S(p_2)$, $p_1 = \frac{k}{2} + z$, $p_2 = \frac{k}{2} - z$).

N-point functions and Wilson short-distance expansion. For N-point functions F_N, Theorem 3.6 is still valid, z' denoting a set of $N - 3$ relative energy-momenta among $p_3, \ldots, p_N$.

A more refined phase-space analysis [IM4], omitted here, along the same lines as above, allows one, moreover, to establish uniform factorization properties (at k, z' fixed, $z \to \infty$) which are essentially equivalent to Wilson-Zimmerman short-distance expansion. We state the result at first order ($\omega(k, z)$ denotes as above the product $S(p_1)S(p_2)$ of 2-point functions).

Theorem 3.8.

$$F_{N+2}(k, z, z') = \frac{\omega(0, z)}{\omega(k, z)}F_4(0, z, 0)\Lambda_N(k, z') + R_{N+2}(k, z, z') \quad (3.132)$$

with

$$|R_{N+2}(k, z, z')| < \frac{\text{cst}(k, z')}{1 + |z|} \qquad (3.133)$$

and

$$\Lambda_N(k, z') = \delta_{2,N} + (1\omega R_{N+2})(k, z')$$
$$= \lim_{\rho \to \infty} \frac{(1\omega F_{N+2})_\rho(k, z')}{(1\omega F_4)_\rho(0, 0)}, \tag{3.134}$$

where $(1\omega A)(k, z') = \int \omega(k, z) A(k, z, z') \mathrm{d}z$ and $(1\omega A)_\rho$ is similarly defined in the theory with ultraviolet cut-off ρ.

In view of (3.133), the remainder R_{N+2} decays faster than F_{N+2} as $z \to \infty$ (with the difference of one power of $|z|$), and the behaviour of F_{N+2} is given by the term $F_4(0, z, 0)$ for all $N \geq 2$ as $z \to \infty$, or (after Fourier transformation) as the dual variable $x_1 - x_2$ tends toward zero.

§5. Bosonic models: complements

5.1 Ultraviolet limit in (massive) $P(\varphi)_2$ models

We only treat here the $\rho \to \infty$ limit of the quantity

$$Z_{\Delta, \rho} = \int e^{-\lambda \int_\Delta :\varphi^4(x):_\rho \, \mathrm{d}x} \mathrm{d}\nu_\rho(\varphi), \tag{3.135}$$

where Δ is a unit square. As already mentioned in Sect. 3.2, the term $C_\rho(0, 0)$ behaves like $c_1\rho$ as $\rho \to \infty$, so that $|Z_{\rho, \Delta}| < e^{c_2 \lambda \rho^2}$ as a simple consequence of the relation $\int \mathrm{d}\nu_\rho(\varphi) = 1$ and of the inequality (3.83), with $c_2 > 0$ independent of ρ. We show below, by an adaptation of Nelson's original ideas, that, as a matter of fact, $Z_{\rho, \Delta}$ has a finite limit as $\rho \to \infty$ (which is nonzero).

We use the momentum-space decomposition $\tilde{C}_\rho(p) = \sum_{i=1}^\rho \tilde{C}^{(i)}(p)$, introduce auxiliary real variables v_i, $i = 1, \ldots, \rho$, that will vary between 0 and 1 and define

$$C_\rho(v) = \sum_{i=1}^\rho v_i C^{(i)}, \tag{3.136}$$

where $v = \{v_i\}$. A corresponding definition of $Z_{\Delta, \rho}(v)$ follows. Leaving Δ implicit below, a Taylor expansion of $Z_\rho \equiv Z_\rho(1)$ of order N_i (specified below) around $v_i = 0$, with respect to each variable v_i, yields:

$$Z_\rho = \prod_{i=1}^\rho \left[\sum_{n_i=0}^{N_i-1} \frac{1}{n_i!} \frac{\mathrm{d}^{n_i}}{\mathrm{d}v_i^{n_i}} \cdots |_{v_i=0} + \int_0^1 \mathrm{d}v_i \frac{(1 - v_i)^{N_i-1}}{(N_i - 1)!} \frac{\mathrm{d}^{N_i}}{\mathrm{d}v_i^{N_i}} \right] Z_\rho(v). \tag{3.137}$$

As will appear, it will be convenient to choose a large enough N_i, $N_i \to \infty$ as $i \to \infty$, but not too large, for example, $N_i \sim i^2$.

Bound on a given term in the expansion (3.137). We first consider a given term in the expansion (3.137), corresponding to given values of all n_i ($\leq N_i$), $i = 1, \ldots, \rho$. By carrying out the derivatives (which apply either to the propagators $C_\rho(v)$ in $:\varphi^4:_\rho(v)$ in the exponential, or the measure $d\nu_\rho(\varphi; v)$ of covariance $C_\rho(v)$), this term is written as a sum of contributions whose number can be bounded by $\prod_i \left[i^{O'(1)n_i} \right]$ and where $O'(1)$ is a fixed constant. Each contribution is of the form

$$\int [\ldots] e^{-\lambda \int_\Delta :\varphi^4(x):_\rho(v) \, dx} \, d\nu_\rho(\varphi; v), \tag{3.138}$$

where $[\ldots]$ is a product of coupling constants, of fields φ at various points, integrated in Δ, and of propagators. It can be bounded in modulus by:

$$\left[\int |[\ldots]|^2 d\nu \right]^{1/2} \times \left[\int e^{-\lambda \int_\Delta :\varphi^4(x):_\rho(v) \, dx} \, d\nu_\rho(\varphi; v) \right]^{1/2}. \tag{3.139}$$

The factor $\int |[\ldots]|^2 d\nu$ can be bounded by a sum, over contraction schemes, of terms that can be treated by methods analogous to those used to get bounds on Feynman integrals in Sect. 2.2. For each graph, one gets a bound of the form $\prod_i \left[\lambda^{1/2} M^{-i/4} \right]^{2n_i}$ (using the fact that each connected component has at least two vertices in view of Wick products). The number of terms is bounded by $\prod_i [i^{O''(1)n_i}]$ where $O''(1)$ is again a fixed constant.

On the other hand, the second factor in (3.139) is bounded by $e^{O'''(1)j^2}$, where j is the lowest index such that $n_i < N_i$ for all $i > j$: the same argument as that giving the original bound in $e^{cst\rho^2}$ on Z_ρ, using here the fact that all v_i of index $i > j$ are fixed at zero so that the sum in (3.136) is restricted to slices $i \leq j$. Since $n_j = N_j$ (if $j \geq 1$), j^2 can be replaced in this bound by n_j if N_j is of the order of j^2.

The final result is a bound, on any given term in the expansion, of the form:

$$\prod_{i=1}^\rho \left[\lambda^{1/2} i^{O(1)} M^{-i/4} \right]^{n_i} \left(\leq \prod_{i=1}^\rho \left[cst \, \lambda^{1/2} M^{-i/8} \right]^{n_i} \right).$$

Bound on $Z_{\Delta,\rho}$ in the $\rho \to \infty$ limit and related results. Z_ρ, which is the sum over all possible sets $\{n_i\}$, is in turn uniformly bounded with respect to ρ (for a small enough λ) by

$$\prod_{i\geq 1}\left\{1+\sum_{n\geq 1}\left[\text{cst }\lambda^{1/2}M^{-i/8}\right]^n\right\} \leq e^{\sum_i[\text{cst}' \lambda^{1/2}M^{-i/8}]}(<\infty),$$

where the conditions $n_i \leq N_i$ for each i, and $i \leq \rho$, have been removed.

A slightly more careful analysis yields a lower bound different from zero, and in fact close to 1, at small λ. The existence of the (nonzero) $\rho \to \infty$ limit Z of Z_ρ is also established (e.g from related bounds on $Z_\rho - Z_{\rho-1}$).

5.2 Phase-space analysis in more general models: massive φ_3^4 and "infrared φ_4^4" (outline)

As indicated in Sect. 1.1, we treat below (massive) φ_3^4 for definiteness, in a way that can directly be adapted to other models. Infrared φ_4^4 will be briefly mentioned at the end.

N-point functions $S_{\Lambda,\rho}(x_1,\ldots,x_N)$ are again of the form $N_{\Lambda,\rho}(x_1,\ldots,x_N)/Z_\Lambda$ with

$$N_{\Lambda,\rho}(x_1,\ldots,x_N)$$
$$= \int \varphi(x_1)\ldots\varphi(x_N)e^{-\lambda\int_\Lambda \varphi^4(x)\,dx+\alpha_\rho\int_\Lambda \varphi^2(x)\,dx}d\nu_\rho(\varphi),$$

$$(3.140)$$

where $d\nu_\rho$ is the measure of covariance C_ρ; $\tilde{C}_\rho$ is of the form $[1/p^2+1]$ $e^{-M^{-2\rho}(p^2+1)}$ and α_ρ will be specified later. The coupling λ can be chosen independent of ρ in φ_3^4.

In contrast to GN$_2$, and as was already the case for $P(\varphi)_2$, it is not possible to get rigorous results in bosonic models from an expansion of the exponential in λ. It is here convenient, in view of phase-space analysis, to introduce a decomposition of C_ρ as a sum of slice propagators $C^{(i)}$, defined as previously, and also new variables $\varphi^{(i)}$ and measures $d\nu^{(i)}$ of covariance $C^{(i)}$, $i = 1,\ldots,\rho$. A new expression of $N_{\Lambda,\rho}$ is then obtained by the following manipulations: each φ in (3.140) is replaced by $\sum_{i=1}^\rho \varphi^{(i)}$ and $d\nu_\rho$ is replaced by the product of the measures $d\nu^{(i)}(\varphi^{(i)})$. However, this expression is not yet satisfactory, in particular because of renormalization problems, and will be modified inductively from slice ρ to 1. We describe the first step of this induction, in which case new variables and measures introduced are $\varphi^{(\rho)}$, $\varphi_{\rho-1}$ and $d\nu^{(\rho)}$, $d\nu_{\rho-1}$ (of covariance $C^{(\rho)}$ and $C_{\rho-1} = \sum_{i=1}^{\rho-1} C^{(i)}$, respectively); φ and $d\nu_\rho$ are then replaced in (3.140) by $\varphi^{(\rho)} + \varphi_{\rho-1}$ and $d\nu^{(\rho)}(\varphi^{(\rho)})d\nu_{\rho-1}(\varphi_{\rho-1})$.

A dependence of $N_{\Lambda,\rho}$ on new auxiliary variables is now introduced: variables $s^{(\rho)}$ corresponding to a cluster expansion in slice ρ with respect to a paving of Λ with cubes of size $M^{-\rho}$, and variables $v_\Delta^{(\rho)}$, one for each cube Δ of this paving. Fixing a variable v_Δ at zero will then amount to decouple locally (in the cube Δ) slice ρ from the slice below ρ. (E.g. φ^4 is now replaced by $(\varphi^{(\rho)}+v^{(\rho)}\varphi_{\rho-1})^4+[1-(v^{(\rho)})^4]\varphi_{\rho-1}^4$ where $v^{(\rho)}(x) = v_\Delta^{(\rho)}$ if $x \in \Delta$. If $v^{(\rho)} = 1$, this expression gives $\varphi^4 = (\varphi^{(\rho)} + \varphi_{(\rho-1)})^4$ whereas at $v^{(\rho)} = 0$ one obtains $\varphi^{(\rho)^4} + \varphi_{\rho-1}^4$, which will yield decoupling of slices ρ and $\rho - 1$.) A Taylor expansion will now be made around the origin in the variables $s^{(\rho)}$ and $v^{(\rho)}$, up to order 1 in $s^{(\rho)}$ and, for example, to order 4 in $v^{(\rho)}$; the order in $v^{(\rho)}$ has to be sufficient to accommodate renormalization procedures. More precisely, one first carries out a suitable a la Mayer procedure, and renormalization then amounts to define $\delta\alpha_\rho$ (hence, $\alpha_{\rho-1} = \alpha_\rho + \delta\alpha_\rho$) in terms of 2-point functions of slice ρ. The procedure is pursued inductively and yields the renormalized mass, close to $1 + \alpha_1$, of the order of $1 + O(\lambda)$. (A further renormalization, which amounts to modify the coefficient of p^2 in the propagator has also to be made.)

As already mentioned in Sect. 1.1, an important technical difficulty is encountered at each step of the induction. It is treated in [FMRS2] by a "domination procedure" taking into account the sign of the term $-\lambda\varphi^4$ in the interaction. This procedure is somewhat complicated and requires various unpleasant technical aspects. A different one, directly adapted from [AIM,IM5], which is somewhat simpler and has a somewhat more general character, is based on the introduction of an auxiliary "ultralocal" field σ, namely, a random real valued function $x \to \sigma(x)$ in Euclidean space-time, with a probability measure $d\nu'(\sigma)$ of covariance $\int \sigma(x)\sigma(y)d\nu'(\sigma) = \delta(x - y)$, that is, with propagator 1 in energy-momentum space. By simple manipulations, Eq. (3.140) is replaced by

$$N_{\Lambda,\rho} = \int \varphi(x_1)\ldots\varphi(x_N)e^{i\sqrt{2}\int_\Lambda \left[\sqrt{\lambda}\varphi^2(x)-\frac{\alpha_\rho}{2\sqrt{\lambda}}\right]\sigma(x)\,dx}d\nu\,(\varphi)d\nu'(\sigma).$$

$$(3.141)$$

The procedure is then analogous to the above except that, besides variables $\varphi^{(i)}$, it is convenient to introduce a family $\vec{\sigma} = \{\sigma^{(i)}\}$ of (non-independent) functions $\sigma^{(i)}$ (one for each i) and a measure $d\nu(\vec{\sigma})$ of covariance $\int \sigma^{(i)}(x)\sigma^{(j)}d\nu(\vec{\sigma}) = \delta(x - y)$. Their introduction is not intended to make an actual energy-momentum analysis (in contrast to the introduction of the $\varphi^{(i)}$) but will help to exhibit suitable factorization properties between slices. The procedure is again inductive. At, for example, the first step, the exponent $\int dx\,[\sqrt{\lambda}\varphi^2(x) - \frac{\alpha_\rho}{2\sqrt{\lambda}}]\sigma(x)$ is replaced by $\int dx\big[\sqrt{\lambda}\{(\varphi^{(\rho)})^2\sigma^{(\rho)} + 2\varphi^{(\rho)}\varphi_{\rho-1}\sigma_{\rho-1} + \varphi_{\rho-1}^2\sigma_{\rho-1}\} - (1/2\sqrt{\lambda})\alpha_\rho\sigma_{\rho-1}\big]$.

The introduction of the variable $v^{(\rho)}$ is made at that stage by replacing $d\nu(\vec{\sigma})$, where here $\vec{\sigma} = (\sigma^{(\rho)}, \sigma_{\rho-1})$, by $d\nu(\vec{\sigma}, v^{(\rho)})$ with $\int \sigma^{(\rho)}(x)\sigma_{\rho-1}(y)$ $d\nu(\vec{\sigma}, v^{(\rho)}) = v^{(\rho)}\delta(x - y)$ (and by replacing the term $\varphi^{(\rho)}\varphi_{\rho-1}\sigma_{\rho-1}$ in the above exponent with $\varphi^{(\rho)}\varphi_{\rho-1}v^{(\rho)}\sigma_{\rho-1}$).

The technical problem arising from the occurrence of large numbers of "low" fields $\varphi_{\rho-1}$ is treated by "elimination": these low fields are regrouped by pairs at the same point modulo gradients, and terms in $\varphi_{\rho-1}(x)^2$ can be replaced by $\frac{1}{\Delta}\int_\Delta \varphi_{\rho-1}(y)^2 dy$ (modulo gradients), and in turn by $\frac{1}{\Delta}\int_\Delta \sigma_{\rho-1}(y)dy$, up to "good" terms. "Power counting" arguments will then allow one to treat the contributions obtained. Details are omitted.

The treatment of infrared φ_4^4 is similar to above. Slices refer now to approximate regions $M^{-i} \lesssim |p| \lesssim M^{-(i-1)}$, with boxes of size M^i for each $i = 1, \ldots, \rho$. The inductive procedure is made from slice 1 to ρ. (The $\rho \to \infty$ limit is here the infrared limit). A dependence of the coupling λ on the slice i is then needed and the renormalization procedure is made also, at each step of the induction, on 4-point functions of the given index i, so that quantities $\delta\lambda_i$ and effective couplings λ_i are also defined.

CHAPTER IV

Particle Analysis in
Constructive Field Theory

§1. Introduction

We start in Sect. 1.1 with some indications on the perturbative approach and will then give a more precise introduction to the constructive (and semi-axiomatic) analysis in Sect. 1.2. With regard to notations used in this chapter, see Sect. 3.2 of the general Introduction.

1.1 The perturbative approach

As explained in Sect. 2.1 of Chapter III, (connected) N-point functions appear in the perturbative approach as formal infinite sums of possibly renormalized Feynman integrals, with possibly renormalized coupling constants at each vertex. (For simplicity, we leave aside the effective expansions mentioned in Sect. 2.2 of Chapter III.) We consider these series, initially, in Euclidean energy-momentum space where, in a theory with positive masses (e.g. φ_2^4 with $m > 0$), each individual integral is analytic. (The factors $p_i^2 + m^2$ or $k_\ell^2 + m^2$ in denominators of the propagators $\tilde{C}(p_i)$ or $\tilde{C}(k_\ell)$ do not vanish: $p_i^2 \geq 0$, or $k_\ell^2 \geq 0$, in the Euclidean region.) These integrals can be analytically continued away from Euclidean space for example, by distorting integration contours in complex space, in order to avoid the poles of internal propagators away from Euclidean space. Singularities will occur when distortions of contours are no longer possible. We give below some results on the analytic structure obtained and explain how information on the analytic structure of N-point functions (and related properties) can in turn be derived, at least from a heuristic viewpoint.

Analytic structure of individual Feynman integrals. The mass m is different from the physical mass μ. However, for reasons explained in the second part of this subsection, one should here consider Feynman integrals whose propagators have poles at the physical mass, that is, at $p^2 (\equiv p^2_{(0)} - \vec{p}^2) = \mu^2$ (in Minkowskian notations) rather than $p^2 = m^2$. Results mentioned below, obtained in the 1960s (see [Ph2] and references therein) and the 1970s (see [KK] and references therein), apply to usual Feynman integrals whenever they are convergent and also, to a large extent, to renormalized integrals otherwise. For simplicity, we discuss below mainly the first situation.

Following works of Bros[BL,B3] on more general "$\mathcal{G}$-convolution" integrals (see Sect. 6.1 in Chapter II), we first note that Feynman integrals can always be analytically continued, at least in the primitive axiomatic domain introduced in Chapter II. By using the analyticity of the kernels at each vertex (which are here constants), they can be analytically continued in a much larger multisheeted domain in complex space: starting from Euclidean energy-momentum space, various paths of analytic continuation around possible singularities can be considered to that purpose. For each given N, their physical boundary values in Minkowski space, whose formal sum will give the chronological function, can be obtained from the directions of the same cell in each real region. They can be written in the same way as in Euclidean space except that propagators are now of the form $1/\left[k_\ell^2 - \mu^2 + i\varepsilon\right]$ with Minkowskian notations. For each connected graph G, the boundary value $I_G(p_1, \ldots, p_N)$ is shown to be analytic outside the $+\alpha$-Landau surfaces of G and of graphs G' obtained from G by contraction, and the following more precise information is obtained. The analytic essential support, or microsupport $ES_P(I_G)$, at any physical point P is shown to be the set of points $x = (x_1, \ldots, x_N)$ (defined modulo global translations) such that there exists a classical multiple scattering diagram $\mathcal{D}_+$ associated with G, with external energy-momenta P_i (or $-P_i$ if i is initial with the sign conventions of Chapter II), and such that each x_i lies on the external trajectory in the past of the interaction vertex involving line i if i is initial, or in its future if i is final. If external propagators $\tilde{C}(p_i)$ are factored out, the same result is obtained except that x_i must lie at the interaction vertex involving the external line i. This result entails that singularities are the $+\alpha$-Landau surfaces already mentioned and plus $i\varepsilon$ rules at $+\alpha$-Landau points also follow (either in the off-shell or on-shell contexts).

Singularities of Feynman integrals in other sheets are more generally modified Landau surfaces (see Chapter I-3.1) and these integrals are shown to be always holonomic, with "regular singularities": see [KK,KKO].

Various discontinuity formulae of interest have been established, including the well-known Cutkosky formula of the 1960s, which expresses the local discontinuity around $L_+(G)$ as a convolution integral associated to G over *on-shell* internal energy-momenta. However, some of the discontinuity formulae (relevant, for example, in the derivation of generalized optical theorems) are not, in general, proved to our knowledge. For results or conjectures, applying also to somewhat more general Feynman-type integrals, see Sect. 5.

N-point functions. Results on N-point functions (and in turn on the S matrix) cannot be directly derived from results on individual integrals, even from a formal heuristic viewpoint. A first problem, already mentioned, is that propagators initially have poles at the bare mass m (or renormalized mass m_{ren}), different from the physical mass. As a matter of fact, partial resummations lead to replace these propagators by new functions that will indeed have a pole at the physical mass. In, for example, a super-renormalizable theory like $P(\varphi)_2$, the actual 2-point function is obtained by taking the sum of the terms

$$= \tilde{C}(p) + \tilde{C}(p)K(p;\lambda)\tilde{C}(p) + \cdots = \left[p^2 - m^2 - K(p;\lambda)\right]^{-1},$$

where K is the (formal) sum of Feynman integrals associated to "1-particle irreducible" $(1-p.i.)$ graphs with two external lines, from which external propagators have been factored out. $(1 - p.i.$ graphs are those that cannot be cut into two parts, with each one of the external lines in each part, by cutting only one internal line.) The pole of the 2-point function is now obtained at $p^2 - m^2 - K(p;\lambda) = 0$. We note that K is, as a matter of fact, a function of p^2 and λ and includes a factor λ. The dependence on λ is left implicit below. Under adequate conditions, proved later in constructive theory at small coupling, there is in particular one (and only one) pole. Its position determines the basic physical mass $\mu = \mu(\lambda)$ $(\mu(\lambda) \to m$ as $\lambda \to 0)$.

The 2-point function itself will have other singularities, besides the pole just mentioned, in particular due to the singularities of K at the 2-particle threshold, or the 3-particle threshold in an even theory, and higher thresholds. These singularities will not play an important role in the analysis of the 2-particle region of a given process (below the 3-particle threshold, or the 4-particle threshold in an even theory), but cannot be ignored otherwise. Using the fact that $K\left(\mu^2\right) = \mu^2 - m^2$, one possible approach is then to rewrite $p^2 - m^2 - K\left(p^2\right)$, in the denominator of the 2-point function, in the form

$$(p^2 - \mu^2)\left(1 - \frac{K(p^2) - K(\mu^2)}{p^2 - \mu^2}\right)$$

and to make a corresponding expansion of the factor $\left[1 - \frac{K(p^2) - K(\mu^2)}{p^2 - \mu^2}\right]^{-1}$. In this expansion, each factor $(K(p^2) - K(\mu^2))/(p^2 - \mu^2)$ is a sum of "subtracted" Feynman integrals, divided by $p^2 - \mu^2$. (This procedure does not introduce multiple poles at $p^2 = \mu^2$ since $K(p^2) - K(\mu^2)$ vanishes at $p^2 = \mu^2$.)

Results on the analytic structure of N-point functions (and related ones) can then be extrapolated from previous results on individual integrals, if there are no other problems generated by the infinite sum. (We come back below to this question.) In this case, one reobtains the microsupport (and hence causality) property of chronological N-point functions, namely, Conjecture 2.2 of Chapter II-7.2 and related results. Other properties (e.g. modified Landau singularities in other sheets) are derived similarly. Discontinuity formulae of N-point functions can in principle also be checked by formal resummations of discontinuities of individual integrals.

We note that, although each individual integral is holonomic, this property cannot be expected for the infinite sum. In, for example, a $2 \to 2$ process, all integrals associated to graphs with n sets of two internal lines joining successive vertices (as in example (i) of Sect. 3.1 in Chapter I) are singular at the 2-particle threshold $s = 4\mu^2$ (for $n \geq 1$). Each term is equal to $\lambda^{n+1}\varphi(s)^n$ where $\varphi(s)$, obtained at $n = 1$, behaves locally near $\sigma = 0$, $\sigma = s - 4\mu^2$, like $\sigma^{(d-3)/2}$ if d is even, but like $\sigma^\beta \ell n \, \sigma$ if d is odd. The sum, equal to

$$\lambda + \lambda^2\varphi(s) + \lambda^3\varphi(s)^2 + \cdots = \lambda/[1 - \lambda \, \varphi(s)]^{-1}, \qquad (4.1)$$

is nonholonomic in the latter case. (Although only a partial sum of Feynman integrals has been considered here, nonholonomy is established in a rigorous way in axiomatic or constructive theories at the 2-particle threshold if d is odd, and is expected at most other singularities at d whether even or odd.)

On the other hand, results indicated above can no longer be expected to hold as they stand in various situations in which new singularities are generated by the infinite sums. As confirmed in Sect. 4 in a more complete way, such singularities can occur even in simple situations, as can be seen from partial resummations analogous to the above at $d = 2$ or $d = 3$. At, for example, $d = 2$ and in an even theory, the partial sum (4.1) will give a contribution to the 4-point function with a pole on the real s-axis close to (and less than) 2μ at small coupling, namely, the zero of $1 - \lambda \varphi(s)$. Since $\varphi(s)$ behaves like cst $\sigma^{-1/2}$ at $d = 2$ near $\sigma = 0$, such a zero is always obtained however small the coupling λ is. (It is closer and closer to 2μ as $\lambda \to 0$). This pole lies in the second sheet for some models, like φ_2^4, but lies in the physical sheet in other $P(\varphi)_2$ models. This partial analysis will be fully confirmed, and made more complete and precise, in Sect. 4 in axiomatic and constructive theories. A pole will correspond in the physical sheet to a 2-particle bound state in accordance with the ideas presented at the end of Sect. 1.2 in Chapter II. A similar, though more delicate, analysis applies equally for theories that are not even, and provides a similar pole of the 2-point (and 4-point) function close to 2μ, which corresponds again to a bound state in some models, at $d = 2$ and $d = 3$: see Sect. 4.2. In theories in which such bound states (and possibly further ones) occur, they must be taken into account in an adequate way and will contribute to new Landau singularities.

1.2 Constructive and semi-axiomatic approaches

General results. We first give a general presentation of results and will then give a description of contents including some further details. Theories defined in Chapter III satisfy the Euclidean axioms, as explained in Chapter III-3 for $P(\varphi)_2$ models and as can also be established for φ_3^4, GN_2 and other theories. Hence, via the OS analytic continuation from imaginary (= Euclidean) to real times, they do yield Wightman field theories in Minkowski space-time. However, no result is obtained on particle analysis: mass spectrum, asymptotic completeness, structure of Green functions and of the S matrix in terms of particles. Results in this direction can be obtained thus far in constructive theory in a different way, through structure equations (first established in Euclidean space) expressing N-point functions in terms of "irreducible" kernels that are shown to satisfy better analyticity properties in complex energy-momentum space. The analysis is then made via analytic continuation of these equations from Euclidean to Minkowski energy-momentum space. It can again be considered as a development of the program proposed by Symanzik [Sy1] and is related to the earlier approach of Bros presented in Chapter II-6 in the axiomatic framework, with the following important differences. Irreducible kernels were defined in Chapter II-6 in terms of N-point functions via integral equations, and their analyticity properties were derived from asymptotic completeness. Here, these kernels will be defined and shown to satisfy suitable analyticity properties in a direct way, and structure equations will also be directly established in view of the underlying structure of the theories considered. These equations are used in turn to derive results on particle analysis, including information on the mass spectrum and on asymptotic completeness. The methods will apply, for example, to the weakly coupled $P(\varphi)_2$, φ_3^4, GN_2 models introduced in Chapter III, and to other models defined in similar ways. We shall also consider semi-axiomatic approaches in which irreducible kernels (playing the role of the potential in nonrelativistic theory) would be the basic quantities, in which case their analyticity properties and relevant structure equations are the starting point.

Before pursuing, we note, as already mentioned in Chapter III, that some special models in space-time dimension 2 can be treated by different, more explicit methods, with results including, in particular, asymptotic completeness (see [Ruij] and references given there). This approach and results, which have their own interest, will be omitted and we restrict our attention to developments along the lines indicated above. Besides earlier results that have exhibited a basic lowest physical mass μ, a number of results were obtained in the second part of the 1970s on $P(\varphi)_2$ models in the low-energy 2-particle region, starting with works of Spencer and Zirilli ([Sp,SpZ]) and further works on possible bound states [DE,Ko], with also some results in the 3-particle region for P even in the early 1980s [CD]. However, as they stand, methods used in these works are restricted to $P(\varphi)_2$ models and to the low-energy regions that have been mentioned. A class of more general "r-particle irreducible" kernels has been directly defined by the methods of [Sp] (see [GJ] and references therein) in view of the treatment of higher energy regions, but this class is probably too limited. We shall here present new methods [I5,IM1-3,BI3,IG] developed in the second part of the 1980s that are more convenient and have provided more general results both in the 2-particle region, for $P(\varphi)_2$ and other theories in dimension 2 or 3, including nonsuper-renormalizable ones, and (with some limitations) in more general energy regions.

Irreducible kernels and related equations to be considered will depend on the energy region one wishes to explore in Minkowski space and also, as explained below, on the character of the theory from the viewpoint of renormalization. In all theories to be considered, a basic (lowest) physical mass $\mu > 0$, which depends on the (possibly renormalized) coupling λ and corresponds to a pole of the 2-point function at $s = \mu^2$ away from Euclidean space, is first obtained. This pole is easily exhibited through the use of a $1 - p.i.$ (1-particle irreducible) 2-point function.

The mass μ is close (at small coupling) to the bare mass m in $P(\varphi)_2$, or to the renormalized mass in other theories. Further results are then obtained in the low-energy 2-particle region, up to $s = (3\mu)^2 - \varepsilon$ (or $(4\mu)^2 - \varepsilon$ if the theory is even) in Minkowski space where $\varepsilon = \varepsilon(\lambda) \to 0$ as $\lambda \to 0$ (see further comments on the loss of ε below). The analysis is made in even $P(\varphi)_2$ models via the use of the 4-point kernel G which is $2 - p.i.$ in the channel $(1, 2; 3, 4)$ and is linked to the 4-point (connected, amputated) function F via the Bethe-Salpeter (BS) equation $F = G + GoF$, where o is Feynman-type 2-particle convolution as in Chapter II-6. It also extends, with some modifications, to noneven $P(\varphi)_2$ models and to other possibly nonsuper-renormalizable theories, like GN_2. For the latter, somewhat different BS-type kernels and equations, of the type initially proposed from axiomatic or semi-axiomatic viewpoints in [B1-4,BL,Sy2,BD] and introduced in a precise way in the constructive framework in [IM2,3], have to be used in view of ultraviolet problems.

Although one might expect the absence of 2-particle bound states at small coupling, such bound states, involved in the mass spectrum, do occur in some of the theories in dimension 2 or 3, however small the coupling is, as analyzed already in [SpZ,DE,Ko] and in a more complete way in [BI3]. The mass μ_B of the bound state is less than 2μ, and is closer and closer to 2μ as $\lambda \to 0$. It corresponds to a pole of F at $s = \mu_B^2$ in the physical sheet, due to specific kinematical factors in dimension 2 or 3, as already mentioned in Sect. 1.1 (and in accordance with ideas presented at the end of Sect. 1.2 of Chapter II). Other results include, as in Chapter II-6, the 2-sheeted (d even) or multisheeted (d odd) structure of F around the 2-particle threshold singularity $s = 4\mu^2$, and asymptotic completeness in the 2-particle region (up to $s = (3\mu)^2 - \varepsilon$, or $(4\mu)^2 - \varepsilon$ in an even theory). The loss of ε is natural and one cannot expect better results at that stage of the analysis. For instance, 3-particle (or 4-particle) bound states at a mass close to and below 3μ (or 4μ) cannot *a priori* be excluded. Results previously achieved cannot be expected to be valid without modification up to $s = (3\mu)^2$ (or $(4\mu)^2$) in such a case. A more complete study, involving higher order irreducible kernels, is then required if one wishes to analyze the situation in the neighborhood of $s = (3\mu)^2$ (or $(4\mu)^2$).

Methods developed in [IM1-4] allow one to define general irreducible kernels that will be needed, following [I5], in particle analysis in higher and higher energy regions (up to $s = (r\mu)^2 - \varepsilon$, for higher and higher values of r), and to exibit in a natural way corresponding structure equations of the type initially analyzed in [I5] from a semi-heuristic viewpoint. The latter will be expansions, with convergence properties at small coupling, of connected functions in terms of Feynman-type integrals with irreducible kernels at each vertex (and possibly modified 2-point functions on internal lines). These expansions generalize the series expansion $F = G + GoG + \cdots$ of F in terms of G obtained from the BS equation. Analyticity properties, including various discontinuity formulae, established or conjectured on individual integrals in view of the analyticity of irreducible kernels involved, yield related properties of connected functions, including unitarity-type and asymptotic completeness relations, generalized optical theorems, macrocausal properties. We note that the analysis applies so far for smaller and smaller couplings as the energy increases. On the other hand, divergences of the expansions away from Euclidean space, already at the origin of possible 2-particle bound states in cases previously mentioned in dimensions 2 or 3, may occur even at small coupling. The analysis thus remains formal to some extent and results should apply, as they stand, to theories without bound states, a more complicated analysis being otherwise needed.

Description of contents. We now give more details on these various topics and present the organization of this Chapter. We start in Sect. 2 with $P(\varphi)_2$ models. General irreducible kernels (with various degrees of irreducibility in various channels) can then be defined at small coupling via expansions analogous to those of connected N-point functions in Chapter III-3, except that (i) a larger number of explicit propagators are exhibited from the outset via "higher order" cluster expansions and (ii) the sums are now restricted to graphs with corresponding irreducibility in the graphical sense. These kernels are intrinsic, that is, independent of technical details in the cluster expansion (size of the paving), and are from the perturbative viewpoint, sums of Feynman integrals of graphs with the same irreducibility properties. They are also shown, from their graphical definition, to satisfy a better exponential decay in Euclidean space-time than connected N-point functions, and are therefore analytic, via the Laplace transform theorem, in larger strips around Euclidean space in complex energy-momentum space. Structure equations in Euclidean space follow, on the other hand, from factorization properties of the cluster expansions of order > 1, via suitable regroupings of terms in the expansions of connected functions. (The souvenir of the cluster expansion has disappeared in these regroupings.)

Cluster expansions of order > 1 are presented in Sect. 2.1. Section 2.2 first treats the 2-point function, shown to be equal to $\tilde{C}(p) + \tilde{C}(p)K(p;\lambda)\tilde{C}(p) + \cdots = \left[p^2 - m^2 - K(p;\lambda)\right]^{-1}$ (in Minkowskian notations) where the $1 - p.i.$ 2-point kernel K is shown to be analytic in p up to $s = (2m)^2 - \varepsilon(\lambda)$ and bounded by $cst|\lambda|$. As a consequence, the

2-point function is itself analytic in a similar region apart from its pole at $s = \mu^2$ (the zero of $p^2 - m^2 - K(p; \lambda)$). Section 2.2 then introduces more specifically the BS kernel G, starting from a cluster expansion of order 4.

Phase-space analysis is required, as in Chapter III, for the treatment of other theories, like φ_3^4 or GN_2. In models like φ_3^4, the BS kernel G can still be defined and the BS equation can still be used for particle analysis in the 2-particle region. This will no longer be true in nonsuperrenormalizable theories. Two different approaches will be presented in Sects. 3.1 and 3.2 respectively. For definiteness, we consider in either case the massive GN_2 model. General ideas and methods of Sect. 3.1 can in principle be extended to more general renormalizable or nonrenormalizable theories that might be defined in related ways via phase-space analysis. On the other hand, Sect. 3.2, at least as it stands, applies more specifically to GN_2, namely, to an asymptotically free, renormalizable theory with only 2- and 4-point relevant or marginal operators. The approach of Sect. 3.1 is closer from a technical viewpoint to that of Sect. 2 for subsequent particle analysis. It is based on the introduction of a $2 - p.i.$ BS-type kernel G_M which is not intrinsic, depends in fact on technical details of phase-space analysis (e.g. the choice of M) and does not have a simple perturbative content. However, it is defined in a natural way in the constructive framework, starting from a cluster expansion of order 4 in the lowest momentum slice (and usual cluster expansions of order 1 in higher slices). Graphical 2-particle irreducibility is understood in the definition of G_M with respect to propagator lines of slice 1 only. This is sufficient to show that G_M is still analytic up to $s = (4\mu)^2 - \varepsilon$. G_M is now linked to the actual function F (of the theory without cut-off) via a regularized BS equation $F = G_M + G_M o_M F$, of the form introduced in the axiomatic framework (see Chapter II-6), with a fixed ultraviolet cut-off (depending on M) in o_M that eliminates divergence problems in the integral.

The approach of Sect. 3.2 is based instead on the actual BS kernel B, or the renormalized BS kernel B_{ren}, denoted G, which differ by a constant in GN_2. They can still be defined through phase-space analysis starting from cluster expansions of order 4 in all momentum slices (irreducibility being now considered with respect to propagator lines of all slices), and are again analytic up to $s = (4\mu)^2 - \varepsilon$. B has a perturbative content only in the theory with ultraviolet cut-off and is then the (formal) sum of nonrenormalized Feynman integrals of $2 - p.i.$ graphs with bare couplings λ_ρ. G is perturbatively (in the theory without cut-off) the sum of renormalized Feynman integrals of $2 - p.i.$ graphs with renormalized coupling. From the further study of large momentum properties (in Euclidean directions) of B, it is shown that the BS equation $F = B + BoF$ still makes sense, but the convergence of the integral BoF (due to factors in $1/\ln$) is too weak to allow its direct exploitation. Subsequent particle analysis in this approach will then be made via "subtracted" BS equations, involving differences of values of B or of G, introduced from a heuristic or semi-axiomatic viewpoint in [Sy2,BD1]. Refined large momentum properties presented and established for these differences in [IM3], will be needed.

An alternative, related method will start from the expansion $F = G + (GoG)_{\rm ren} + (GoGoG)_{\rm ren} + \cdots$ of F in terms of G, which has again convergence properties at small coupling (in contrast to the formal expansion $F = B + BoB + \cdots$) and is also presented in Sect. 3.2.

Results in the 2-particle region are then presented in Sect. 4. Sects. 4.1 and 4.2 present the method based on the BS or regularized BS equation for even and noneven theories respectively, with some complications in the latter case. Methods based on the BS or renormalized BS kernel in GN_2 are presented in Sect. 4.3. Although somewhat more complicated, they have their own conceptual interest and are also more appropriate in a semi-axiomatic approach in which one would start from irreducible kernels as basic quantities. In all cases, the analysis of possible poles near $s = 4\mu^2$ in dimension 2 or 3 is made in the simplest and most complete way via the introduction of the (intrinsic) irreducible kernel U already introduced in Chapter II but here defined in terms of G (or G_M). It extends similarly to hypothetical theories in dimension 4 where poles near $s = 4\mu^2$ no longer occur at small coupling (the kinematical factor $\sigma^{-1/2}$ being then replaced by $\sigma^{1/2}$).

Section 5 presents general methods and results (as also conjectures) in higher energy regions, on the basis of irreducible kernels satisfying possibly regularized structure equations. The structure equations adapted to the study of a given energy region are presented in Sect. 5.1. Discontinuity formulae on individual terms, conjectured on the basis of partial results, are presented in Sect. 5.2 and are shown in Sect. 5.3, at least formally, to yield desired discontinuity formulae of Green functions and related results.

§2. Irreducible kernels in super-renormalizable models

2.1 Cluster expansions of order $\nu > 1$

Cluster expansions which are sufficient if the aim is to define (Euclidean) N-point functions in the $\Lambda \to \infty$ limit have been presented in Chapter III. Cluster expansions of order ν, $\nu > 1$, are introduced similarly. The pairwise cluster expansion is now obtained from a Taylor expansion of order ν, again with an integral remainder, with respect to the variables $s_{\Delta,\Delta'}$. The result is similar to that of Chapter III, except that there may be up to ν propagator lines between two squares. As in Chapter III, this cluster expansion is only formal in bosonic models like $P(\varphi)_2$. The inductive cluster expansion of order ν, which can be applied in these models, is obtained by an extension of the method presented in Chapter III, with some changes outlined at the end of this subsection. In either case, one is led again to Eqs. (3.61)–(3.63), with different terms $I(X; \{x_r\}; G)$ and a sum over connected graphs G linking squares of X with at most ν lines between any pair of squares. Moreover, the following factorization property, important for our later purposes, applies. (It has an empty content at $\nu = 1$.)

Factorization property. If (X, G) can be divided into two parts (X', G'), (X'', G'') by cutting a set S of $q < \nu$ propagator lines of G, then $I(X; \{x_r\}; G)$ is equal to the following convolution product with propagators $C(u_l, v_l)$ on each line of S:

$$I(X; \{x_r\}; G) = \int_{\substack{u_l \in \Delta'_l \\ v_l \in \Delta''_l}} \cdots \int \prod_{l \in S} du_l \, dv_l \, C(u_l, v_l)$$
$$\times \hat{I}\left(X'; \{x_r\}', \{u_l\}; G'\right) \hat{I}\left\{X''; \{x_r\}'', \{v_l\}; G''\right\}, \quad (4.2)$$

where $\Delta_l^{(')}, \Delta_l^{('')}$ are the squares of X' and X'' respectively, linked by line l and each term $\hat{I}$ is defined in a way similar to I, except that there are further operators $\delta/\delta\varphi(u_l)$ or $\delta/\delta\varphi(v_l)$ in the integrand for each line l of S.

This factorization is due, in the pairwise cluster expansion, to the fact that variables $s_{\Delta',\Delta''}$ have been fixed at zero for any pair (Δ', Δ'') of squares such that $\Delta' \subset X'$, $\Delta'' \subset X''$ (since there are less than ν derivatives). It is also checked in the inductive cluster expansion now described.

Inductive cluster expansion of order ν. Starting from a given square Δ_1 as in Sect. 3.2 of Chapter III, I_Λ is written in the form (3.73)–(3.75), where it is now convenient to denote $\Delta_{1,1}$ as the square denoted Δ_2 in Chapter III. $I_{\Delta_1, \Delta_{1,1}}(., s_1)$ is itself written as $I_{\Delta_1, \Delta_{1,1}}(., 0)$ plus a complement. This yields, leaving the set of points x_r implicit,

$$I_\Lambda = I_\Lambda(0) + \sum_{\Delta_{1,1} \in \Lambda \setminus \Delta_1} I_{\Delta_1, \Delta_{1,1}}(0)$$

$$+ \sum_{\substack{\Delta_{1,1} \in \Lambda \setminus \Delta_1 \\ \Delta_{1,2} \in \Lambda \setminus \Delta_1}} \int ds_1 (1 - s_1) I_{\Delta_1, \Delta_{1,1}, \Delta_{1,2}}(s_1), \qquad (4.3)$$

where $I_{\Delta_1, \Delta_{1,1}, \Delta_{1,2}}$ is defined in a way similar to $I_{\Delta_1, \Delta_{1,1}}$ but now with two propagators $C(u_1, v_1)$ and $C(u_2, v_2)$ joining $\Delta_1, \Delta_{1,1}$ and $\Delta_1, \Delta_{1,2}$ respectively; $\Delta_{1,1}$ and $\Delta_{1,2}$ are both different from Δ_1 but may coincide. The procedure is pursued, in a way that depends on the term considered in the last sum. Namely, it is stopped for a given set of squares $\Delta_{1,1}, \ldots, \Delta_{1,\sigma} \in \Lambda \setminus \Delta_1$ as soon as ν squares $\Delta_{1,i}$ coincide. It is continued otherwise. One thus obtains a sum of terms $\frac{1}{\sigma!} I_{\Delta_1, \Delta_{1,1}, \ldots, \Delta_{1,\sigma}}(0)$, in which at most $\nu - 1$ squares $\Delta_{1,i}$ coincide, and of terms $\int_0^1 ds_1 \frac{(1-s_1)^{\sigma-1}}{(\sigma-1)!} I_{\Delta_1, \ldots, \Delta_{1,\sigma}}(s_1)$ in which $\Delta_{1,\sigma}$, denoted below Δ_2, coincides with $\nu - 1$ previous squares $\Delta_{1,i}$. In the first case, Δ_1 is decoupled from $\Lambda \setminus \Delta_1$ in the measure $d\mu$ (s_1 fixed at zero), although it is linked to the squares $\Delta_{1,1}, \ldots, \Delta_{1,\sigma}$ by propagators. The way to treat such terms is indicated below. The factors $I_{\Delta_1, \ldots, \Delta_{1,\sigma}}(s_1)$ involved in the second terms are treated as follow. A new variable s_2 is introduced, with $C(x, y; s_1, s_2)$ equal to $C(x, y; s_1)$ if x, y both belong to $\Delta_1 \cup \Delta_2$ or to $\Lambda \setminus (\Delta_1 \cup \Delta_2)$ and to $s_2 \, C(x, y; s_1)$ otherwise. Taylor expansions in s_2 are then considered. One obtains in that way factors $I_{\Delta_1, \Delta_{1,1}, \ldots, \Delta_{1,\sigma}, \Delta_{2,1}, \Delta_{2,2}, \ldots}(s_1, s_2)$, with $\Delta_{2,1}, \Delta_{2,2}, \ldots \neq \Delta_1, \Delta_2$, propagators $C(u_l, v_l)$ joining as before Δ_1 to $\Delta_{1,1}, \ldots$ and new propagators $s_1 \, C$ joining $\Delta_1 \cup \Delta_2$, hence, either Δ_1 or Δ_2, to $\Delta_{2,1}, \Delta_{2,2}, \ldots$. More precisely, one has terms in which s_2 is fixed at zero, while the dependence on s_2 (with integration over s_2) is kept whenever ν lines (with propagators C or $s_1 \, C$) join $\Delta_1 \cup \Delta_2$ to a common square, to be denoted Δ_3. In the second case, a new variable s_3 is introduced and so forth.

When one s_k is fixed at zero, $\Delta_1 \cup \Delta_2 \ldots \cup \Delta_k$ and its complement in Λ are decoupled in the measure (although some propagators may still join some of the squares $\Delta_1, \ldots, \Delta_k$ to other squares in Λ). Then a new square Δ_{k+1} is chosen in $\Lambda \setminus (\Delta_1 \cup \ldots \cup \Delta_k)$ and the same procedure is now applied in $\Lambda \setminus (\Delta_1 \cup \ldots \cup \Delta_k)$ with either s_{k+1} fixed at zero at some stage or Δ_{k+1} joined to one or more squares of Λ in a way such that the set formed by Δ_{k+1} and those squares cannot be separated into two parts by cutting less than ν propagator lines.

The final result is an expansion of the form (3.61)–(3.63). As, already mentioned, connected graphs G linking squares of X have, at most, ν lines between any pair of squares and for each G, leaving the points x_r implicit

$$I(X,G) = \sum_P I(X,G,P), \tag{4.4}$$

where the sum runs over a set of procedures P that give rise to the same G. If we again restrict our attention, for definiteness, to the case when external lines are attached to distinct points x_r, each $I(X,G,P)$ is of the form

$$I(X,G,P) = \prod_i \int_0^1 ds_i \, \frac{(1-s_i)^{r_i-1}}{(r_i-1)!} M(s_1,\ldots,s_\sigma)\, J(X,G,P;s), \tag{4.5}$$

$$J(X,G,P;s) = \int_{\substack{u_l \in \Delta_l \\ v_l \in \Delta'_l}} \prod_{l \in G} du_l \, dv_l \, C(u_l,v_l) \frac{\delta}{\delta\varphi(u_l)} \frac{\delta}{\delta\varphi(v_l)}$$

$$\times \left\{ \prod_{r, x_r \in X} :-4\lambda\varphi^3(x_r): \right\} e^{-\lambda \int_X :\varphi^4(x): \, dx}$$

$$d\mu\left(\varphi; s_1, s_2, \ldots, s'_1 = s'_2 = \cdots = 0\right), \tag{4.6}$$

where variables s_i have been relabelled (variables fixed at zero are denoted $s'_1,\ldots$). M in (4.5) is a polynomial in the variables s_i. M and $d\mu$ depend on X, G, P and satisfy factorization properties. As a consequence, $I(X,G,P)$ satisfies itself a factorization property of the type (4.2) with, however, factorized terms that still depend on the overall procedure P and not only on the procedures P', P'' determined by P in X', X'': see example given below. But, as will appear on this example, the *sum* $I(X,G)$ over procedures P in (4.4) does factorize according to (4.2).

Example. Let $\nu = 2, (X,G) = $ ⊟—⊟⊟⊟ (with labels Δ″, Δ, Δ′) and $\Delta_1 = \Delta$. Then 3 procedures P_1, P_2, P_3 are encountered:

$$
\begin{aligned}
P_1: &\quad \Delta_{1,1} = \Delta'', \ \Delta_{1,2} = \Delta_{1,3} = \Delta'(=\Delta_2), \ \Delta'' = \Delta_3 \\
P_2: &\quad \Delta_{1,1} = \Delta', \ \Delta_{1,2} = \Delta'', \ \Delta_{1,3} = \Delta'(=\Delta_2), \ \Delta'' = \Delta_3 \\
P_3: &\quad \Delta_{1,1} = \Delta_{1,2} = \Delta'(=\Delta_2), \ \Delta_{2,1} = \Delta''(=\Delta_3)
\end{aligned}
$$

with, in either case, the same pair (P', P'') determined by P with respect to $X' = \{\Delta \cup \Delta'\}$, $X'' = \{\Delta''\}$ $(P': \Delta_{1,1} = \Delta_{1,2} = \Delta' = \Delta_2, P''$ trivial), and

$$I(X, G, P_1) = I(X, G, P_2) = \int_{\substack{u \in \Delta \\ v \in \Delta''}} du\, dv\, C(u, v)$$

$$\times \hat{I}\left(\Delta''; v\right) \int_0^1 ds\, \frac{(1-s)^2}{2!}\, \hat{J}\left(\begin{array}{c}\Delta\quad\Delta'\\ \square\!\square\!\square\end{array}; u; s\right), \quad (4.7)$$

$$I(X, G, P_3) = \int_{\substack{u \in \Delta \\ v \in \Delta''}} du\, dv\, C(u, v)\hat{I}\left(\Delta''; v\right)$$

$$\times \int_0^1 ds\, (1-s)s\hat{J}\left(\begin{array}{c}\Delta\quad\Delta'\\ \square\!\square\!\square\end{array}; u; s\right). \quad (4.8)$$

The factor s in the integrand of (4.8) is due to the fact that the propagator issued from $\Delta \cup \Delta'$ to Δ'' in P_3 is a propagator $s\, C(u, v)$ (where s is the variable, not fixed at zero, introduced at the first stage). The sum of the three contributions above gives the factorized form:

$$\sum_{i=1}^3 I(X, G, P_i) = \int_{\substack{u \in \Delta \\ v \in \Delta''}} du\, dv\, C(u, v)\hat{I}\left(\Delta''; v\right)$$

$$\times \left[\int_0^1 ds\, (1-s)\hat{J}\left(\begin{array}{c}\Delta\quad\Delta'\\ \square\!\square\!\square\end{array}; u; s\right)\right]. \quad (4.9)$$

2.2 The 2-particle irreducible Bethe-Salpeter kernel and the BS equation in $P(\varphi)_2$ models

We consider more specifically below φ_2^4 as in Sect. 3 of Chapter III, but results can be easily adapted to any even $P(\varphi)_2$ model. (For noneven P, see Sect. 4.2). For each given ν, an expansion of, for example, the 4-point function H_Λ^c follows from the cluster expansion and the a la Mayer procedure in the same way as in Chapter III. Lemma 3.1 of Sect. 3.2 of Chapter III still applies to the new terms $I(X, G)$ involved, and the existence of the $\Lambda \to \infty$ limit can be established in the same way as before with only minor complications (and for smaller and smaller couplings as ν increases). The interest of considering values $\nu > 1$ is not this new derivation, but the possibility of introducing irreducible kernels and related structure equations. We restrict our attention for simplicity, in this section, to irreducible kernels of interest in the low-energy region, but more general definitions and results follow in the same way, as explained in Sect. 5. For purposes of this section, we fix either $\nu = 3$ or, as will be later convenient in even theories, $\nu = 4$. The connected 4-point

Euclidean function $S_\Lambda^c(x_1, \ldots, x_4)$, so far at given Λ, can be written in the form (see Eq. (3.57)):

$$
\begin{aligned}
S_\Lambda^c(x_1, \ldots, x_4) = \int_{\substack{z_i \subset \Lambda \\ i=1,\ldots,4}} dz_1 \ldots dz_4 \prod_{i=1}^4 C(x_i - z_i) \\
\times \Big[\lambda\, \delta(z_1 - z_2)\, \delta(z_1 - z_3)\, \delta(z_1 - z_4) \\
+ \cdots + H_\Lambda^c(z_1, \ldots, z_4) \Big],
\end{aligned} \tag{4.10}
$$

where the dots correspond to various contributions to $N_\Lambda'(z_1, \ldots, z_4)/Z_\Lambda$ with one or two δ-functions between some of the points $(z_1, \ldots, z_4)$.

A corresponding expansion of S_Λ^c follows from those of the function H_Λ^c and related ones in the dots. It gives, by direct graphical inspection (using the factorization property of Sect. 2.1):

$$
\begin{aligned}
S_\Lambda^c(x_1, \ldots, x_4) = \int_{\substack{u_i \in \Lambda \\ i=1,\ldots,4}} \left[\prod_{i=1}^4 S_\Lambda(x_i, u_i) \right] \\
\times F_\Lambda(u_1, \ldots, u_4)\, du_1 \ldots du_4,
\end{aligned} \tag{4.11}
$$

where $S_\Lambda(x_1, x_2)$ is the 2-point function and where F_Λ, the amputated, connected 4-point function (in the sense of Sect. 6 of Chapter II, but here in Euclidean space-time), admits an expansion analogous to that of S_Λ^c except that it is restricted to graphs **G** that are 1-particle irreducible in all channels of the form $(i; j, k, l)$. We then define the $2 - p.i.$ Bethe-Salpeter kernel $G_\Lambda(u_1, \ldots, u_4)$ as the sum over all graphs that are, moreover, 2-particle irreducible in the channel $(1, 2; 3, 4)$. We note that, in the definition of irreducibility used here, Mayer lines cannot be cut.

The one-particle irreducible 2-point function $K(u_1, u_2)$ is defined similarly as follows. The 2-point function can be written:

$$
\begin{aligned}
S_\Lambda(x_1, x_2) = \int_{\substack{u_i \in \Lambda \\ i=1,2}} du_1\, du_2 \prod_{i=1,2} C(x_i, u_i) \Big[\lambda \delta(u_1 - u_2) \langle :\varphi^2: \rangle_\Lambda \\
+ H_\Lambda'^c(u_1, u_2) \Big] \\
= \int_{u_i \subset \Lambda} du_1\, du_2 \prod_{i=1,2} C(x_i, u_i)\, \hat{S}_\Lambda(u_1, u_2).
\end{aligned} \tag{4.12}
$$

Then, $K_\Lambda(u_1, u_2)$ is the sum over all 1-particle irreducible graphs in the expansion of $\hat{S}_\Lambda(u_1, u_2)$.

The following equations hold by direct inspection for 2- and 4-point functions respectively:

$$S_\Lambda = C + CK_\Lambda S_\Lambda, \tag{4.13}$$

$$F_\Lambda = G_\Lambda + G_\Lambda o_\Lambda F_\Lambda, \tag{4.14}$$

where $CK_\Lambda S_\Lambda\,(x_1, x_2)$ is the double convolution integral

$$\int_{u_1, u_2 \in \Lambda} C(x_1, u_1) K_\Lambda(u_1, u_2) S_\Lambda(u_2, x_2)\, du_1\, du_2$$

and $G_\Lambda o_{(\Lambda)} F_\Lambda$ is a Feynman-type convolution integral with 2-point functions on internal lines:

$$G_\Lambda o_{(\Lambda)} F_\Lambda\,(u_1, \ldots, u_4) = \int_{\substack{u_i' \in \Lambda \\ i=1,\ldots,4}} G_\Lambda\,(u_1, u_2, u_1', u_2')\, S_\Lambda\,(u_1', u_3')\, S_\Lambda\,(u_2', u_4')$$

$$\times\, F_\Lambda\,(u_3', u_4', u_3, u_4) \prod_{i=1}^{4} du_i', \tag{4.15}$$

that is, Eq. (4.14) is the usual Bethe-Salpeter equation apart from the dependence in Λ. In view of (4.10), (4.11), K_Λ and G_Λ admit decompositions of the form:

$$K_\Lambda\,(u_1, u_2) = \lambda\,\delta\,(u_1 - u_2)\,\langle: \varphi^2 :\rangle_\Lambda + K_\Lambda'\,(u_1, u_2; \lambda), \tag{4.16}$$

$$G_\Lambda\,(u_1, \ldots, u_4) = \lambda\,\delta\,(u_1 - u_2)\,\delta\,(u_1 - u_3)\,\delta\,(u_1 - u_4) + \cdots$$

$$+\, G_\Lambda'\,(u_1, \ldots, u_4; \lambda), \tag{4.17}$$

where the dots refer again to terms with one or two δ-functions between the points $u_1, \ldots, u_4$, multiplied by corresponding functions $G_{\Lambda,1}', G_{\Lambda,2}', G_{\Lambda,3}'$. The following result can then be established:

Theorem 4.1. $\forall \varepsilon > 0$, $\exists \lambda_\varepsilon$, cst_ε (independent of Λ) such that K_Λ' and the functions G_Λ' converge to well-defined functions in the $\Lambda \to \infty$ limit, $\forall \lambda$ such that $|\lambda| < \lambda_\varepsilon$. Moreover, the following bounds hold ($\forall \Lambda$ finite or infinite), $\forall \lambda$, $|\lambda| < \lambda_\varepsilon$:

$$|K_\Lambda'\,(u_1, u_2)| < \mathrm{cst}(\varepsilon)|\lambda|^2(1 + |\ln|u_1 - u_2||)^3 e^{-3(m-\varepsilon)|u_1 - u_2|}, \tag{4.18}$$

$$|G_\Lambda'\,(u_1, \ldots, u_4)| < \mathrm{cst}(\varepsilon)|\lambda|^4(1 + |\ln \ln f|u_i - u_j||)^6$$

$$e^{-(m-\varepsilon)L_3(u_1, u_2; u_3, u_4)}, \tag{4.19}$$

where L_3 is the shortest length of all possible connected graphs, joining $u_1, \ldots, u_4$ and possibly intermediate points, that are, moreover, 3-particle irreducible in the channel $(1, 2; 3, 4)$, and 2-particle irreducible in

channels $(i; j, k, l)$. (Similar results apply to other terms $G'_{\Lambda,1}, \ldots, G'_{\Lambda,3}$).
The terms F, K, G obtained in the $\Lambda \to \infty$ limit are Euclidean invariant.

Finally, $CK_\Lambda S_\Lambda$ and $G_\Lambda o_{(\Lambda)} F_\Lambda$ converge in the $\Lambda \to \infty$ limit to CKS
and GoF respectively, and F and G satisfy the BS equation

$$F = G + GoF. \tag{4.20}$$

F, considered in Euclidean energy-momentum space, is also equal at
small coupling to its Neumann series

$$F = G + GoG + GoGoG + \cdots, \tag{4.21}$$

which is convergent at small coupling in Euclidean energy-momentum
space (where F and G are analytic functions).

Note. We denote here F and G as the functions defined in (Euclidean)
space-time and also, in the last part, their Fourier transforms in (Euclidean) energy-momentum space.

Proof. The convergence, for example, of G'_Λ in the $\Lambda \to \infty$ limit is established by methods similar to those used previously for H^c_Λ. The fact
that a number of terms involved in the original expansion of H^c_Λ have
been eliminated from that of G'_Λ does *not* simplify but instead compli-
cates the proof: regroupings of terms used in Sect. 3 in Chapter III are
no longer valid and more refined ones are needed. We here omit details.
Euclidean invariance is obtained by the same argument as in Chapter III.
Finally, GoF is well defined in view of the exponential fall-off properties
of G, F and of the 2-point functions S involved in o, and of the fact that
singularities at short distances are, at most, logarithmic (apart from
δ-functions that can be integrated explicitly). The fact that $G_\Lambda o_{(\Lambda)} F_\Lambda$
converges to GoF can be obtained by decomposing $GoF - G_\Lambda o_\Lambda F_\Lambda$ as
$(G - G_\Lambda) oF + G_\Lambda (o - o_\Lambda) F + G_\Lambda o_\Lambda (F - F_\Lambda)$: uniform bounds that include exponential fall-off factors in the distance between $u_1, \ldots, u_4$ and
the boundary $\partial\Lambda$ of Λ are again obtained.

The convergence of the Neumann series (4.21) follows, for example, in
(Euclidean) energy-momentum space from bounds in $\mathrm{cst}|\lambda|$ on G. The
latter follow from previous results.

Remark. The kernel G that has been defined above satisfies, as we have
seen, the BS equation. It is then expected to be from a perturbative viewpoint the sum of Feynman integrals of graphs with the same irreducibility properties: in fact this sum also satisfies (for the same graphical
reasons) the BS equation (and the BS kernel was initially introduced in

that way). This perturbative content is not *a priori* transparent for reasons similar to those mentioned in the remark that concludes Sect. 3.2 in Chapter III, but it can be directly checked from a more detailed analysis.

We conclude with the following analyticity properties in complex energy-momentum space.

Theorem 4.2. Momentum-Space Analyticity
$\forall \varepsilon > 0, \exists \lambda_\varepsilon > 0$ such that, for $|\lambda| < \lambda_\varepsilon$:

> (i) The 2-point function S is analytic in the region $s < (3m)^2 - \varepsilon$ apart from a pole at $s = (\mu(\lambda))^2$; μ is arbitrarily close to m if λ is sufficiently small.
>
> (ii) $G(k; z, z')$, where $k = p_1 + p_2 = p_3 + p_4$, $z = \frac{p_1 - p_2}{2}$, $z' = \frac{p_3 - p_4}{2}$, is analytic (and bounded) in a region of the form $k \in \Delta$, $z \in D$, $z' \in D$, where Δ is a complex neighborhood of the set $k = (k_0, 0)$, $k_0 \leq 4(m - \varepsilon)$ and $D = \{(z_0, z_1); |\mathrm{Re}\, z_0| < 2(m - \varepsilon), |\mathrm{Im}\, z_1| < \varepsilon\}$.

Proof. By projection on the time axis, the exponential fall-off factors of the bounds (4.18)–(4.19) yield fall-off factors $\mathrm{e}^{-3(m-\varepsilon)|(u_1-u_2)_0|}$, respectively $\mathrm{e}^{-(m-\varepsilon)\mathrm{d}(u_1,\ldots,u_4)}$, where $\mathrm{d}(x_1,\ldots,x_4) = 4\left|\frac{(x_1)_0 + (x_2)_0}{2} - \frac{(x_3)_0 + (x_4)_0}{2}\right| + |(x_1)_0 - (x_2)_0| + |(x_3)_0 - (x_4)_0|$. A small part of the fall-off factors of (4.18)–(4.19) can also be used to get fall-off factors in space variables.

Property (ii) then follows directly from the Laplace transform theorem in a simple way, as already done in [SpZ] (where the fall-off in $\exp -(m - \varepsilon)d(u_1,\ldots,u_4)$ was first derived by another method).

From direct inspection, the 2-point function is equal in energy-momentum space to $\tilde{C}(p) + \tilde{C}(p)K(p)\tilde{C}(p) + \cdots = [p^2 - m^2 - K(p)]^{-1}$, where by the Laplace transform theorem K is analytic in the region $\mathrm{Re}\, p_0 < 3m - \varepsilon'$, and is, moreover, uniformly bounded there by $\mathrm{cst}(\varepsilon')|\lambda|$. Property (i) follows from invariance properties; μ^2, the zero of $p^2 - m^2 - K$, is arbitrarily close to m^2 for sufficiently small λ, for example, via the implicit function theorem.

§3. Irreducible kernels in nonsuper-renormalizable theories

3.1 Irreducible kernels satisfying regularized equations

As mentioned in Sect. 1, we consider for definiteness the Gross-Neveu model. Bethe-Salpeter type irreducible kernels satisfying regularized BS equations can be introduced as follows. We outline below the method described in [IM2]. Somewhat different methods along the same general lines are also possible. We first come back to the phase-space expansion of $H^c_{(\Lambda,\rho)}$ described in Sect. 4 of Chapter III, and modify it as follows. Cluster expansions ("of order 1" as in Chapter III), Mayer procedures and renormalization are first applied in slices ρ to 2 only. In the last slice, all 2-point insertions are first resummed in the propagator, at sufficiently small λ_ρ, hence small $\delta m_i, \delta\zeta_i$. This gives an effective propagator $C_{(1)}$ in slice 1 of the form:

$$C_{(1)}(p) = \frac{C^{(1)}(p)}{1 - \varphi_1 C^{(1)}(p)}$$

$$= \mathrm{e}^{-V_1(p)}\left[\not{p}\left(1 - \sum_{i=2}^{\rho}\delta\zeta_i\mathrm{e}^{-V_1}\right) + m_\rho + \sum_{i=2}^{\rho}\delta m_i\mathrm{e}^{-V_1} + \cdots\right]^{-1}, (4.22)$$

where $\varphi_1(p)$ is the sum of 2-point insertions and $V_1 = \left(p^2 + m_\rho^2\right)M^{-4}$ as already indicated. Methods of Chapter III give, on the other hand, control on the remainder in the bracket of the right-hand side of (4.22). It is also shown that $1 - \sum_{i=2}^{\rho}\delta\zeta_i\mathrm{e}^{-V_1}$ and $m_\rho + \sum_{i=2}^{\rho}\delta m_i\mathrm{e}^{-V_1}$ are close to 1 and $m_\rho + \sum|\delta m_i|$ respectively. By choosing a small enough λ_1 (i.e. D large enough: see Chapter III), $C_{(1)}(p)$ will then have a unique pole, away from Euclidean space, at a mass (which is not yet the physical mass) close to m_0.

A cluster expansion of order 4, relative to $C_{(1)}$, and a Mayer procedure, are then applied in slice 1. Beside recovering results already given in Euclidean space in Chapter III in the $\rho \to \infty$ limit, the aim is then (i) to establish the analytic structure of the 2-point function in complex energy-momentum space (analyticity up to $(4\mu)^2 - \varepsilon$ apart from the pole whose position defines the physical mass) and (ii) to define a 2-particle irreducible kernel G_M satisfying a regularized BS equation:

$$F = G_M + F o_M G_M. \qquad (4.23)$$

The analysis is close to that given for $P(\varphi)_2$ models with, however, some complications: in particular, 2-point functions are not directly exhibited in the graphical expansion of the 4-point connected (nonamputated) function. This is due to the fact that "particle analysis" is made only in slice one. We have thus to reconstruct a BS-like equation. The result will be the regularized equation (4.23), where o_M is the convolution with modified 2-point functions that have now a sufficient decrease in Euclidean directions in energy-momentum space. Their pole and the residue at the pole are the same as those of the actual 2-point function (as required to derive asymptotic completeness in Sect. 4). On the other hand, we note that the quantities such as C_ρ and φ are matrices and not scalars. However, they commute (since by invariance properties they can be expressed in terms of the matrices $\not{p}$ and 1).

In all the following, irreducibility properties are to be understood with respect to the propagator lines of slice 1: a graph is r-particle irreducible in a given channel if it cannot be divided into two parts (with respect to that channel) by cutting r propagator lines of slice 1 or less. Leaving, unless explicitly mentioned, the cut-off ρ (that will later tend to infinity) implicit, the phase-space expansions of the functions H^c give a corresponding phase-space expansion of the 4-point connected function $S^c(x_1, x_2, x_3, x_4)$ which, by direct graphical inspection, can be written:

$$S^c(x_1, \ldots, x_4) = \int \left[\prod_{k=1}^{2} S'(x_k, u_k) \right]$$

$$\times F'(u_1, \ldots, u_4) \left[\prod_{k=3}^{4} S'(u_k, x_k) \right] du_1 \ldots du_4, \quad (4.24)$$

with (in energy-momentum space):

$$S'(p) = C_\rho(p) + C_\rho (\varphi_1 + \varphi_2) S''(p), \quad (4.25)$$

$$S''(p) = C_{(1)}(p) + C_{(1)} \varphi_2 C_{(1)} + C_{(1)} \varphi_2 C_{(1)} \varphi_2 C_{(1)} + \cdots \quad (4.26)$$

$$= \frac{C_{(1)}(p)}{1 - \varphi_2 C_{(1)}(p)} = \frac{C^{(1)}(p)}{1 - (\varphi_1 + \varphi_2) C^{(1)}(p)} = \frac{e^{-V_1}}{\not{p} + m_\rho - \varphi e^{-V_1}},$$

where φ_2 is the 2-point 1-particle irreducible function and $\varphi = \varphi_1 + \varphi_2$. F' is defined by an expansion analogous to that of S^c except that it is restricted to graphs that are 1-particle irreducible in all channels $(i; j, k, l)$.

We note from (4.25) that:

$$S'(p) = \frac{e^{-V_\rho}}{\not{p} + m_\rho - \varphi e^{-V_1}}. \quad (4.27)$$

This shows that S' is symmetric like S'' so that in (4.24) the factor S' on both sides is the same.

The 2-point function. The 2-point function is equal, by direct inspection, to

$$S = C_\rho + C_\rho \varphi C_\rho + C_\rho \varphi C_{(1)} \varphi C_\rho + C_\rho \varphi C_{(1)} \varphi_2 C_{(1)} \varphi C_\rho + \cdots = C_\rho + S' \varphi C_\rho$$

$$(4.28)$$

and satisfies the relation:

$$S = AS', \tag{4.29}$$

$$A = 1 + \varphi C_\rho \left[1 - e^{(V_\rho - V_1)} \right]. \tag{4.30}$$

For any $\varepsilon > 0$ and $|D| > D_0$, D_0 being sufficiently large (depending on ε), φ is known by a combination of previous methods to be analytic and bounded in modulus by $m_0 + \mathrm{cst}(\varepsilon) |\lambda_1|$ in the region $s < (3m_0)^2 - \varepsilon$; in fact, $\varphi_1 = m_0 + O(\lambda_1)$ and $\varphi_2 = O(\lambda_1)$. On the other hand,

$$C_\rho\big(1 - e^{(V_\rho - V_1)}\big) \equiv (\not{p} + m_\rho)^{-1} \left[1 - e^{(p^2 + m_\rho^2)(M^{-2\rho} - M^{-2})} \right]$$

is analytic, without pole, and bounded in that region. More precisely, one sees easily that, for a sufficiently small λ_1 and a large enough M (depending on m_0), the functions A and A^{-1} are analytic and close to 1. Thus, S' and S have a common (unique) pole at the physical mass μ defined by the equation:

$$p^2 \left[1 - ae^{-V_1} \right]^2 + \left[m_\rho - be^{-V_1} \right]^2 = 0, \tag{4.31}$$

where we have put:

$$\varphi(p) = a(p^2)\not{p} + b(p^2), \tag{4.32}$$

and, by an easy calculation, S is shown to be of the form:

$$S(p) = \frac{Z(-\not{p} + \mu)}{p^2 + \mu^2} + R(p), \tag{4.33}$$

where R is analytic in p^2 up to $(4\mu)^2 - \varepsilon$ in Minkowskian space (and is bounded in Euclidean directions), and where Z, the residue of S at the pole, coincides with that of S'.

The BS-type kernel G_M. In view of (4.24) and (4.29), the connected amputated 4-point function F is equal to:

$$F(p_1, \dots, p_4) = \left[\prod_{k=1}^{4} A(p_k) \right]^{-1} F'(p_1, \dots, p_4). \tag{4.34}$$

F', and hence F, are known by previous methods to be analytic and bounded in Euclidean space (and in a strip around it). Similarly, λ_{ren}, that is, the value of F at zero energy-momentum, is itself shown to be equal to λ_1 at first order in λ_1 and is then different from zero at small λ_1 (nontriviality), as already indicated in Chapter III.

Let G'_M now be defined as F', except that the sum is restricted to graphs that are 3-particle irreducible in the channel $(1, 2; 3, 4)$ under consideration and let G_M be defined as:

$$G_M (p_1, \ldots, p_4) = \left[\prod_{k=1}^{4} A(p_k) \right]^{-1} G'_M (p_1, \ldots, p_4). \qquad (4.35)$$

By direct inspection, G'_M satisfies the equation

$$F' = G'_M + G'_M o''_M F', \qquad (4.36)$$

where o''_M is defined with functions S'' (see (4.26)) on internal lines. This shows in turn that:

$$F = G_M + G_M o_M F \qquad (4.37)$$

with o_M now defined with functions $A S'' A = S A e^{V_\rho - V_1}$ on internal lines, that is, functions $S A e^{-V_1}$ in the $\rho \to \infty$ limit.

Since $A e^{V_\rho - V_1} = 1$ at $p^2 = -\mu^2$ (in Euclidean notations) and when $\not{p}$ is replaced by $-\mu$, the residue of $S A e^{V_\rho - V_1}$ does coincide with the residue $Z(-\not{p} + \mu)$ of S at the pole.

Irreducibility properties of G'_M (and in turn of G_M in view of (4.35)), in the sense of exponential fall-off in Euclidean space-time and corresponding analyticity in complex momentum-space up to $(4\mu)^2 - \varepsilon$, are reestablished by a combination of previous methods. The analysis includes the existence of the $\rho \to \infty$ limits and the derivation of the regularized BS equation in that limit.

3.2 The Bethe-Salpeter and renormalized BS kernels in the Gross-Neveu model

The kernel B that will satisfy the actual Bethe-Salpeter equation $F = B + BoF$, where o denotes 2-particle convolution with 2-point functions on internal lines, can be defined as follows in the weakly coupled Gross-Neveu model. The same analysis as in Sect. 4 of Chapter III is carried out on the connected 4-point function but with cluster expansions of order 4 in all slices. As in Sect. 2.2, this gives somewhat more complicated expansions but convergence properties are established in a similar way. The expansion of the 4-point connected, amputated function F (with at first the cut-off ρ that will tend to infinity) is obtained by restricting the sum to graphs that are $1 - p.i.$ in all $(i; j, k, l)$ channels, and B is in turn defined by restricting the sum, before renormalization, to graphs that are $2 - p.i.$ in the $(1, 2; 3, 4)$ channel under consideration. Here, irreducibility is to be understood, in contrast to Sect. 3.1, with respect to propagator lines of all slices (e.g. a $2 - p.i.$ graph in the channel $(1, 2; 3, 4)$ cannot be cut into two parts, with external lines 1, 2 and 3, 4 in each side, by cutting two propagator lines of any slice, or less). The expansion is first shown to be convergent (before renormalization) again for values of the bare coupling that tend toward zero like $M^{-(\rho+\varepsilon)/2}$ as $\rho \to \infty$, and the BS equation is established by direct graphical inspection for these values of λ. Results below will allow extensions (for a large enough D) to the usual value

$$\lambda_\rho = \left[-\beta_2(\ln M)\rho + \frac{\beta_3}{\beta_2}\ln \rho + D\right]^{-1}. \qquad (4.38)$$

As a matter of fact, the renormalization procedure of Sect. 4.2 of Chapter III is applied in a similar way to B. It gives the following results:

(i) for all diagrams that are not reduced to a single vertex, the effective coupling constants in each slice i are the same λ_i as before: this is due to the fact, analogous to that mentioned in the remark given below in the perturbative framework, that 2-particle irreducibility properties are not modified by replacing some vertices by 4-point subgraphs.

(ii) the situation is different for the trivial diagram with a single vertex. In this case, the flow of the coupling constant is modified: the effective coupling in each slice i is given by

$$\lambda'_{i-1} = \lambda'_i + \sum[2 - p.i. \text{ graphs}], \qquad (4.39)$$

where $[\quad]$ means the value at $k = z = z' = 0$ and where the graphs involved in $[\quad]$ have effective coupling λ_j in all slices $j \geq i$.

Since there is a ratio γ of $2 - p.i.$ graphs (in the channel $(1, 2; 3, 4)$) of the form ⟩⟨ or ◁ among all graphs of this form, Eq. (4.39) gives:

$$\lambda'_{i-1} = \lambda'_i + \gamma (\beta_2 \ln M) \lambda_i^2 + \text{cst } \lambda_i^3 + \mathcal{O}\left(\lambda_i^4\right), \qquad (4.40)$$

which finally yields

$$\lambda'_{\text{ren}} = \gamma\lambda_{\text{ren}} + \mathcal{O}\left(\lambda_{\text{ren}}^3\right). \qquad (4.41)$$

The usual methods then show that the contribution of the graphs of (i) to B is still well defined for λ_ρ of the form (4.38) and has a well-defined limit B' as $\rho \to \infty$. B is thus equal to

$$B = \lambda'_{\text{ren}} + B'. \tag{4.42}$$

B' is equal from the perturbative viewpoint to the sum of renormalized $2 - p.i.$ graphs not reduced to a single vertex: this can be seen by completing the "useful" renormalization by "useless" ones. The relation between B and the "renormalized" BS kernel B_{ren}, denoted G (which is perturbatively the sum of all renormalized $2 - p.i.$ graphs, and is also well defined nonperturbatively in the weakly coupled GN_2 model as $B' + \lambda_{\text{ren}}$) is thus:

$$G(\equiv B_{\text{ren}}) = B' + \lambda_{\text{ren}} = B + (1 - \gamma)\lambda_{\text{ren}} + \mathcal{O}\left(\lambda_{\text{ren}}^3\right). \tag{4.43}$$

Remark. The equality

$$B = B_{\text{ren}} + \text{cst}, \tag{4.44}$$

can be seen as follows from a perturbative viewpoint in "a la φ_4^4" theories whose renormalization parts are at most 4-point functions such as the GN_2 model. The 2-particle irreducibility of a graph in the channel (1, 2; 3, 4) is not modified if a vertex is replaced by any 4-point subgraph (or conversely), except for the trivial graph with a single vertex. The formal (simple, though not quite trivial) algebraic argument that shows that the sum of nonrenormalized Feynman integrals of graphs with bare coupling is equal to the sum of renormalized integrals with renormalized coupling then yields (4.44). The constant in the right-hand side of (4.44) is perturbatively the sum of nonrenormalized Feynman integrals of $2 - p.i.$ graphs with bare coupling, at $k = z = z' = 0$. (These arguments must, strictly speaking, be first considered in the theory with cut-off ρ since $\lambda_\rho = 0$ at ρ infinite. In a way similar to above, the constant in the right-hand side of (4.44) can be re-expressed in terms of renormalized quantities: the fact that it is finite in the $\rho \to \infty$ limit then becomes transparent, at least for each term of the expansion in the perturbative argument.)

Large-momentum properties of the BS kernel. In order to establish the BS equation $F = B + BoF$ and further results in the $\rho \to \infty$ limit, the analysis of the behaviour of B in energy-momentum space at large Euclidean values is needed, in particular, to ensure convergence of the integral BoF; in contrast to $P(\varphi)_2$, this convergence does not follow from the decrease (here in $1/|p|$ as $|p| \to \infty$) of the 2-point functions involved in the convolution.

This analysis is analogous to that carried out for F in Sect. 4.3 of Chapter III, but important simplifications occur, at least in some situations, due to the 2-particle irreducibility of diagrams contributing to B.

We restrict here our attention to situations in which $k = p_1 + p_2$ is fixed, while $z = \frac{p_1 - p_2}{2}$ or $z' = \frac{p_3 - p_4}{2}$ or both will tend to infinity in Euclidean space. External lines 1, 2, respectively 3, 4, will belong by convention to slices $i(z), i(z')$ respectively. The new important facts, in comparison with the analysis of Chapter III are as follows.

Let, for example, $i(z) \geq i(z')$. Most contributing diagrams are then "attached" to slice $i(z)$: exponential factors $M^{-(i(z)-l)}$ at least are obtained if $l < i(z)$, where l is the highest internal line of the diagram. The only possible exceptions are diagrams in which the external line 1 and one of the external lines 3, 4 arrive at the same vertex while 2 and the remaining line among 3, 4 also arrive at a common vertex. This fact follows from energy-momentum conservation arguments if 1, or 2, arrives at a vertex involving no other external line. On the other hand, leaving aside the trivial diagram with only one vertex involving $1, \ldots, 4$, 2-particle irreducibility excludes cases in which 1 and 2 would arrive at the same vertex. (The case when 3 or 4 arrives at this vertex is also excluded since contributing diagrams are also 1-particle irreducible.)

The exceptional diagrams that have been mentioned are independent of $k = p_1 + p_2$ (in energy-momentum space). In cases when z' is not close to z, for example, z large, z' fixed or $|z'| < |z|/3$, exceptional diagrams are removed by energy-momentum conservation arguments: the energy-momentum $p_1 + p_3$ or $p_1 + p_4$ is equal to $z \pm z'$. If $|z'| < |z|/3$, then $|z \pm z'|$ remains of the order of $|z|$.

We then state:

Theorem 4.3. There exists $\lambda_0 > 0$ such that, if $\lambda_{\text{ren}} < \lambda_0$,

$$|B(k, z, z')| \leq f(k) \, \text{Inf}\left[\lambda_{\text{ren}}, \frac{1}{1 + \ln(1 + |z|)}\right] \quad \text{if} \quad |z| \geq 3|z'|. \quad (4.45)$$

Proof. The result follows from remarks above on relevant diagrams and leads to the fact that the effective coupling of a vertex involving 1 or 2 is $\leq \lambda_{i(z)}$, which behaves like $1/\ln|z|$ at large z (and is also less than λ_{ren}) in view of the analysis of Chapter III. QED

The main content of this result for our purposes is that B behaves like $1/\ln|z|$ as $|z| \to \infty$ (z' fixed), a fact that is at the origin of the convergence of the integral

$$FoB\,(k; z, z') = \int F(k, z, \zeta)\omega(k, z)B\,(k, \zeta, z')\,\mathrm{d}^2\zeta, \qquad (4.46)$$

where ω is, as previously, the product of 2-point functions $S(\frac{k}{2}+\zeta)S(\frac{k}{2}-\zeta)$. In fact, each 2-point function S decreases like $1/|\zeta|$ as $|\zeta| \to \infty$, and convergence thus follows from the decrease of B in $1/\ln|\zeta|$ and that of F in $1/(\ln|\zeta|)^\alpha$, $\alpha > 0$ (established in Sect. 4.3 of Chapter III). QED

A similar result applies by symmetry if $|z'| \geq 3|z|$ and a complementary result, omitted here, also applies in the intermediate region. On the other hand, we also mention that, if the trivial diagram with one vertex (whose value is $\lambda_{i(z)}$) is removed, relevant diagrams have at least 2 vertices involving 1 and 2 separately so that $B\,(k, z, z') - \lambda_{i(z)}$ decreases at least like $1/(\ln|z|)^2$ at large z:

$$\left|B\,(k, z, z') - \lambda_{i(z)}\right| \leq f(k)\mathrm{Inf}\left[\lambda_{\mathrm{ren}}^2, \frac{1}{\ln(1+|z|)^2}\right] \quad \text{if} \quad |z| \geq 3|z'|. \tag{4.47}$$

Although Theorem 4.3 ensures convergence of BoF and allows one to show the relation $F = B + BoF$, the convergence in BoF is too weak to allow the further exploitation of this equation, through extensions of Fredholm theory. From the viewpoint of Neumann series, this is due to the fact that the dependence on n of the bounds obtained on the terms $Bo\ldots oB$ (n factors B) does not allow one to show the convergence of the series $F = B + BoB + BoBoB + \cdots$ even at small coupling λ_{ren} (and in Euclidean space). As a consequence, the analysis of Chapter II or that to be given in Sect. 4.1 can no longer be directly applied. (The analysis of Chapter II does apply in the theory with any given cut-off ρ, starting from the BS equation $F_\rho = B_\rho + F_\rho o_\rho B_\rho$, but this is of no use in the $\rho \to \infty$ limit: there might be an increasing number of poles of F_ρ in the second sheet that might accumulate on the real s-axis in that limit.) Two possible adaptations of the methods will be described in Sect. 4.3. They will make recourse to the following further results that we first give in the remainder of this section.

Theorem 4.4. For any $\varepsilon > 0$, there exists $\lambda_0(\varepsilon) > 0$ such that, if $\lambda_{\mathrm{ren}} < \lambda_0(\varepsilon)$

$$\left|B\,(k, z, z') - B\,(0, z, z')\right| < g(k)\frac{\lambda_{\mathrm{ren}}^2}{(1+|z|)^{1-\varepsilon}}, \tag{4.48}$$

$$\left| B\left(k, z, z'\right) - B(k, z, 0) \right| < g'(k)\lambda_{\text{ren}}^2 \left(\frac{1 + |z'|}{1 + |z|} \right)^{1-\varepsilon} \qquad (4.49)$$

for $|z| \geq |z'|$.

Remarks.

(1) A better result, in which ε is removed, can be established, but a more refined analysis is needed. Theorem 4.4 is sufficient for our later purposes.

(2) Theorem 4.4 applies equally to $G = B_{\text{ren}}$ since B and G differ only by a constant.

Proof of Theorem 4.4.

(a) Taking the difference $B(k, z, z') - B(0, z, z')$ has the following effects:

(i) it ensures that all relevant diagrams are now "attached" to slice $i(z)$, hence factors in $M^{-(i(z)-j)}$ are always obtained if j is the lowest line of the diagram (taking into account the factors $M^{-(l-j)}$).

(ii) at the cost of possible powers of $|k|$ (arising from the action of gradients), it ensures further factors M^{-j} in the usual way applied in the renormalization procedures.

The combination of effects (i) and (ii) provides factors in $M^{-i(z)}$ for all diagrams. If a part in $M^{-\varepsilon i(z)}$ is kept for internal summations, a decrease in $M^{-(1-\varepsilon)i(z)}$ is left. Equation (4.48) follows. (A factor λ_{ren}^2 can always be obtained since there are at least two couplings in all diagrams. On the other hand, the restriction to $\lambda_{\text{ren}} < \lambda_0(\varepsilon)$ is due to the fact that $\sum_i M^{-\varepsilon i} \approx \text{cst}/\varepsilon$.)

(b) We establish (4.49) for $|z'| \leq |z|/3$. (Otherwise, the result follows easily from bounds on $B(k, z, z')$ and $B(k, z, 0)$ separately.) Under this condition, the exceptional diagrams mentioned previously can be disregarded as already remarked, so that factors $M^{-(i(z)-j)}$ are obtained for all diagrams. Taking the difference $B(k, z, z') - B(k, z, 0)$ now yields factors in $M^{-(j-i(z'))}$ if $i(z') < j$. The combination gives a factor $M^{-(i(z)-i(z'))}$. A part in $M^{-\varepsilon(i(z)-i(z'))}$ is kept for internal summations as in (a). QED

The following result on the 2-point function $S(p)$ will also be needed.

Theorem 4.5. There exists λ_0 such that, for any λ, $|\lambda| < \lambda_0$, $S(p)$ decreases like $1/|p|$ as p tends to infinity in Euclidean directions. Moreover,

$$|S(k + z) - S(z)| < f(k)/(1 + |z|)^2. \qquad (4.50)$$

That is, the difference $S(k + z) - S(z)$ decreases like $1/|z|^2$ rather than $1/|z|$ as $|z| \to \infty$. The proof of this result, by methods analogous to the above, is omitted. It yields the following corollary on the product $\omega(k, z) = S\left(\frac{k}{2} + z\right) S\left(\frac{k}{2} - z\right)$.

Corollary 4.1.

$$|\omega(k, z) - \omega(0, z)| < g(k)/(1 + |z|)^3. \qquad (4.51)$$

These results will be used in Sect. 4.3 through one of the following related approaches. The following notations will be used: given kernels $A_1, A_2, A_3, \ldots$, the integral $A_1 o A_2 = \int A_1(k, z, \zeta)\omega(k, \zeta)A_2(k, \zeta, z')d\zeta$ will be also rewritten $A_1 \omega A_2$. $A_1 \omega A_2 \omega A_3$, similarly, denotes the double convolution integral over intermediate variables ζ_1, ζ_2, and so forth. $A_1 \overset{\circ}{\omega} A_2$ denotes the above convolution integral with $\omega(k, \zeta)$ replaced by $\omega(0, \zeta)$. Finally, an index zero on the top, the left or the right side of A means that k, z, or z' respectively is fixed at zero: for example, $\overset{\circ}{A}(k, z, z') \equiv A(0, z, z')$, $A_0(k, z, z') \equiv A(k, z, 0)$.

Expansion of F in terms of the renormalized BS kernel. This expansion, which reads

$$F = G + (GoG)_{\text{ren}} + (GoGoG)_{\text{ren}} + \cdots \qquad (4.52)$$

is an analogue of (4.21) but G is here the renormalized BS kernel. This expansion will again be shown to have convergence properties and to be satisfied at small coupling (see more precise statements in Sect. 4.3). Terms $(Go \ldots oG)_{\text{ren}}$ are defined, initially in Euclidean space, as renormalized Feynman integrals, except that constants at each vertex are replaced by G. The renormalization does not apply "inside" the kernels G. Forests introduced in Sect. 2.2 of Chapter III reduce here to possible sets of brackets with no overlap. For example, if we put by convention $[A] = -A(0, 0, 0)$,

$$(GoG)_{\text{ren}} = GoG + [GoG], \qquad (4.53)$$

$$(GoGoG)_{\text{ren}} = GoGoG + Go[GoG] + [GoG]oG$$
$$+ [GoGoG] + [[GoG]oG] + [Go[GoG]]. \qquad (4.54)$$

The expansion (4.52) (which is not true for general theories) can be checked formally in the perturbative approach in models, like GN_2, with only 2- and 4-point relevant or marginal operators. The definition (4.53)–(4.54) is only formal at ρ infinite: individual terms in the right-hand side are expected to be infinite. Simple, *a priori* formal algebraic manipulations will transform these right-hand sides into expressions (Lemma 4.1 below) in which individual terms will be finite in view of Theorem 4.4 and Corollary 4.1. These expressions can then be used to define $(Go\ldots oG)_{\mathrm{ren}}$. Alternatively, if an ultraviolet cut-off is first introduced in each operation o (i.e. o is replaced by o_ρ), initial contributions to $(Go_\rho\ldots o_\rho G)_{\mathrm{ren}}$ are well defined, and the above algebraic procedure is legitimate and yields expressions from which the existence of the $\rho \to \infty$ limit follows.

Lemma 4.1.

(a) $k = 0$. (The index 0 on the top is left implicit in the right-hand side of (4.55) for the simplicity of the notation.)

$$
\begin{aligned}
(Go\ldots oG)_{\mathrm{ren}}^{(n)}(0, z, z') = {} & (G - G_0)\omega(G - G_0)\ldots\omega(G - G_0) \\
& + (G_0)\omega(G - G_0)\omega\ldots\omega(G - G_0) \\
& + \sum (G - {}_0G)\omega\ldots\omega(G - {}_0G)\omega(G_0)\omega(G - G_0)\ldots\omega(G - G_0) \\
& + (G - {}_0G)\omega\ldots\omega(G - {}_0G)\omega(G_0)
\end{aligned}
\tag{4.55}
$$

with a total number of n factors (${}_0G, G_0, G - {}_0G$ or $G - G_0$) in each term.

(An analogous expression in which the roles of G_0 and ${}_0G$ are exchanged also holds.)

(b) $k \neq 0$.

$$
\begin{aligned}
(Go\ldots oG)_{\mathrm{ren}}^{(n)} = {} & \sum \ldots \mathring{F}^{(n_i)}(\omega - \mathring{\omega})\mathring{F}^{(n_{i+1})} \\
& \ldots \mathring{F}^{(\cdot)}\omega(G - \mathring{G})\omega(G - \mathring{G})\ldots\omega\mathring{F}^{(\cdot)}\ldots,
\end{aligned}
\tag{4.56}
$$

where

$$
F^{(\cdot)} \equiv (Go\ldots oG)_{\mathrm{ren}}^{(\cdot)}.
$$

In (4.56), the sum runs over all possible multiple convolutions of factors $\mathring{F}^{(\cdot)}$ and $G - \mathring{G}$ in arbitrary order and number (and arbitrary numbers n_i in the factors $\mathring{F}^{(\cdot)}$), except that the total number of factors $G - \mathring{G}$ and of factors G inside each $\mathring{F}^{(\cdot)}$ must be equal to n. Two successive factors are separated by ω, except that two successive factors $\mathring{F}^{(n_i)}, \mathring{F}^{(n_{i+1})}$ are separated by $\omega - \mathring{\omega}$.

Subtracted BS equation. An alternative approach in the GN_2 model will start from the following subtracted BS equation, which can be checked formally from the BS equation $F = B + BoF$ and can also be established rigorously at small coupling. In particular, the existence of individual terms and the convergence of the series in (4.58) and (4.59) is again a consequence of results stated in Theorem 4.4 and Corollary 4.1 (see more precise statements in Sect. 4.3).

Lemma 4.2.
$$[1 - \mathring{F}(\omega - \mathring{\omega}) - (1 + \mathring{F}\mathring{\omega})(B - \mathring{B})\omega]F \\ = \mathring{F} + (1 + \mathring{F}\mathring{\omega})(B - \mathring{B}) \qquad (4.57)$$

By formal expansion of $[\ldots]^{-1}$, where $[\quad]$ denotes the bracket in the left-hand side of (4.57), one obtains in turn:

$$F = \sum \ldots \mathring{F}(\omega - \mathring{\omega})\mathring{F}(\omega - \mathring{\omega}) \ldots \\ \ldots \mathring{F}\omega(B - \mathring{B})\omega(B - \mathring{B}) \ldots, \qquad (4.58)$$

where the sum runs over all convolution products of factors $\mathring{F}$, $B - \mathring{B}$, in arbitrary number and order; successive factors are separated by ω's except that two factors $\mathring{F}$ are separated by $\omega - \mathring{\omega}$.

Replacing $B - \mathring{B}$ by $G - \mathring{G}$ in (4.57)–(4.58) yields corresponding expressions of F in terms of G. For example, (4.58) gives:

$$F = \sum \ldots \mathring{F}(\omega - \mathring{\omega})\mathring{F}(\omega - \mathring{\omega}) \ldots \\ \ldots \mathring{F}\omega(G - \mathring{G})\omega(G - \mathring{G}) \ldots. \qquad (4.59)$$

These expressions in terms of G can be reobtained from Lemma 4.1 (b) and the relation $\mathring{F} = \sum \mathring{F}^{(n)}$.

§4. Two-particle structure in weakly coupled field theories

4.1 Two-particle bound states and asymptotic completeness in even theories

We consider any even theory in dimension 2, 3 or 4, depending on a possibly renormalized coupling parameter λ, in which the 4-point connected, amputated function F is linked to a BS or BS-type kernel via a possibly regularized BS equation of the form (2.2). More precisely G, or G_α, and $(s - \mu^2) S_2(p)$ are assumed, as established earlier in even $\lambda P(\varphi)_2$, $GN_2, \ldots$, to be analytic in complex domains around Euclidean space of the form $(k, z, z') \in \Delta \times \mathcal{D} \times \mathcal{D}$ and $k \in \Delta'$, respectively, where Δ and Δ' go, in particular, up to $s = (4\mu)^2 - \varepsilon$ and $(3\mu)^2 - \varepsilon$, respectively, in Minkowski space, and to be bounded there by $c_1|\lambda|$ and c_2, respectively, where c_1 and c_2 are given constants; $\varepsilon > 0$ can be chosen arbitrarily small, but corresponding constraints may follow on the maximal possible values of the coupling. At small coupling, it is first seen that poles of F arising from the BS or regularized BS equation, if they exist, either lie "far away" in the unphysical sheet(s) or are close to the 2-particle threshold. This is due to the convergence, at small coupling, of the Neumann series $F = G + GoG + \cdots$ of F in term of G, with possibly o, G replaced by o_α, G_α or o_M, G_M in this subsection (the index α or M will be left implicit). Individual terms $Go \cdots oG$ are defined away from Euclidean space by analytic continuation in a 2-sheeted (d even) or multisheeted (d odd) domain around $s = 4\mu^2$. To that purpose, locally distorted integration contours $\Gamma(k)$ (initially Euclidean space) are introduced as in Sect. 6 of Chapter II so as to avoid the pole singularities of the two 2-point functions involved in o, the singularities at $s = 4\mu^2$ being due to the pinching of this contour between these poles as $s \to 4\mu^2$.

More precisely, we first state the following result.

Theorem 4.6. Given any bounded (compact) region D, in the 2-sheeted (d even) or multi-sheeted (d odd) domain of F, whose distance to $s = 4\mu^2$ is strictly positive, $\exists \lambda_0 > 0$ such that F has no pole in D if $|\lambda| < \lambda_0$.

Proof. The terms $Go \cdots oG$ (n factors G) of the Neumann series of F, defined away from Euclidean space by analytic continuation as recalled above, satisfy in D bounds of the form:

$$\left| (Go \cdots oG)(k, z, z') \right| < c(c_1'|\lambda|)^n, \qquad (4.60)$$

where c, c_1' depend on c_1 and on D but not on λ. These bounds are obtained easily since the distorted contours of integration stay away from the poles of 2-point functions. (On the other hand, convergence of the integrals is ensured in view of the decay properties at infinity, in Euclidean space, of 2-point functions on internal lines involved in o or o_M.) The dependence on D is due to the facts that the contour must be more and more distorted as D becomes larger and larger and that it must approach more and more the pole singularities of 2-point functions as the distance of D to $s = 4\mu^2$ tends toward zero.

The Neumann series is absolutely convergent, and F is bounded (thus has no pole) at a small enough λ. QED

Further investigations can be restricted to values of s in a complex neighborhood V of $s = 4\mu^2$. To that purpose, the following preliminary lemma on 2-particle convolution will be useful. In the latter, the index M, if present, is again left implicit in o, and the operation $*$ is, as in

Sect. 6 of Chapter II, on-shell convolution at s real $> 4\mu^2$ or is obtained by analytic continuation for complex values of s around $4\mu^2$, $g = 1/2$ if d is even, $g(s) = (i/2\pi)\ell n\ \sigma$, $\sigma = 4\mu^2 - s$, if d is odd, and the operation ∇ is defined by the formula

$$\nabla = o - g(s) *. \qquad (4.61)$$

Lemma 4.3 will indicate that, in contrast to o and $*$, which are singular at threshold, ∇ is fully regular at threshold, i.e. o is equal to $g(\sigma)*$, up to a regular background. A corollary of Lemma 4.3 (which can be extended to a complex neighborhood of the real s-axis, up to $(4\mu)^2 - \varepsilon$) is the discontinuity formula $o_+ - o_- = *_+$ mentioned (and used) in Chapter II since $\nabla_+ = \nabla_-$ (Lemma 4.3), and $(g(s)*)_+ - (g(s)*)_- = *_+$ as checked by direct inspection. Conversely, if the discontinuity formula $o_+ - o_- = *_+$ is already known, the same argument shows that $\nabla_+ - \nabla_-$ vanishes, that is, that ∇ is uniform around $s = 4\mu^2$. Lemma 4.3 is a more precise version of that result.

Lemma 4.3. Given $A_1(k, z, z')$ and $A_2(k, z, z')$ analytic in $\Delta \times \mathcal{D} \times \mathcal{D} \cap \{s \in V\}$ and bounded there in modulus by constants $|A_1|$ and $|A_2|$ respectively, $A_1 \nabla A_2$ is analytic and bounded by $c'|A_1||A_2|$ in that region, where c' is a given constant, independent of A_1 and A_2 and depending only on the given bounds on $|(s - \mu^2) S_2(p)|$.

Proof.

(a) Geometrical proof (outline). The proof can be obtained from geometrical results of [Ph2,BP]. At d even, the distorted contour $\Gamma(k)$ can in fact be written as:

$$\Gamma(k) = \underline{\Gamma} + \frac{1}{2}\tilde{e}(k), \qquad (4.62)$$

where $\underline{\Gamma}$ is a fixed cycle independent of k whose support stays at a fixed (> 0) distance of the poles of 2-point functions and $\tilde{e}(k)$ is a cycle whose support surrounds the mass-shell, such that integration over $\tilde{e}(k)$ reduces to $*$ by a residue formula. From its definition, ∇ thus amounts to integration over $\underline{\Gamma}$ and Lemma 4.3 follows from the fact that the support of $\underline{\Gamma}$ is not pinched as $s \to 4\mu^2$.

The situation is more complicated at d odd, in which case (4.62) is no longer valid. A proof can still be obtained by a suitable use of various partial geometrical results of [Ph2]. We omit it here and give instead an alternative proof below, applying at least, as it stands, at $d = 2, 3$.

(b) Proof making use of results on the Feynman integral for $d = 2, 3$. We below evaluate $A_1 o A_2$ as follows. It will be assumed, as is the case in the application in view of Lorentz invariance, that the dependence of $A_1(k, z, \zeta)$ or $A_2(k, \zeta, z')$ with respect to ζ reduces to a dependence on ζ_0, ζ_1^2 at $d = 2$, or if $d = 3$ on ζ_0, $\boldsymbol{\zeta}^2$ and a supplementary angle variable θ [angle between $\mathbf{z}$ and $\boldsymbol{\zeta}$, or $\boldsymbol{\zeta}$ and $\mathbf{z}'$; k is taken of the form $(k_0, \mathbf{0})$]. We note on the other hand that ω can be written (with Minkowskian variables: k_0, ζ_0 imaginary in Euclidean space) as:

$$\omega(k, \zeta) = \left[\left(\frac{k}{2} + \zeta \right)^2 - \mu^2 \right]^{-1} \left[\left(\frac{k}{2} - \zeta \right)^2 - \mu^2 \right]^{-1} \alpha(k, \zeta)$$

$$(4.63)$$

with $\alpha(k, \zeta)$ analytic and bounded in $\Delta \times \mathcal{D}$. A_1, A_2 and α will then be decomposed as $A_{1|} + A_1'$, $_| A_2 + A_2'$, $\alpha' + \underline{\alpha}$ where, for example, $A_{1|}(k, z, \zeta) = A_1(k, z, \zeta_0 = 0, \boldsymbol{\zeta}^2 = -\frac{\sigma}{4}, \theta)$, and similar decompositions of A_2 and α. The values $\zeta_0 = 0$, $\boldsymbol{\zeta}^2 = -\frac{\sigma}{4}$ are those that make the intermediate energy-momenta $\frac{k}{2} \pm \zeta$ *on-shell*.

The basic geometrical idea to be used is the following. The differences $A_1' = A_1 - A_{1|}$, $A_2' = A_2 -_| A_2$ or $\alpha' = \alpha - \underline{\alpha}$ generate factors that will cancel one of the poles produced by the Feynman propagators: it is then easily seen that the integration contour in corresponding contributions is no longer pinched and can thus be moved so as to avoid the remaining pole singularity. Such contributions will therefore be analytic in the neighborhood of $\sigma = 0$. The contribution corresponding to the terms $A_{1|}$, $_| A_2$, $\underline{\alpha}$, which do *not* depend on ζ_0, $\boldsymbol{\zeta}^2$, is, on the other hand, treated by direct computation of the Feynman integral

$$I(k) = \int \left[\left(\frac{k}{2} + \zeta \right)^2 - \mu^2 \right]^{-1} \left[\left(\frac{k}{2} - \zeta \right)^2 - \mu^2 \right]^{-1} d\zeta, \quad (4.64)$$

first in Euclidean space (k_0, ζ_0 imaginary) and then away from it by analytic continuation, with indeed a singularity in $\sigma^{-1/2}(d = 2)$ or $\ln \sigma$ $(d = 3)$ in the complex neighborhood of $\sigma = 0$.

This contribution is then shown explicitly, in view of the definition of $*$, to be equal to $g(\sigma) A_1 * A_2$ modulo an analytic remainder on which desired bounds are obtained. QED

We next state:

Lemma 4.4. $\exists \lambda_0 > 0$ such that $\forall \lambda, |\lambda| < \lambda_0$, the equation

$$U = G + U\nabla G$$

defines the kernel $U(k, z, z')$ as an analytic function in $\Delta \times \mathcal{D} \times \mathcal{D}$, $s \in V$, bounded by $cst|\lambda|$, where the constant depends only on the bounds on $|G|$ and of those on $|(p^2 - \mu^2) S_2(p)|$. Moreover, the relations

$$F = U + g(s)F * U = U + g(s)U * F \qquad (4.65)$$

and

$$F = U + g(s)U * U + g(s)^2 U * F * U \qquad (4.66)$$

hold for s in a 2-sheeted (d even) or multisheeted (d odd) domain constructed over $V \backslash \{s = 4\mu^2\}$.

Proof. The local existence of U as a meromorphic function, uniform around $\sigma = 0$, follows from the local analyticity of G and the uniformity of the operation ∇ (e.g. by Lemma 1.1 of Chapter I-6). To show that U is, moreover, locally analytic, it is sufficient to note that, in view of Lemma 4.3 and the bounds on $|G|$, the terms $|G\nabla \cdots \nabla G|$ (n factors G) are bounded by $cst(cst|\lambda|)^{n-1}$, so that the Neumann series $U = G + G\nabla G + \cdots$ is absolutely convergent and defines U as a bounded (therefore analytic) function locally.

Eq. (4.65) follows from the equalities $F = G + FoG$, $G = U - G\nabla U$ and Lemma 1.1 of Chapter I-6. Eq. (4.66) follows easily. (We have used the easy equalities $FoG = GoF$, $G\nabla U = U\nabla G$.)

Applications to Theories in Dimensions 4, 2 and 3. We now draw conclusions in dimensions 4, 2 and 3 respectively.

(a) $d = 4$. Theorem 4.6 still holds when the restriction on the distance of D to $s = 4\mu^2$ is removed. The series $\sum GoG \cdots oG$ remains, in fact, convergent at small coupling for s in V, as can be seen by decomposing, near $\sigma = 0$, o as $\nabla + \frac{1}{2}*$ and noting that the factor $\sigma^{(d-3)/2} = \sigma^{1/2}$ generated by $*$ gives no problem. Alternatively, the fact that F is bounded (and thus has no pole) near $\sigma = 0$ can be derived from the convergence of the Neumann series $U + gU * U + g^2 U * U * U + \cdots$ of F in terms of U (due to the same reason).

(b) $d = 2$ [scalar models such as even $P(\varphi)_2$]. The mass-shell is trivial at $d = 2$ (the relation $p_1 + p_2 = k$, p_1, p_2 on-shell, determines the pair p_1, p_2). Values of F and U when p_3, p_4 (respectively p_1, p_2) are on-shell but not necessarily p_1, p_2 (respectively p_3, p_4), depend only on s and z (respectively s, z') and will be denoted $F_|(s, z)$ and $U_|(s, z)$ [respectively $_|F(s, z')$, $_|U(s, z')$]. The full mass-shell restrictions of F and U depend only on s and will be denoted $f(s)$ and $u(s)$.

The composition product $F * U$ involved in Eq. (4.65) reduces to

$$g(s)[F * U](k, z, z') = \frac{1}{a(s)} F_|(k, z)_|U(k, z'), \qquad (4.67)$$

where the factor $1/a(s)$ is the product of $\frac{1}{2}Z^2$ with a kinematical factor generated by $*$ of the form $\text{cst}/\left(\sqrt{s}\, \sigma^{1/2}\right)$. Hence, $a(s)$ is of the form $\text{cst}\,\sqrt{s}\,\sigma^{1/2}$. One then easily obtains:

Theorem 4.7. The following equations, in which U, $U_|$, $_|U$, and u are analytic for $s\,(= 4\mu^2 - \sigma)$ in V, hold, on-shell and off-shell:

(i) $$f(s) = u(s) + (1/a(s))f(s)u(s), \qquad (4.68)$$

or equivalently

$$f(s) = \frac{a(s)u(s)}{a(s) - u(s)} \qquad (4.69)$$

(ii) $$F(k, z, z') = U(k, z, z') + \frac{U_|(k, z)_|U(k, z')}{a(s) - u(s)}. \qquad (4.70)$$

Proof. Equation (4.68) follows directly from (4.65) (by restriction to the mass-shell) and (4.67). Equation (4.70) follows, in turn, from Eq. (4.66), rewritten in the form:

$$F = U + \frac{1}{a(s)}U_| \cdot_| U + \frac{1}{a(s)^2}U_| \cdot_| f \cdot_| U.$$

QED

The relations (4.69) and (4.70) completely characterize the local structure of f and F, for s in V, including possible poles (= zeroes at $a(s) - u(s)$), their residues and the behaviour at threshold. Concerning the latter, it is easily seen that F remains bounded as $\sigma \to 0$ for any given small λ if u is different from zero at $\sigma = 0$ (as is the case at small $\lambda \neq 0$: see below) and $F_|$, $_|F$ or f tend to zero like $\sigma^{1/2}$, as $\sigma \to 0$.

The analysis of possible poles can be made as follows. U is equal at first order in λ to G. On the other hand, G is known, for example, by methods of previous sections, to be equal to a first known term in λ (theories with a φ^4 term) or λ^2 (theories without φ^4 term) plus a remainder bounded by higher powers of λ (times a constant). Since $a(s) = (\text{cst}\sqrt{s})\sigma^{1/2}$, results announced in Sect. 1 follow easily: at small λ and for s in V, one and only one pole at $s = \mu_B^2$, $\mu_B < 2\mu$, $|\mu_B - 2\mu| = \text{cst } \lambda^2 + O(\lambda^3)$ in the theories with a $\pm\lambda\varphi^4$ term in the interaction. This pole lies either in the physical sheet $(-\lambda\varphi^4)$ or in the second sheet $(\lambda\varphi^4)$. One also determines the residue at its pole $s = \mu_B^2$ of the (bound state)–(bound state) 2-point function: $r(\lambda) = \lim\limits_{s\to\mu_B^2} \left(-s + \mu_B^2\right)$ $\sim$

cst λ. The case with no φ^4 term is also treated easily (one 2-particle bound state. The lowest order contribution to G is in λ^2, with a sign that gives a pole in the physical sheet, again at $s = \mu_B^2 < 4\mu^2$. Note that in this case $|\mu_B - 2\mu| = \text{cst } \lambda^4 + O(\lambda^5)$).

(c) $d = 2$ (spinorial case: GN_2 model). All equations now involve matrices (whose coefficients are functions of k, z, z', ...). For example, for the GN_2 model, 4×4 matrices, some of which arise from the 2×2 matrices associated with 2-point functions, or their residues at the pole $k^2 = \mu^2$.

The results are, however, essentially unchanged.

Poles of F are now the zeroes of $\det(a(s)\mathbb{1} - m(s)u(s))$, where $m(s)$ is the 4×4 matrix obtained from the 2×2 residue matrices. The matrix m has elements that are almost all equal to zero at threshold so that conclusions at small coupling are easy.

The detailed analysis requires the consideration of different channels ($\psi\psi$ or $\psi\bar{\psi}$, spin and colour indices) and is omitted.

(d) $d = 3$ (outline). The results are similar: F is, for example, decomposed as $F' + F''$ where F' is the "$l = 0$ partial wave component" of F, namely $F' = \frac{1}{2\pi} \int F \, d\theta$ if F is expressed (using invariance properties) in terms of variables s, z_0, $\vec{z}^2$, z_0', $\vec{z}'^2$ and $\cos\theta$, θ being the angle of $\vec{z}$, $\vec{z}'$ (i.e. the "scattering angle" of the channel). Its complement F'' is shown to be locally bounded. (Kinematical problems at the origin of possible poles in dimensions 2 or 3 are eliminated for F'' through the occurrence of a further factor σ.) F' is again given by formula (4.70) with U replaced by U'. The factor $a(s)$ now behaves like $\text{cst}/\ell n\,\sigma$ as $\sigma \to 0$: hence, there is in general (at small coupling) either no pole for s in V or one pole in the physical sheet at $s = \mu_B^2 < 4\mu^2$, with $|\mu_B - 2\mu| \approx \text{cst } e^{-\text{cst}/\lambda}$. There is no pole for φ_3^4. There will be one pole in the noneven $\left(\lambda\varphi^4 + \lambda'\varphi^3\right)_3$ models for some values of λ, λ' : see Sect. 4.2.

4.2 Noneven theories

The situation is somewhat different in noneven theories. The analysis of Sect. 4.1 applies *a priori* to the connected, amputated function F_1 which is also $1 - p.i.$ in the channel $(1, 2; 3, 4)$: F_1 is the kernel linked, via the (possibly regularized) BS equation, to the $2 - p.i.$ kernel G (or G_M), here analytic up to $s = (3\mu)^2 - \varepsilon$ in Minkowski space. The analysis then yields an analytic structure of F_1 in that region analogous to that of F in Sect. 4.1, with a possible pole in the physical sheet at $s = \mu_1^2$, $\mu_1 < 2\mu$, μ_1 close to 2μ at small coupling, in dimension 2 or 3. However, this does not directly yield corresponding results on F itself, which is linked to F_1 via the relation

$$F(k, z, z') = F_1(k, z, z') + \underset{}{\text{—①—}}(k, z)S_2(k)\,\underset{}{\text{—①—}}(k, z'),\qquad (4.71)$$

where $\underset{}{\text{—①—}}$, $\underset{}{\text{—①—}}$ are similar 3-point $1 - p.i.$ kernels. The analysis of the structure of the various terms involved in the right-hand side of (4.71) shows not only that F has, as $S_2(k)$, a pole at $s = \mu^2$ (not present in F_1), but also that other possible poles of F and F_1 are different: for example, the pole of F_1 at $s = \mu_1^2$ is cancelled by a pole in the second term in the right-hand side of (4.71), whereas a new pole of F at $s = \mu_B^2$, $\mu_B \neq \mu_1$ will appear. On the other hand, S_2 and F will have poles at the same position (namely at $s = \mu^2$ and, in the example above, at $s = \mu_B^2$). This is in fact a generic situation. For example, the fact that a pole of F_1 at a value s_1 of s is not expected to be a pole of F can be seen as follows. The pole of F_1 will yield, through Eqs. (4.72)–(4.74) below, corresponding poles of $\underset{}{\text{—①—}}$, $\underset{}{\text{—①—}}$ and of the 2-point $1 - p.i.$ kernel $\underset{}{\text{—①—}}$ at the same value s_1 of s. The pole of $\underset{}{\text{—①—}}$ yields a zero of

$S_2(k)$ (in view of the relation $S_2(k) = \left[s - m^2 - \underset{}{\text{—①—}}\right]^{-1}$ in $P(\varphi)_2$: see Sect. 2.2) which cancels one of the poles of $\underset{}{\text{—①—}}$ or $\underset{}{\text{—①—}}$ in the second term of the right-hand side of (4.71). Moreover, the remaining pole is cancelled, generically, with that of F_1 as a consequence of factorization properties of the residue of F_1 at the pole and corresponding results on $\underset{}{\text{—①—}}$, $\underset{}{\text{—①—}}$ and $\underset{}{\text{—①—}}$ derived from Eqs. (4.73)–(4.75).

To carry out the analysis, it is useful to consider, besides the BS equation written diagrammatically in the form

$$\underset{}{\text{—①—}} = \underset{}{\text{—②—}} + \underset{}{\text{—②—①—}} \qquad (4.72)$$

the related set of relations

$$\underset{}{\text{—①—}} = \underset{}{\text{—②—}} + \underset{}{\text{—①—②—}}, \qquad (4.73)$$

$$-\text{①} \;=\; -\text{②} \;+\; -\text{②}\;\text{①}\;,\tag{4.74}$$

$$-\text{①}- \;=\; -\text{②}- \;+\; -\text{①}\;\text{②}-\;,\tag{4.75}$$

where all $2-p.i.$ kernels are analytic up to $(3\mu)^2-\varepsilon$ and satisfy bounds analogous to those of Sect. 4.1.

At weak coupling, results of Sect. 4.1, away from the 2-particle threshold, or also close to $s=4\mu^2$ in hypothetical theories at $d=4$, apply similarly (up to $s=(3\mu)^2-\varepsilon$), namely,

Theorem 4.8. Given any bounded region D in the 2-sheeted or multi-sheeted domain of F whose distance to $s=4\mu^2$ is >0, $\exists\,\lambda_0$ such that F and S_2 have in D one and only one pole, at $s=\mu^2$, if $|\lambda|<\lambda_0$. The same result applies also at $d=4$ when the condition on $\mathrm{dist}(D,\,s=4\mu^2)$ is removed.

Proof. The previous result of Sect. 4.1 applies to F_1 and in turn to the terms $\text{①}-$ and $-\text{①}$. At $d=4$, previous results on o are used. The term $-\text{①}-$, and in turn S_2, are also controlled. QED

The neighborhood of $s=4\mu^2$ at $d=2$ or $d=3$ can be treated at small coupling in two different ways. The first one follows the lines indicated above. For example, at $d=2$, Eq. (4.70) does apply to F_1 with U linked to G as in Sect. 4.1. Using the relation $o=\nabla+g*$, one obtains

$$\text{①}- \;=\; \text{⑧}- \;+\; \frac{U_|(k,z)n(s)}{a(s)-u(s)},\tag{4.76}$$

where $\text{⑧}- \;=\; \text{②}- \;+\; {\scriptstyle UV}\;\text{②}-$ and n is its mass-shell restriction ($n\equiv {}_|\text{⑧}-$). In view of the regularity of ∇ (Lemma 4.3 of Sect. 4.1), N is locally analytic. A similar result applies to $-\text{①}$, and in turn

$$=\; d(s) \;+\; \frac{n(s)^2}{a(s)-u(s)},\tag{4.77}$$

where $d=-\text{②}- \;+\; -\text{②}\;{}_\nabla\;\text{⑧}-$ is also locally analytic. The local expression of $S_2=\left[s-m^2-\text{①}-\right]^{-1}$, in, for example, $P(\varphi)_2$, follows and a local expression of F is in turn derived from (4.71). One obtains

Theorem 4.9. In a complex neighborhood V of $s=4\mu^2$,

$$S_2 \;=\; -\frac{(a(s)-u(s))/d'(s)}{a(s)-\hat u(s)},\tag{4.78}$$

$$F = \hat{U} + \frac{\hat{U}_|(k,z)_| \hat{U}(k,z')}{a(s) - \hat{u}(s)}, \qquad (4.79)$$

where $d'(s) = s - m^2 + d(s)$, $\hat{U} = U + [\,\text{─Ⓝ─}(k,z)\ \text{─Ⓝ─}(k,z')]/d'(s)$ and $\hat{u} = u - (n^2/d')$ is the mass-shell restriction of $\hat{U}$; N, d' and in turn $\hat{U}$ are analytic for s in V, at small coupling.

As a matter of fact, ─Ⓝ─, ─Ⓝ─ and $d'(s)$ do not vanish and are equal to lowest order in λ to ─②─, ─②─ and $s - m^2 - \text{─②─}$ respectively; in particular, $d'(s)$ remains close to $3\mu^2$ at small λ. The common pole of S_2 and F in V is given by the zero of $a(s) - \hat{u}(s)$. As in Sect. 4.1, it lies either in the first or second sheet at $d = 2$ depending on a sign, that is, on the model.

Method based on a modified BS kernel. Results of Theorem 4.9 on F can be equally recovered by the following more direct method. We define the "modified" BS kernel $\hat{G}$ via the equation

$$\hat{G} = G - \frac{\text{─②─}(k,z) \cdot \text{─②─}(k,z')}{\text{─②─}}, \qquad (4.80)$$

where ─②─ is the contribution to $\text{─①─} = -1/S_2$ from which 2-particle reducible graphs have been taken out ($\text{─②─} = s - m^2 - \text{─②─}$ in $P(\varphi)_2$).

This kernel is, like G, analytic at small coupling up to $s = (3\mu)^2 - \varepsilon$, apart, however, from a pole (the zero of ─②─) at a value $s = \hat{\mu}^2$ where $\hat{\mu}$ is close to μ, and thus does not lie in V, at small λ. On the other hand, one checks that $\hat{G}$ is now linked to F itself via the BS equation

$$F = \hat{G} + \hat{G} \circ F, \qquad (4.81)$$

so that Eq. (4.79) is reobtained with $\hat{U}$ directly defined in terms of $\hat{G}$ via the equation $\hat{U} = \hat{G} + \hat{U}\nabla\hat{G}$. (It can be seen directly that it does coincide with the kernel $\hat{U}$ defined previously.)

Applications. Previous expressions of S_2 and F, obtained by either one of the two methods that have been presented, yield corresponding informations on possible poles near $s = 4\mu^2$ at $d = 2$ and results can again be extended as in Sect. 4.1 in dimension $d = 3$. If we consider the models $\lambda\varphi^4 + \lambda'\varphi^3$ that do exist (and are nontrivial) at $d = 2$ or $d = 3$, and put $\lambda'^2 = \alpha\lambda$, the kernel $\hat{u}$, at $\sigma = 0$ and at lowest order in λ, is (as $\hat{G}$) a sum of contributions of the form $-\lambda$ (graph $\asymp$ in $\overline{\underline{}②\overline{}}$ = G), λ'^2/μ^2 (graphs of the form $_3^1\overline{\bigstar}$ in $\overline{\underline{}②\overline{}}$) and $-\lambda'^2/3\mu^2$ (contribution arising from the last term in the right-hand side of (4.80), $\hat{S}_2 = -1/\overparen{-②-}$ being equal to $-1/3\mu^2$ at lowest order). There is one pole (= the zero of $a(s) - \hat{u}(s)$) in the physical sheet, if α is such that the sum is > 0, one pole in the second sheet ($d = 2$) or no pole ($d = 3$) if it is < 0. (If α is such that the sum is zero, a calculation to the next order is needed.)

4.3 Gross-Neveu model and semi-axiomatic approaches

Results of Sect. 4.1 do apply, as has been mentioned, to the GN_2 model: one starts to that purpose from the regularized BS equation $F = G_M + F\, o_M G_M$ introduced in Sect. 3.1. One may alternatively start from the BS or renormalized BS kernels introduced in Sect. 3.2 in either one of the two approaches that have been described at the end of that section. In the first one, the bounds (4.48) and (4.49) with B replaced by $B_{\rm ren} = G$ and (4.51), together with Lemma 4.1 of Sect. 3.2, the analyticity properties of G and S_2 and the properties of the operation o allow one to show that each term $(Go\ldots oG)_{\rm ren}$ (n factors G) is well defined and analytic in a 2-sheeted domain around $s = 4\mu^2$, and satisfies bounds in $c_1(c_2|\lambda|)^{n-1}$ if one stays at a minimal > 0 distance of $s = 4\mu^2$. Corresponding results on F follow. The neighborhood of $s = 4\mu^2$ can be treated by an adaptation of the U-kernel method: the kernel U can be directly defined at small coupling in terms of G through expansions in terms of convolutions involving $G - \mathring{G}$, $\omega - \mathring{\omega}$. In the second approach, one may start at weak coupling from the expansion (4.59). Bounds on $\mathring{F}$ (in Euclidean space) and properties of $G - \mathring{G}$, $\omega - \mathring{\omega}$ and the operation o allow one again to derive desired results, first away from $s = 4\mu^2$ and then in its neighborhood (via a definition of U in terms of G analogous to above).

Results obtained in either case include the 2-sheeted structure of F, the analysis of possible poles and 2-particle asymptotic completeness.

Semi-axiomatic approaches. They extend similarly in semi-axiomatic approaches in which B or G are the basic quantities, starting from assumptions similar to those established in the GN_2 model, and from basic formulae defining F in terms of B or G. Two possible situations can be considered at $d = 2$ or $d = 4$. At $d = 2$, assumptions include desired analyticity properties of B, or G, and of S_2 and the large momentum properties (4.48), (4.49) and (4.50). In a related hypothetical "à la φ_4^4" scalar theory at $d = 4$, $S_2(k)$ is assumed to decrease like $1/|k|^2$ at large k (in Euclidean directions), the factor $(1 + |z|)^{-2}$ in (4.50) is replaced by $(1 + |z|)^{-3}$ and the factor $1/(1 + |z|)^3$ in (4.51) is in turn replaced by $(1 + |z|)^{-5}$. Results similar to the above are then obtained at weak coupling. In the first approach, F is defined in terms of G via Eq. (4.52). In the second one, it can be defined in terms of B or G via Eqs. (4.58) or (4.59), but one has, from the axiomatic viewpoint here considered, to first reconstruct $\overset{\circ}{F}$ itself in terms of B (or G). This can be achieved at weak coupling through the following equation which can be formally derived from the BS equation (and is rigorous in weakly coupled GN$_2$: see [IM4])

$$F(0, z, z') = R_{(r)}(0, z, z') + {}_{(\ell)}R(0, z, 0)\Lambda_{(r)}(0, z')$$
$$+ F(0, 0, 0){}_{(\ell)}\Lambda(0, z)\Lambda_{(r)}(0, z'), \qquad (4.82)$$

where, at $k = 0$, $R_{(r)} = [1 - (B - B_0)\omega]^{-1}(B - B_0)$ and admits the expansion $(B - B_0) + (B - B_0)\omega(B - B_0) + (B - B_0)\omega(B - B_0)\omega(B - B_0) + \cdots$, ${}_{(\ell)}R$ is defined similarly in terms of $B -_0 B$, $\Lambda_{(r)}(0, z') = 1 + (1\omega R_{(r)})(0, z, z')$ (which is independent of z) and ${}_{(\ell)}\Lambda(0, z) = 1 + ({}_{(\ell)}R\omega 1)(0, z, z')$ (independent of z').

An analogous expression in which the two first terms are replaced by ${}_{(\ell)}R(0, z, z') +_{(\ell)} \Lambda(0, z)R_{(r)}(0, z, z')$ can alternatively be used.

The analogue of (4.82) in terms of G is obtained by replacing all differences, $B - B_0$, $B -_0 B$ by $G - G_0$ and $G -_0 G$, respectively, and $F(0, 0, 0)$ by $G(0, 0, 0)$ [as a particular aspect of Eq. (4.52): at $k = z = z' = 0$ all terms vanish except $G(0, 0, 0)$]. The latter equation does allow one to define $\overset{\circ}{F}$ in terms of G at weak coupling and to show bounds in cst λ. If one starts from B, the analysis is completely analogous if $F(0, 0, 0)$ is by definition equal to cst λ (as is the case in models where λ is the renormalized coupling). Otherwise a supplementary assumption is needed to control it.

Note. In contrast to the situation in the super-renormalizable case, where the series $F = G + G \circ G + \cdots$ is convergent at small coupling in the whole Euclidean energy-momentum space, the series $F = G + (G \circ G)_{\text{ren}} + \cdots$ or the related series (4.58) and (4.59) arising from the subtracted BS equation (with a given "subtraction point," such as the origin, in the definition of renormalization or in the subtracted BS equations) will not be shown to be convergent (and will not allow the definition of F, in the semi-axiomatic framework) in the whole Euclidean space, even at weak coupling. Results are established only for bounded values of the channel energy $k_0 = (p_1 + p_2)_0$ (with $\mathbf{k} = \mathbf{p}_1 + \mathbf{p}_2$, fixed at zero) in view of the dependence on k of constants in relevant bounds as k_0 becomes larger and larger. (More precisely, values of k_0 can be chosen arbitrarily large but for smaller and smaller values of the coupling.) This is sufficient for the study of 2-particle structure, which involves local properties in complex energy-momentum space. A more complete study would involve the use of more general bounds for arbitrary subtraction points (either proved in the GN_2 model, or assumed).

§5. Many-particle structure analysis: general results and conjectures

Note: relevant indices of irreducibility will most of the time be left implicit in diagrammatical notarions.

5.1 Structure equations

We consider again theories with a lowest basic physical mass $\mu > 0$ (whose limit, as the coupling λ or λ_{ren} tends to zero, remains strictly positive). In view of the analysis of the energy region $s < [(r+1)\mu]^2 - \varepsilon$ (in a given channel) in Minkowski space, we shall consider, for each connected, amputated N-point function F, an expansion of the form

$$F = \sum_G F_G, \qquad (4.83)$$

where the sum will run over a class of graphs G, depending on r. Each F_G, which also depends on r, will be a Feynman-type integral with irreducible kernels at each vertex and (possibly regularized) 2-point functions on internal lines, up to modifications explained later in the general case. Irreducible kernels involved generalize those of Sects. 2 and 3.1, that is, are nonintrinsic in nonsuper-renormalizable theories but can again be defined in a natural way in the constructive framework. We shall not attempt to generalize the formalism of Sects. 3.2 and 4.3 based on "true" or renormalized irreducible kernels.

In super-renormalizable theories like $P(\varphi)_2$, each irreducible kernel is formally, from the perturbative viewpoint, the sum of Feynman amplitudes of graphs with corresponding graphical properties, and Eq. (4.83) will be from that viewpoint a partial summation of the perturbative series. It corresponds to the best possible regrouping of Feynman integrals that have, in the region $s < [(r + 1)\mu]^2$, a common analytic and monodromic structure. In the constructive framework, cluster expansions of sufficiently high order (depending on r) are used and all irreducible kernels are again well defined, initially in Euclidean space, by expansions restricted to graphs with desired graphical properties with respect to propagator lines (of the lowest momentum slice in the general case). These expansions are, as in Sect. 3.1, convergent at sufficiently small coupling. Moreover, in view of their graphical definition, bounds including an exponential fall-off factor in $\exp -(\mu - \varepsilon)L_{\mathrm{irr}}(x_1, \ldots, x_N)$ are obtained in Euclidean space-time, where $L_{\mathrm{irr}}(x_1, \ldots, x_N)$ is here the minimal length of all connected graphs joining $x_1, \ldots, x_N$ and possibly intermediate points and satisfying corresponding irreducibility properties. Analyticity in complex energy-momentum space follows (again by the Laplace transform theorem). Structure equations will be, on the other hand, satisfied by graphical inspection and the series (4.83) are again convergent, initially in Euclidean space, at small coupling. This is due to the fact that, while well adapted (via subsequent analytic continuation away from Euclidean space) to the analysis of given energy regions in Minkowski space, they are more limited than the perturbative series: cluster expansions of a possibly high, but fixed, order have been used. This allows one both to define irreducible kernels through convergent expansions and to obtain the convergence of the series (4.83). We note, however, that in both cases convergence is obtained for values of couplings that are smaller and smaller as r increases.

The regroupings of terms leading to the expansions of interest, for each given value of r, are the simplest ones such that, in view of their analyticity properties, the irreducible kernels involved are in fact analytic for practical purposes. Their singularities should *not* generate singularities of the integrals F_G in the region of interest (in Minkowski space), the only singularities being generated by the pole singularities of internal lines. In the simplest case ($r = 3$, in an even theory), the expansion of the 4-point function F of interest is that already mentioned in terms of the BS (or BS-type) kernel G (or G_M). This kernel is $2 - p.i.$ in the channel $(1, 2; 3, 4)$ and $1 - p.i.$ in channels $(i; j, k, \ell)$: following the terminology introduced below, it is "totally $2 - p.i.$" in the channel $(1, 2; 3, 4)$. Related expansions, for $r = 2$ in noneven theories, follow for the kernels F_1, $\overline{}\!\!\text{①}\!\!-$, $-\text{①}\!\!\overline{}$ and in turn for F. We shall first describe below the situation for the 6-point function at $r = 4$ in even theories, that is, in the 3-particle region (below the 5-particle threshold), a case in which important simplifications still occur, and will then give indications in more general cases. We first give relevant definitions.

Preliminary definitions. By definition, a connected graph is $r - p.i.$, respectively, totally $r - p.i.$ in a channel g (i.e. a partition of the set of external lines into two subsets) if it cannot be divided with respect to g into two successive connected, respectively connected *or* nonconnected, multiparticle scattering subgraphs G_1, G_2 (with at least one orientation of all lines of G) by cutting $r' \leq r$ intermediate lines: the latter must include internal lines of G but may also include some external lines of G as explained in the example below. G is (totally) r-particle reducible in the channel if it can be divided into two parts by cutting r intermediate lines.

Example. The graph ⟨graph⟩ is infinitely irreducible in the channel

$(1, 2, 3; 4, 5, 6)$ (i.e. is not reducible in this channel). It is totally 1, 2 and $3 - p.i.$, but is *not* totally $4 - p.i.$ in that channel in view of the possible cutting:

into the two subgraphs $G_1 = $ ⟨graph⟩ , $G_2 = $ ⟨graph⟩ .

Its index of total reducibility in this channel is 4. In the channel $(1, 2, 4; 3, 5, 6)$, the index of reducibility or of total reducibility is 2.

Remarks.

(1) Total r-particle irreducibility in a given channel g entails irreducibility properties in various channels. For instance, total 3-particle irreducibility in the channel $(123; 456)$ entails:

- 3-particle irreducibility in the channel $(1, 2, 3; 4, 5, 6)$
- 2-particle irreducibility in any channel $(ij; k456)$ or $(123k'; \cdot)$
- 1-particle irreducibility in any $1 \to 5$ channel and in any crossed channel such as $(1, 2, 4; 3, 5, 6)$.

  From the viewpoint of analyticity properties, it follows from the above definitions and from analyticity properties associated with irreducibility that a totally $r - p.i.$ kernel (in a given channel g) is, in particular, analytic in a neighborhood of the physical region (and of a further domain of the complex mass-shell) in the region $s < [(r + 1)\mu]^2 - \varepsilon$. We emphasize that this property is *not* satisfied for usual indices or irreducibility, rather than total irreducibility: the 3-particle irreducible function in the channel $(1, 2, 3; 4, 5, 6)$ has singularities associated with graphs with one internal line and triangle graphs in the region $s < (4\mu)^2$.

(2) Indices in each channel are not independent. For instance, if we again consider a 6-point function and the channels $g_1 = (1, 2, 3, 4; 5, 6)$, $g_2 = (1, 2, 3; 4, 5, 6)$, and if ν_{g_1} is an index of total reducibility, possible indices ν_{g_2} of total irreducibility must satisfy, as easily checked, $\nu_{g_2} \leq \nu_{g_1}$. If indices are not consistent, then by convention the irreducible kernel is identically zero.

$3 \rightarrow 3$ *processes in the 3-particle region (even theories).* The relevant expansion is written below for the 6-point function F_1 which is $1-p.i.$ in the $3 \rightarrow 3$ channel $(1, 2, 3; 4, 5, 6)$ considered ($F_1 = F - $).

The analysis, following ideas to be presented in the general case, leads here to introduce the 4-point and 6-point kernels denoted as and , which are "totally" $2-p.i.$ and $3-p.i.$ respectively, in the channel corresponding to lines on the left and right, respectively. The class of graphs G involved in (4.83) includes in this case

- all graphs with only $2 \rightarrow 2$ vertices, of the form , , , .

- all graphs in which one or more of the $2 \rightarrow 2$ vertices above are replaced by sequences of $2 \rightarrow 2$ vertices.

- all graphs composed of sequences of the graphs above and of $3 \rightarrow 3$ vertices.

One thus reobtains the expansion already described (from a different viewpoint) in Sect. 6.3 of Chapter II.

Remarks.

(1) The kernel L introduced in Chapter II-6 has the same expansion, except that all terms must start and end with a 6-point totally $3-p.i.$ kernel . As already mentioned at the end of Chapter II-6, this is the origin of the fact that subchannel cuts and crossed channel poles do not occur for L.

(2) Apart from the terms of the form and where denotes either or any sequence ,

⟂⟂⟂ ,..., whose sums are equal to ⟂Ⓕ⟂Ⓕ⟂ and

⟂Ⓕ⟂Ⓕ⟂ respectively, the singularities of all other terms (in the region considered) are restricted on-shell, in the physical sheet, to $s = (3\mu)^2$. In nonphysical sheets, singularities of these other terms (which can be conjectured to be Landau, or more precisely, modified Landau singularities) should become effective away from $s = (3\mu)^2$, probably in more and more remote sheets according to the complication of the graph G.

More general situations. The following properties will be satisfied in more general situations, in even or noneven theories. We do not attempt here to give a complete description, but only indicate main features and give some examples that clarify the origin of the rules. The kernel $F^{(r)}$, totally $r - p.i.$ in the channel g considered, is one of the terms of the expansions, associated with the trivial graph G with only one vertex and no internal line. For reasons that appear on the example given below, the definition of the terms F_G will have to be extended in general:

- F_G will be (a finite) sum of Feynman-type integrals $F_{G,\alpha}$ in which irreducible kernels involved will also have in some cases, beside their indices of total irreducibility, specified degrees of (total) *reducibility* in some channels (consistent with their irreducibility indices). These kernels are again defined graphically.
- The (possibly regularized) 2-point function associated to each internal line may have to be replaced by a modified function with a further index $\nu > 1$ of irreducibility: this modified function will still have a pole at $k^2 = \mu^2$ (where k is the energy-momentum of the line in Minkowskian notations) but, apart from it, will be analytic up to $k^2 = [(\nu + 1)\mu]^2 - \varepsilon$. In, for example, $P(\varphi)_2$, the usual 2-point function, equal to $\left[s - m^2 - \text{—①—}\,\right]^{-1} \equiv$

$$\left[s - \mu^2 - (\;\text{—①— — —①—}_{|\mu^2}\;)\right]^{-1}, \text{ will be replaced in relevant}$$

cases by $S_2^{(\nu)} = \left[s - \mu^2 - (\;\text{—Ⓥ— — —Ⓥ—}_{|\mu^2}\;)\right]^{-1}$ where

—Ⓥ— is the $\nu - p.i.$ 2-point kernel. There may be inclusions, on an internal line of index $\nu > 1$, of one or more 2-point subgraphs (such as —⟂⟂— , —⟂⟂⟂— ,... if $\nu = 2$) arising from the expansion of —①— in terms of kernels with higher irreducibility. These inclusions give a multiplicative factor equal to the value of the subgraph, minus its value at $k^2 = \mu^2$, divided by $k^2 - \mu^2 - ($ $- $ $\big|_{\mu^2})$.

General requirements on the expansion are as follows:

(a) The sum runs over connected graphs G such that, given any internal line ℓ of G, there is at least one way of dividing G, with respect to the channel g considered, into two (connected or nonconnected) successive subgraphs by cutting a set of, at most r, intermediate lines that contains ℓ.

(b) For each graph G, there is a finite sum of terms corresponding to various ways of attributing indices of (total) irreducibility, and also in some cases of reducibility, to each bubble b, such that any way of dividing G with respect to g that goes now "across" one or more bubbles, [i.e. crosses lines of one or more subgraphs G_b, if each bubble b is replaced by a subgraph G_b compatible with irreducibility, and reducibility properties of b] includes at least $r + 1$ intermediate lines.

(c) There must also be at least $r + 1$ intermediate lines for cuttings of (original) internal lines of G if one of them (or more) is counted for $\nu_\ell + 1$ instead of one, where ν_ℓ is its irreducibility index.

The second and third rules are in agreement with the general principle, according to which singularities of irreducible kernels involved (either explicitly or implicitly in the 2-point functions of internal lines) should not contribute to the generation of singularities of the integrals in the region considered. The first rule is intended to ensure that pole singularities of internal lines do contribute to generate singularities of the integral; otherwise, simpler structure equations should be available for the analysis in the region of interest.

Regroupings of terms in the expansions of (connected, amputated) N-point functions, in a way such that these rules be satisfied, lead to the simple expansions mentioned previously in the 2-particle region or in the 3-particle region in an even theory. However, the fact that each term must be counted once, and only once, leads to some complications in more general situations. For example, if the theory is no longer even, such complications already occur for the 6-point function at $r = 3$. For the subsequent analysis of the 3-particle region, new terms then have to be considered, such as

, , , ,

where indices of irreducibility (or also reducibility) have been omitted either in the bubbles or on the internal lines.

Although the first rule is still satisfied, sets of more than r intermediate lines dividing the graph G cannot be avoided, in contrast to previous cases. On the other hand, some internal lines have to include indices of irreducibility > 1. This is the case in the first term shown above: internal lines must have an index of irreducibility 2. (Otherwise, one could obtain the number 3 by cutting the two internal lines, one of them being counted for two. In such a case, the singularity at $k_\ell^2 = 4\mu^2$ would contribute to singularities of the integral in the region $s < (4\mu)^2 - \varepsilon$.)

Examples in which several terms $F_{G,\alpha}$ are needed for a given graph G occur, for instance, in a $2 \to 2$ process at $r = 4$ $\left(s < (5\mu)^2 - \varepsilon\right)$ in the channel $g = (1, 2; 3, 4)$. Consider, for example, terms of the form

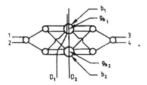

Each internal line ℓ of G is such that G can indeed be divided with respect to g into two subgraphs with a set of at most 4 lines that contains ℓ as required by condition (a). On the other hand, the index $\nu(b_1, g_{b_1})$ of (total) irreducibility of b_1 in the channel g_{b_1} must be ≥ 1 : otherwise, the division D_1, which crosses b_1, could have less than 5 intermediate lines, in contradiction with condition (b). Similarly, $\nu(b_2, g_{b_2}) \geq 1$. But one cannot simply put $\nu(b_1, g_{b_1}) = \nu(b_2, g_{b_2}) = 1$: otherwise, the division D_2 (which crosses b_1 and b_2) might have less than 5 intermediate lines. One is thus led to consider a term $F_{G,\alpha}$ with pure indices of total irreducibility $\nu(b_1, g_{b_1}) = \nu(b_2, g_{b_2}) = 2$, but also other ones, namely, a term with a degree of (total) irreducibility of 1 and a further degree of *reducibility* equal to two for (b_1, g_{b_1}) (possibility of cutting 2 lines), and as above, a degree of (total) irreducibility equal to two for (b_2, g_{b_2}) (3 lines at least cut), or a similar term in which the roles of b_1 and b_2 are exchanged.

Some conjectures. We now conclude this section with the following conjectures, which are only partly established, on individual integrals F_G or $F_{G,\alpha}$, and related remarks.

Conjecture 4.1 (analytic properties of irreducible kernels). Irreducible kernels are analytic in the primitive domain. Moreover, if a kernel is

$r_g - p.i.$ in each channel g, corresponding energy-cuts start at $[(r_g+1)\mu]^2$. (More precisely, if the total energy-momentum k_g of the channel is of the form $((k_g)_0, \vec{0})$, the cut in the complex $(k_g)_0$-plane is $(k_g)_0 = (r_g+1)\mu+\rho$, $\rho \geq 0$.)

This conjecture goes somewhat beyond results that can be *a priori* established at weak coupling from the definition of irreducible kernels and should be a particular consequence of a further analysis in higher energy regions (e.g. through expansions analogous to (4.83) of the irreducible kernel considered, assuming that no pole, associated with divergences of the expansions, will occur).

We next state:

Conjecture 4.2 (analyticity and holonomicity properties of integrals F_G). Given a channel g and r, each term F_G or $F_{G,\alpha}$ in the expansion (4.83) of F has, at $s < [(r + 1)\mu]^2$, the same analytic and monodromic structure as the (possibly regularized) Feynman integral of G, in a neighborhood of the physical region (and in a further domain). It is in particular holonomic, with regular singularities in that region (in the sense of [SKK]), and its singularities are Landau singularities of G and of graphs obtained by contraction: more precisely, $+\alpha$-Landau singularities in the physical region and other branches of (modified) Landau singularities in other sheets.

This property should be a consequence of analyticity properties of irreducible kernels and of condition (b) above on the terms F_G or $F_{G,\alpha}$. We recall that Feynman integrals are indeed holonomic with regular singularities, as established in [KKO] (and references therein).

Conditions that have been imposed on the expansions (4.83) of F entail that these expansions are *minimal* ones in terms of holonomic contributions. Further regroupings of terms may lead in various cases to expansions in terms of holonomic contributions with a smaller number of terms, but there is apparently no general rule of this type.

The expansion (4.83) is still not unique. (It depends on the regularization and other details.) It has convergence properties (up to limitations mentioned previously) at small coupling. Other expansions, involving, for example, generalizations of the kernel U, with possibly better properties are not known so far.

5.2 Discontinuity formulae of Feynman-type integrals

We consider a (regularized) Feynman-type integral F_G or $F_{G,\alpha}$ involved in the expansion of F. Conjecture 4.2 of Section 5.1 means, in particular, that in a neighborhood of the physical region (and of a further domain of the complex mass-shell) singularities encountered at $s < [(r+1)\mu]^2$ are Landau singularities. Conjecture 4.2 also entails that there is, at sufficiently small values of s below all thresholds, for example, at $s < (2\mu)^2$ for a term of the form $\equiv\!\bigcirc\!\bigcirc\!\equiv$, or $s < (3\mu)^2$ for a term of the form $\equiv\!\bigcirc\!\equiv\!\bigcirc\!\equiv$, a common analytic function which, by plus $i\varepsilon$ and minus $i\varepsilon$ analytic continuations around all normal thresholds and other singularities encountered gives boundary values denoted $(F_G)_+$ and $(F_G)_-$, respectively. The latter can be defined in a field theory context as integrals over well-defined contours $\Gamma_+(k)$ and $\Gamma_-(k)$. Other analytic continuations and boundary values will be considered later. We first state:

Conjecture 4.3 (graph by graph unitarity). The following discontinuity formula holds at $s < [(r+1)\mu]^2$, in a real neighborhood of the physical region (and a further domain of the complex mass-shell):

$$
\begin{aligned}
\Delta F_G &\equiv (F_G)_+ - (F_G)_- \\
&= \sum_{(G_1, G_2)} (-1)^{N_c(G_2)+1} (F_{G_1})_+ * (F_{G_2})_-,
\end{aligned}
\tag{4.84}
$$

where $* \equiv *_+$ and the sum runs over all ways of dividing the graph G with respect to g, with at most r intermediate lines, into two (connected or nonconnected) subgraphs G_1, G_2, each of which has at least one nontrivial connected component. (Trivial components are those with one incoming and one outgoing line.) $N_c(G_2)$ is the number of nontrivial connected components in G_2. If G_1, respectively G_2, is not connected, $(F_{G_1})_+$, respectively, $(F_{G_2})_-$, is the product of the (nontrivial) connected components (each one depending on its own energy-momenta variables). The operation $*_+$ denotes a convolution integral over on-mass-shell values of the energy-momenta of intermediate lines joining nontrivial components.

Examples (indices of irreducibility or reducibility are left implicit). In all examples below, the channel g corresponds to incoming and outgoing lines on the left and right sides, respectively.

If $\bigcirc$ denotes the BS kernel in an even theory, then in the region $s < (4\mu)^2$,

$$
\underbrace{\left(\overline{\bigcirc\bigcirc\cdots\bigcirc}\right)}_{n} - \underbrace{\left(\overline{\bigcirc\bigcirc\cdots\bigcirc}\right)}_{n} = \sum_{\substack{r \\ 1 \leq r < n}} \underbrace{\overline{\bigcirc\cdots\bigcirc}}_{r}\Big|\underbrace{\overline{\bigcirc\cdots\bigcirc}}_{n-r} .
\tag{4.85}
$$

In the region $s < (5\mu)^2$,

$$\left(\begin{smallmatrix}1\\2\\3\end{smallmatrix}\boxed{}\begin{smallmatrix}4\\5\\6\end{smallmatrix}\right) - \left(\boxed{}\right) = {}_3\boxed{}{}_6 + \boxed{} . \quad (4.86)$$

We note that singularities encountered in the latter case include the normal thresholds associated with the two contracted graphs $\underset{2}{\overset{1}{}}\!\!\!\rightarrowtail\!\!\!\underset{3}{}$ and $\underset{6}{\overset{4}{}}\!\!\!\prec\!\!\!\overset{5}{}$, and the triangle singularity of the graph $\underset{3}{}\!\!\bigtriangledown\!\!\underset{6}{}$. The discontinuity, in the form above, contains only the two terms associated with the two normal thresholds.

Eq. (4.85) can be checked inductively from the relation $o_+ - o_- = *_+$. Eq. (4.84) has been more generally partially established in some cases. (Its perturbative analogue for Feynman integrals is a consequence of works of the 1970s by H. Epstein and V. Glaser, restricted, however, to the perturbative framework.)

Formulae involving more general boundary values. We mainly consider the $3 \to 3$ case in view of the application to generalized optical theorems in connection with Sect. 5.3 of Chapter I, but a large part of the discussion applies more generally. It is useful to give some preliminary definitions and results. As before, the analysis is made with respect to a given channel g, for example, $(1, 2, 3; 4, 5, 6)$ in the $3 \to 3$ case. Sets S of channels to be considered may or may not include it. The function $(F_G)^S$ defined below will later be considered (Conjecture 4.4) as obtained in "good" cases by plus $i\varepsilon$ and minus $i\varepsilon$ analytic continuation around normal thresholds corresponding to channels of S and to other channels respectively. $(F_G)^S$ is initially defined, for each set S of channels, by the formula:

$$(F_G)^S = \sum_L (-1)^l (F_{G_1})_+ * (F_{G_2})_+ * \cdots * (F_{G_{l+1}})_+, \quad (4.87)$$

where $* \equiv *_+$ and the sum runs over all ways $L = (L_1, \ldots, L_l)$ of dividing G with respect to the channel g considered into successive (connected or nonconnected) subgraphs $G_1, \ldots, G_{l+1}, l \geq 1$, such that each division $L_\sigma, \sigma = 1, \ldots, l$, has at most r intermediate lines and is consistent with S. Namely, given any division of G into two *connected* subgraphs $G_{1,\sigma}$, $G_{2,\sigma}$ obtained by cutting lines of L_σ that are *internal* lines of G, the external lines of $G_{1,\sigma}$ and $G_{2,\sigma}$ must be those of a channel of S. For example, the division $\underset{3}{\overset{1}{}}\!\!\boxed{}\!\!\bigcirc\!\!\underset{6}{\overset{4}{}}$ respectively $\underset{3}{\overset{1}{}}\!\!\boxed{}\!\!\underset{6}{\overset{4}{}}$, with respect to the channel $g = (123; 456)$, is consistent with the channel $(12; 3456)$, respectively $(124; 356)$. In view of this definition, $(F_G)^S$ can be also be written:

$$(F_G)^S = \sum_{S' \subset S} (-1)^{|S'|}(F_G)_{S'}, \qquad (4.88)$$

where $(-1)^{|S|}(F_G)_S$ is defined as in (4.87), but with a sum in the right-hand side running only over sets L such that each channel of S is obtained for at least one set of internal lines of one L_σ. As appears in the examples given below, the sum is moreover limited by:

Lemma 4.5. $(F_G)_S \equiv 0$ in the following situations:

(a) S contains a channel g' that cannot be obtained from any division L_σ of G with respect to g. For example,

$$\left(\begin{array}{c} \text{image} \end{array}\right)_{(3,4)} \equiv 0 \;, \qquad (4.89)$$

where S is here composed of only one channel, the channel (34), that is, following notations of Chapter I-5, $(34) \equiv (124; 356)$.

(b) S contains two overlapping channels g_1, g_2.

In case (b), one cannot in fact find divisions $L = (L_1, \ldots, L_l)$ into *successive* subgraphs such that both g_1 and g_2 are obtained from divisions $L_{\sigma_1}, L_{\sigma_2}$ of L. For example,

$$\left(\begin{array}{c} \text{image} \end{array}\right)_{t,(3,4)} \equiv 0, \qquad (4.90)$$

where S is the set of the two channels $t \equiv g = (123; 456)$ and (34).

Condition (a) depends on the graph G. Condition (b) is independent of it. In the $3 \to 3$ case, it restricts sets S such that $(F_G)_S$ is not identically zero to those already mentioned in Sect. 5.3 of Chapter I. On the other hand, various contributions to the terms $(F_G)_{S'}$ in the right-hand side of (4.88) may coincide, apart possibly from multiplicative coefficients. They can then be regrouped together, and in some cases cancel among themselves.

Example. Let $F_G = $.

(a) Let $S_1 = \{(3), (4)\}$, where, as in Chapter I, $(3) = (12; 3456)$ and $(4) = (1234; 56)$. Then,

$$(F_G)^{S_1} = (F_G)_+ - (F_G)_{(3)} - (F_G)_{(4)} + (F_G)_{(3),(4)} \qquad (4.91)$$

with $(F_G)_{(3)} = $ 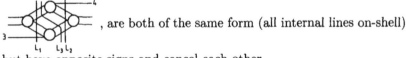 $, (F_G)_{(4)} = $

and $(F_G)_{(3),(4)} = $.

(b) $S_2 = \{t, (34)\}$:

$$(F_G)^{S_2} = (F_G)_+ - (F_G)_t - (F_G)_{(34)}, \qquad (4.92)$$

where

$$(F_G)_t = \quad , (F_G)_{(34)} = \quad .$$

The term $(F_G)_{t,(34)}$ is identically zero (see (4.90)).

(c) $S_3 = \{3, 4, t\}$:

$$(F_G)^{S_3} = (F_G)_+ - (F_G)_{(3)} - (F_G)_{(4)} - (F_G)_t$$
$$+ (F_G)_{(3),t} + (F_G)_{t,(4)}, \qquad (4.93)$$

where, for example, $(F_G)_{t,(3)} = $.

In fact, the terms $(F_G)_{(3),(4)}$ and $(F_G)_{(3),(4),t}$, which correspond respectively to $L = (L_1, L_2)$ and $L = (L_1, L_2, L_3)$, with $L_1 L_2 L_3$ shown in

, are both of the same form (all internal lines on-shell) but have opposite signs and cancel each other.

It is finally useful to define $(\bar{F}_G)^S$ in a way analogous to $(F_G)^S$, but with minus signs $[(\bar{F}_G)^S = (F_G)_-$ if S is the empty set]:

$$(\bar{F}_G)^S = \sum_L (-1)^{l+N_c+1} (F_{G_1})_- * \cdots * (F_{G_{l+1}})_-, \qquad (4.94)$$

where N_c is the total number of nontrivial connected components of the graphs $G_1, \ldots, G_{l+1}$.

The following result can be checked from Conjecture 4.3 (graph by graph unitarity) in a real neighborhood of the physical region, at $s < [(r+1)\mu]^2$:

Lemma 4.6.

$$(F_G)^S = (\bar{F}_G)^{\bar{S}}, \forall S, \tag{4.95}$$

where $\bar{S}$ is the complement of S in the set E of all channels. In particular:

$$(F_G)^E = (F_G)_-. \tag{4.96}$$

Another result, useful below, says that if S is a good set (no pair i, f such that $(if) \in S$, $t \in S$, $(i) \notin S$, $(f) \notin S$ or $(if) \notin S$, $t \notin S$, $(i) \in S$, $(f) \in S$ in the $3 \rightarrow 3$ case: see Chapter I-5), there is no "interference" between divisions occurring in the expansion of $(F_G)^S$ and $(\bar{F}_G)^{\bar{S}}$.

Lemma 4.7. ($3 \rightarrow 3$ case). If S is a good set, then given any set L of divisions $L_1, \ldots, L_l$ giving rise to a nonzero term in the expression (4.87) of $(F_G)^S$ (respectively in the expression of $(\bar{F}_G)^{\bar{S}}$), there exists no division consistent with $\bar{S}$ (respectively with S) whose internal lines all belong to the set of internal lines determined by the division $L_1, \ldots, L_l$.

The content of this lemma can be illustrated on the term F_G of the example above. First consider a "bad" set, for example, $S_1 = \{(3), (4)\}$, and the set $L = (L_1, L_2)$ that gives rise to the nonzero contribution $(F_G)_{(3),(4)}$ to $(F_G)^{S_1}$. Then the division L_3 that is consistent with $t \in \bar{S}_1$, or the analogous division consistent with the channel (34) of $\bar{S}_1$, has internal lines that all belong to those of L. But if we consider a good set, such as $S_3 = \{3, 4, t\}$, then this same contribution is absent as explained below (4.93) and Lemma 4.7 is thus easily checked.

Lemma 4.7 is checked more generally by inspection. We now state:

Conjecture 4.4. For every good set S, $(F_G)^S$ is (on-shell, in the physical region) analytic at $s < [(r+1)\mu]^2$ except on some branches (depending on S) of (modified) Landau surfaces of G and of graphs obtained from G by contraction. Apart from exceptional points, good functions $(F_G)^S$ are, moreover, boundary values of analytic continuations, on the complex mass-shell, of the unique analytic function F_G. Starting from small values of s, analytic continuation is made beneath, respectively above, normal thresholds associated with S, respectively $\bar{S}$ (i.e. with minus $i\varepsilon$ and plus $i\varepsilon$ rules, respectively). Analytic continuation around other singularities is also well specified, in a way consistent with previous rules (e.g. at points where singularities are tangent).

Similar results apply in the neighbourhood of the physical region, off-shell. If the set S corresponds to a cell, $(F_G)^S$ is also the boundary value of F_G from the (off-shell) directions of the cell (which are in this case independent, for each S, of the real point considered).

Proof (partial arguments). The proof should provide more precise and more general information, for example, on on-shell or off-shell analyticity domains. We only give below some arguments that support the plus and minus $i\varepsilon$ rules of analytic continuation, in the physical region, around normal thresholds associated with S or $\bar{S}$. The analysis makes use of results on the micro-support of on-shell convolution integrals in terms of the micro-supports of individual factors, and hence, by duality on possible directions of analyticity with respect to imaginary parts of the variables. External energy-momenta can either be strictly restricted to the mass-shell or be allowed to vary in a real neighborhood. From Conjecture 4.1, the micro-support of each term $(F_{G_k})_+$, and respectively $(F_{G_k})_-$, at any physical point is known to be associated with relative configurations of external trajectories of classical space-time diagrams $(\mathcal{D}_k)_+$, and respectively of opposite diagrams $(\mathcal{D}_k)_-$, whose topological structure is G_k, or with relative configurations of points in space-time (one for each initial and final particle) that lie at corresponding external vertices of such diagrams. Theorems on products and integrals recalled in Sect. 3 of the Appendix then assert that the micro-support of the integral is associated with relative configurations of external trajectories (or external points) of space-time diagrams $\mathcal{D}$ which are collections of subdiagrams $(\mathcal{D}_k)_+$, and respectively $(\mathcal{D}_k)_-$, that "fit together." (I.e. space-time trajectories associated to a given original internal line, considered in either one of the two subdiagrams in which it is involved, must coincide.) Some limiting procedures may also have to be considered in "$u = 0$" cases.

The p-particle normal threshold associated with a channel g_i is the singularity of the basic graph $G_{g_i, p}$ with two vertices, whose sets of external lines are those of the channel g_i and p internal lines between these two vertices. The typical case where analytic continuation around it is blocked occurs when the set of internal lines over which there is on-mass-shell integration includes a set associated with a division of G having p internal lines and consistent with g_i. The graph G can in this case be reduced by contraction to $G_{g_i, p}$, and the micro support of the integral at real points of the threshold includes the two opposite directions corresponding to the (relative) external configurations of space-time diagrams $(\mathcal{D}_{g_i,p})_+$ and $(\mathcal{D}_{g_i,p})_-$ respectively. The two vertices of $G_{g_i,p}$ are represented in these diagrams by points a and b such that $b - a = \lambda k(g_i)$, where $k(g_i)$ is the energy-momentum of the channel, with $\lambda > 0$ or $\lambda < 0$ respectively. By duality, these two directions correspond to directions of analyticity Im $s_{g_i} > 0$ and Im $s_{g_i} < 0$, which are conflicting.

To show that $(F_G)^S$ continues above normal thresholds associated with any channel g_i in $\bar{S}$, one may use the expression (4.87). If S is a good set, Lemma 4.7 ensures in fact that the typical situation just mentioned in which analytic continuation is blocked does not occur for any individual term in this expression. On the other hand, cases in which a diagram $\mathcal{D}_{b_1}$ reduces to $(\mathcal{D}_{g_i,p})_+$ do occur. Let us consider, for example, the term $(F_G)_{(3)}$ in the expression (4.93) of $(F_G)^{S_3}$; $(F_G)_{(3)}$ has been written below (4.91). Its two bubbles on the left and on the right will be denoted b_1 and b_4, and b_2 and b_3 will denote the two bubbles in the middle. We then consider corresponding diagrams with respective space-time vertices $a_1, \ldots, a_4$. Then the situation just mentioned is obtained for $g_i = (34) \in \bar{S}_3$ and $p = 4$, by contracting lines between a_1 and a_2 and between a_3 and a_4 with $a_3 - a_1 \equiv a_4 - a_2 \in V_+$, a condition consistent with the requirement that the two lines between a_2 and a_4 are associated with internal lines of a term $(F_{G_k})_+$. Thus one does obtain by duality a plus $i\varepsilon$ rule of analytic continuation corresponding to Im $s_{g_i} > 0$, unless *external configurations* of other relevant diagrams (different from $\mathcal{D}_{g_i,p}$) lead to some conflict, for example, they coincide with the external configurations of $(\mathcal{D}_{g_i,p})_-$.

Whether such a situation may arise or not requires a further analysis. If we consider, for example, the purely on-shell framework and the term $F_G = \begin{smallmatrix}1\\3\end{smallmatrix}\!\!\!\begin{smallmatrix}\text{---}\bigcirc\text{---}\bigcirc\text{---}\bigcirc\text{---}\\ \text{---}\bigcirc\text{---}\bigcirc\text{---}\end{smallmatrix}\!\!\!\begin{smallmatrix}4\\5\\6\end{smallmatrix}$, $S = \{3, 4, t\}$ and $g_i = (3, 4) \in \bar{S}$, terms such as $\begin{smallmatrix}\text{---}\bigcirc\text{---}\bigcirc\text{---}\\3\text{---}\bigcirc\text{---}\bigcirc\text{---}\end{smallmatrix}{}^{4}$ or $\begin{smallmatrix}\text{---}\bigcirc\!\!\!\text{---}\bigcirc\text{---}\\3\text{---}\bigcirc\text{---}\end{smallmatrix}{}^{4}$ are encountered in the expansion of $(F_G)^S$. These terms are individually singular along a mixed-α branch of the Landau surface of the graph ${}_3\!\!\diagdown\!\!\!\diagup{}^{4}$ which contains in the physical region the 1-particle threshold singularity of the graph $G_{g_i,1} = {}_3\!\!\times{}^{4}$. Moreover, external configurations of corresponding diagrams do coincide with those of the diagrams $(\mathcal{D}_{g_i,1})_-$. Thus, the above argument cannot be used to show that $(F_G)^S$ has a plus $i\varepsilon$ analytic continuation around the 1-particle threshold of the channel $(3, 4)$. However, the problem does not occur in this case since the channel $(3, 4)$ is not a relevant channel. $(F_G)^S$ is in fact equal to $(\bar{F}_G)^{(34)} \equiv (F_G)_-$, which is not singular along the 1-particle threshold of $G_{g_i,1}$. We shall assume that a more complete analysis leads to the same conclusion, except in exceptional situations.

The arguments given above cannot be applied to channels $g_i \in S$, where analytic continuation is indeed blocked in the typical way for some of the individual on-mass-shell integrals occurring in the expression of $(F_G)^S$. However, a fully analogous argument can now be applied to the terms occurring in the expansion of $(\bar{F}_G)^{\tilde{S}}$, which is equal to $(F_G)^S$ by Lemma 4.6, with plus signs replaced by minus signs. It now leads us to expect minus $i\varepsilon$ rules of analytic continuation. QED

According to Conjecture 4.4, differences between terms $(F_G)^S$ and $(F_G)^{S'}$ are (if S and S' are good sets) discontinuities between corresponding boundary values of analytic continuations of a common analytic function F_G. These discontinuities can be evaluated from Eq. (4.88). For example,

$$(F_G)^{S_1} - (F_G)^{S_1 \cup S_2} = \sum_{\substack{S' \subset S_1 \cup S_2 \\ S' \not\subset S_1}} (-1)^{|S'|+1} (F_G)_{S'}, \qquad (4.97)$$

where S_1 and S_2 are disjoint. The terms $(F_G)_S$ represent multiple discontinuities.

We conclude with the following remark on an extension of the previous analysis. Given a set S of channels g_i, and for each i a positive integer $p_i = p(g_i)$, $(F_G)^S_{\{p_i\}}$ can be defined in a way analogous to (4.87) but with the supplementary condition that each set of internal lines of a division L_σ corresponding to a channel $g_i (\in S)$ has at least $p(g_i)$ lines. For example, if S has the unique channel t and if $p(t) = 3$, let F_G be the term $\equiv\!\!\bigcirc\!\!\!\!\!\bigcirc\!\!\!\!\!\bigcirc\!\!\equiv$. Then:

$$\left(\equiv\!\!\bigcirc\!\!\!\!\!\bigcirc\!\!\!\!\!\bigcirc\!\!\equiv\right)^t_{(p=3)} = (F_G)_+ - \equiv\!\!\bigcirc\!\!\!\!\!+\!\!\!\!\!\bigcirc\!\!\!\!\!\bigcirc\!\!\equiv \ . \qquad (4.98)$$

In this case a conjecture analogous to Conjecture 4.4 can be stated, but with analytic continuation above (not beneath) the p-particle normal thresholds associated with S if $p < p(g_i)$. Discontinuity formulae follow in the same way as above.

5.3 *Asymptotic completeness relations, S-matrix discontinuity formulae, and all that*

We now give some examples of applications of the formulae and conjectures in the $3 \rightarrow 3$ case (in any energy region $s < [(r+1)\mu]^2$). In all the following, F^S is identified formally with $\sum_{G,\alpha}(F_{G,\alpha})^S$ in the region considered. Similarly, $F_S = \sum_{G,\alpha}(F_{G,\alpha})_S$. Rules of analytic continuation are derived formally for F^S from those conjectured in Sect. 5.2 for individual terms, at least if there is a common domain of analytic continuation for all terms $(F_{G,\alpha})^S$. If S corresponds to a cell, there are always common directions of analyticity, namely, those of the (complex, off-shell) domain associated with S. We shall admit that this is the case in general, and for the complex mass-shell, apart from exceptional points, such as those already mentioned for F_+ that lie on several Landau surfaces with conflicting $i\varepsilon$ rules in the on-shell context.

The discontinuity formulae of the S matrix and Green functions will follow in all cases from (i) an algebraic transformation of the relevant discontinuity formulae for individual terms, (ii) a formal resummation over G, α and (iii) a further analysis which will be omitted.

Unitarity-type relations. These relations on $\Delta F = F_+ - F_-$ follow from the "graph by graph unitarity" discontinuity formulae presented in Sect. 5.2.

Asymptotic completeness relations. The purpose is now to derive asymptotic completeness relations such as Eqs. (2.96) (which characterize essentially asymptotic completeness). We start from formula (4.97), written in the case of two adjacent cells S_1 and $S_1 \cup t$, $t \notin S_1$. The sum in the right-hand side is taken over all subsets S' containing t and possibly channels of S_1. We then consider the expressions, described below (4.88), of the term $(F_G)_{S'}$ and regroup all terms that have the same first division L_0 consistent with t. L_0 divides G into two connected subgraphs G', G''. For each L_0, one has to consider all possible further divisions of G' and of G'' that are divisions of G with respect to t and are, respectively, consistent with S_1 and $S_1 \cup t$. Let t', and t'' be the respective channels relative to the external lines of G' and G'', determined by the $r'(\leq r)$ lines of L_0. The above divisions of G' and G'' are divisions with respect to t' and t'', which are consistent with the respective channels of sets Σ' and Σ''. In, for example, the $3 \rightarrow 3$ case, Σ' is the set of channels g' that divide the external lines of G' into two subsets, one of which contains two of the three incoming lines of G that are also those of one of the subsets determined by a channel in S_1. It may also contain lines of L. Σ'' is similarly the set of channels g'' that divide the external lines of G'' into two subsets, one of which contains either two of the three outgoing lines of G that are also those of one of the subsets determined by a channel in S_1, or all three outgoing lines of G, plus possibly (in either case) lines of L.

Σ' and Σ'' may include "initial" or "final" channels (one subset composed of two incoming or two outgoing lines of G), and also crossed channels; Σ'' includes also the channel t'' and all channels with one subset composed only of lines of L (incoming lines in G'').

Example. Let F_G and L_1, $L_0 \equiv L_2, L_3$ be the term and divisions

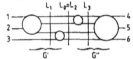

L_1 is consistent with S_1 if S_1 contains the channel $(3) = (12; 3456)$. It corresponds to a crossed channel in Σ'. L_3 is consistent with t and corresponds to a channel in Σ'', one subset of which is composed of two of the four incoming lines of G''.

In view of the definitions of $(F_{G'})^{\Sigma'}$ and $(F_{G''})^{\Sigma''}$, the sum of all terms corresponding to the same division L_0 is thus equal to $(F_{G'})^{\Sigma'} *_+ (F_{G''})^{\Sigma''}$ and the following formula is obtained:

$$(F_G)^{S_1} - (F_G)^{S_1 \cup t} = \sum (F_{G'})^{\Sigma'} * (F_{G''})^{\Sigma''}, \qquad (4.99)$$

where $* \equiv *_+$ and the sum runs over all divisions L_0 of G with respect to t into two *connected* subgraphs G' and G'', these divisions having, moreover, a number $r' \leq r$ of intermediate lines. By formal summation, one thus reobtains Eq. (2.96) in the case $m = n = 3$. The result is *a priori* obtained in a neighbourhood of the physical region and can then be extended by analytic continuation. It can be checked that the sets Σ' and Σ'' introduced here correspond to cells for the $(r' + 3)$-point functions involved and coincide with those introduced (in a different way) in the axiomatic framework.

Generalized optical theorems and other S matrix discontinuity formulae. As an example, we wish to derive below formula (1.59), that is,

$F_t = \begin{smallmatrix}1\\2\\3\end{smallmatrix}\!\!\!-\!\!\!\bigcirc\!\!\!+\!\!\!\boxtimes\!\!\!-\!\!\!\boxed{}\!\!\!-\!\!\!\boxtimes\!\!\!+\!\!\!\bigcirc\!\!\!-\!\!\!\begin{smallmatrix}4\\5\\6\end{smallmatrix}$, and consider to that purpose any graph

G that can be cut with respect to the channel t in one or several ways. $(F_G)_t$ is, according to Sect. 5.2, a sum of terms associated with sets L of divisions $L_1, \ldots, L_l$, that are all consistent with t. Let us divide the sum into a sum over sets L with only one division $L_1(l = 1)$ and a sum over remaining sets, and let us regroup the latter by subclasses that have the same external divisions $L_1, L_1'(= L_l)$. The graph G is divided by L_1 in the first case into two connected successive subgraphs G' and G'' and it is divided by L_1 and L_1' in the second case into three successive subgraphs G', G'' and G''' where G' and G''' are again connected but G'' may be nonconnected, as in the example:

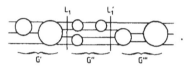

The terms of a class (L_1, L_1') are obtained by considering all possible sets of zero, one or more *intermediate* divisions of G with respect to t. The latter are all possible divisions of G'' (with $\leq r$ intermediate lines) into successive, connected or nonconnected subgraphs with respect to the channel t'' (which divides the set of external lines of G'' into the two subsets determined by the lines of L_1 and L_1'). These divisions are not necessarily consistent with t'', and may be consistent with any other channel. For example, the division L_2 in the example below, which is consistent with t in G, is not consistent with t'' but with the crossed channel $(\alpha\beta\alpha'; \gamma\beta'\gamma')$.

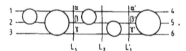

By the definition of $(F_{G''})^E$ (extended possibly to nonconnected graphs G''), the sum of all terms in the class (L_1, L_1') is thus equal to $(F_{G'})_+ * (F_{G''})^E * (F_{G'''})_+$. $(F_{G''})^E$ is also equal by Eq. (4.96) to $(-1)^{N_c(G'')+1} (F_{G''})_-$. Hence, the following formula is obtained (with $* \equiv *_+$):

$$(F_G)_t = \sum_{L_1} (F_{G'})_+ * (F_{G''})_+ + \sum_{L_1, L_1'} (-1)^{N_c(G'')+1}(F_{G'})_+ *$$
$$* (F_{G''})_- * (F_{G'''})_+. \qquad (4.100)$$

The formal resummation then gives Eq. (1.59). In fact, the summation of the terms in the first sum of (4.100) gives ⎯⊙▨⊙⎯ (since G' and G'' are connected), which corresponds to the identity part of ▨-▨ , while the summation of the terms in the second sum gives the remaining contributions corresponding to all nontrivial parts of ▨-▨ (at least one nontrivial connected component). Since G' and G''' have to be connected, one again finds connected terms on the left and right sides of the discontinuity formula, although this is not the case for the minus box.

Other discontinuity (or multiple discontinuity) formulae such as (1.60) and (1.61) and other formulae involved in multiparticle dispersion relations are similarly derived. In, for example, (1.60), the fact that line i (or f) cannot go straight through, corresponds to the fact that the only allowed divisions in the terms $(F_G)_{i,f}$ have to be consistent with i or f but *not* with the crossed channel (if).

Eq. (4.100) can be alternatively derived without regrouping the terms in the class L_1 or (L_1, L_1'). One first obtains (with $* \equiv *_+$):

$$(F_G)_t = \sum_{\substack{G_1,\ldots,G_{n+1} \\ G_1, G_{n+1} \text{ connected} \\ n \geq 1}} (-1)^{n-1}(F_{G_1})_+ * (F_{G_2})_+ \cdots * (F_{G_{n+1}})_+, \quad (4.101)$$

where the sum Σ runs over all ways of dividing G into successive subgraphs $G_1,\ldots,G_{n+1}$, $n = 1, 2, \ldots$; G_1 and G_{n+1} are connected but $G_2,\ldots,G_n$ are connected or nonconnected. (Each division has a set of $\leq r$ intermediate lines.) By formal resummation (and a further analysis), this allows one to express F_t as the formal infinite sum:

$$F_t = \sum_{n>1}(-1)^{n-1} \quad \text{} \quad (4.102)$$

from which Eq. (4.100) is (formally) reobtained.

The same method allows one to reobtain, for example, the discontinuity formula given in Chapter I around the 3-particle threshold $s = (3\mu)^2$ in the $2 \to 2$ case. In fact, for any graph G, $(F_G)_t^{(p=3)}$ is given by a formula analogous to (4.101) but with on-mass-shell convolutions over at least three particles. The analogue of (4.102), with sets of three particles, and the expression of the discontinuity as a double on-shell convolution with the factor $S_{(3)}^{-1}$ between $2 \to 3$ and $3 \to 2$ connected functions, follow.

Alternative forms of this same discontinuity can be obtained by considering the expansion of $(F_G)_t$, and by grouping now together all terms that have the same first division L_1, which divides G into two connected subgraphs G' and G''. Possible further divisions $L_2,\ldots,L_{n+1}$ of G'' consistent with t in G are all possible divisions of G'' with respect to the channel t'' (determined by the lines of L_1 and the outgoing lines of G) consistent with t'' or with any other channel $g_1^{(in)},\ldots,g_\rho^{(in)}$ that divides the set of external lines of G'' into two subsets, one of which is composed of lines of L_1 only. By resummation, one obtains:

$$(F_G)_t = \sum_{L_1}(F_{G_1})_+ *_+ (F_{G_2})^{t,g_1^{(in)},\ldots,g_\rho^{(in)}}, \qquad (4.103)$$

which gives, in turn (formally), an on-shell convolution integral of the $2 \to 3$ connected function and of the corresponding analytic continuation of the $3 \to 2$ function.

Mathematical Appendix:
Distributions, analytic functions,
and microlocal analysis

Results of Sects. 1, 2 and 3 will be found in [BI1, I1, I2] and references therein. For related results in the approach of [SKK], see [SKK, KK] and references therein. More details on the contents of Sect. 4 will be found also in the latter references. Sect. 5 is a summary of [I7].

§1. Microsupport of distributions

We first consider a tempered distribution f defined in the real n-dimensional space $\mathbb{R}^n$ of variables $p = (p_1, \ldots, p_n)$. Variables in the dual space will be denoted $x = (x_1, \ldots, x_n)$, with $px = \sum p_i x_i$. A "generalized" Fourier transform F of f can be defined for each $\gamma > 0$ by the formula

$$F(x, p; \gamma) = \int f(p') e^{-ip' \cdot x - \gamma |x| |p' - p|^2} dp'. \qquad (A.1)$$

It reduces to the usual Fourier transform $\tilde{f}$ of f at $\gamma = 0$. Local analyticity properties of f around a given point p can be characterized in terms of exponential fall-off properties of F in x-space, whereas this is not possible in terms of either the usual Fourier transform $\tilde{f}$ of f or the Fourier transforms of products χf of f with a fixed test function χ. (If χ is, for example, chosen with a compact support around p, this is due to the fact that it has at least C^∞, that is, infinitely differentiable singularities on the boundary of its support.) Given any real point $P = (P_1, \ldots, P_n)$, the ("microlocal," analytic) essential support $\mathrm{ES}_P(f)$ of f at P can be defined as the set of "singular" directions along which $F(x, P; \gamma)$ does *not* decay exponentially in x-space. More precisely, a direction $\hat{x}$ in x-space is by definition outside $\mathrm{ES}_P(f)$ if $F(x, P; \gamma)$ decays exponentially in that direction for all $\gamma > 0$, with a rate of exponential fall-off at least proportional to γ at small γ. Some continuity properties are also required. (Exponential decay can be restricted in the definition to small values of γ: exponential decay for all $\gamma > 0$ follows.)

$\mathrm{ES}_P(f)$ can alternatively be viewed as the cone with apex at the origin in x-space composed of singular directions just defined.

It can be checked that $\mathrm{ES}_P(f) \equiv \mathrm{ES}_P(\chi f)$ for test functions χ that are locally analytic and different from zero at P (and with, for example, a compact support around P). Given a general distribution f defined either in $\mathbb{R}^n$ or on a real analytic manifold M, $\mathrm{ES}_P(f)$ can be defined in a way similar to the above, by introducing a system of real analytic local coordinates of M and test functions χ with sufficiently small support around P: it is independent of χ and of the local coordinate system, and is a well-defined set of directions in the cotangent space at P to M. It is also unchanged if the function $|p' - p|^2$ in the exponential is replaced by a more general function Φ with similar local properties around p.

The essential support $\mathrm{ES}(f)$ is the collection of essential supports at all points p of $\mathbb{R}^n$ or of M. It can be checked that it coincides with the *analytic wave front set*, as also with the *singular spectrum*, introduced independently by different methods by L. Hörmander and by M. Sato, T. Kawaï, and M. Kashiwara, respectively. (The definition applies in the latter case to general hyperfunctions.) This common object is also called *micro-support*, following a terminology proposed by M. Sato which is attractive but has no link with (micro-) locality in field theory.

Remark. The wave front, or C^∞ essential support, $W_P(f)$ of f at P is the set of directions in x-space along which the Fourier transform $\widetilde{f\chi}$ of χf does *not* decay rapidly (= faster than any inverse power of $|x|$) for C^∞ test functions χ, in the limit when the support of χ around P tends to zero. More precisely, a direction $\hat{x}$ is outside $W_P(f)$ if $\widetilde{\chi f}$ decays rapidly in that direction for test functions χ with sufficiently small support (depending on $\hat{x}$). This notion, first introduced explicitly by L. Hörmander, characterizes local analyticity properties, but only modulo undesirable C^∞ backgrounds.

Given $\hat{x} \notin \mathrm{ES}_P(f)$, it can be shown that the generalized Fourier transform of χf is bounded, for all $\gamma \geq 0$, by the product of a rapidly decaying factor $(C_N/1 + |x|^N, \forall N)$ times an exponential fall-off factor of the type described above with a rate depending here also on the local analyticity domain of χ, if χ is now locally analytic at P and is, as above, C^∞ with a sufficiently small support, depending on $\hat{x}$. At $\gamma = 0$, the rapid decay factor remains, so that $\mathrm{ES}_P(f) \subset W_P(f)$.

§2. Local analyticity properties, general decomposition theorems, generalized edge-of-the-wedge theorems

It can be shown, as an extension of results using the usual Fourier-Laplace transformation, that f is analytic at a point P if $\mathrm{ES}_P(f)$ is empty (exponential decay of F in all directions in x-space) and that f is "at P" the boundary value of an analytic function $\underline{f}$ from the (imaginary) directions of an open cone Γ if $\mathrm{ES}_P(f)$ is contained in the closed (convex, salient) dual cone C of Γ. ($\underline{f}$ is here defined in a domain of $\mathbf{C}^n$ or of the complexified manifold $\underline{M}$; f is "at p" the boundary value of $\underline{f}$ from the directions of Γ if, given any slightly smaller cone Γ_ε, there is a real neighborhood $\mathcal{N}_\varepsilon$ of p in which f is the boundary value of $\underline{f}$ from the directions of Γ_ε.)

More generally, $\mathrm{ES}_P(f)$ characterizes by duality the possible (in general nonunique) local decompositions of f as a sum of boundary values of analytic functions. Given any set of closed, convex salient cones C_β such that $\mathrm{ES}_P(f) \subset \cup_\beta C_\beta$, f is locally at p the sum of corresponding boundary values f_β from the directions of the dual cones Γ_β.

Edge-of-the-wedge and generalized edge-of-the-wedge theorems can be obtained locally as a simple corollary of previous results. They apply to a family of distributions f_i, $i = 1, \ldots, m$ such that (i) $\Sigma f_i = 0$ locally and (ii) each f_i is, at p, the boundary value of an analytic function $\underline{f}_i$ from the directions of a cone Γ_i. The local edge-of-the-wedge theorem (either in its original or refined version) is obtained at $m = 2$: if (i) is replaced for convenience by the condition $f_1 = f_2$ locally, it is shown that $\underline{f}_1$ and $\underline{f}_2$ have a common analytic continuation $\underline{f}$, and $f = f_1 = f_2$ is at p the boundary value of $\underline{f}$ from the directions of the convex envelope $(\Gamma_1 \cup \Gamma_2)^c$ of $\Gamma_1 \cup \Gamma_2$ (which is all $\mathrm{I\!R}^n$, i.e. f is analytic at p, if there are opposite directions in Γ_1 and Γ_2, as assumed in the original version).

Proof. From the assumptions, $\mathrm{ES}_P(f_1) \subset C_1$ and $\mathrm{ES}_P(f_2) \subset C_2$ where C_1 and C_2 are dual to Γ_1 and Γ_2 respectively. Since $f_2 = f_1$, $\mathrm{ES}_P(f) \subset C_1 \cap C_2$ where $f \equiv f_1$. In turn, f is at p the boundary value of an analytic function $\underline{f}$ from the directions of the dual cone $\Gamma = (\Gamma_1 \cup \Gamma_2)^c$ of $C_1 \cap C_2$. Finally, $\underline{f}$ is a common analytic continuation of $\underline{f}_1$ and $\underline{f}_2$ as a consequence of the preliminary result, admitted here, according to which an analytic function whose boundary value from the directions of a given cone vanishes locally is identically zero. QED

At $m > 2$, the decomposition theorems yield the existence of common distributions $f'_{i,j}$, $f'_{i,j} = -f'_{j,i}$, boundary values (at p) of analytic functions from the directions of cones (arbitrarily close to) $(\Gamma_i \cup \Gamma_j)^c$, such that $f_i = \sum_{j \neq i} f'_{i,j}$, $i = 1, \ldots, m$.

Two types of extensions of previous results can be considered. On the one hand, given a point P, semi-global notions of essential support can be given. They correspond to specifications of rates of exponential fall-off outside the essential support (decay at least like $e^{-\alpha\gamma|x|}$, for some given $\alpha > 0$, for all γ smaller than some maximal γ_0 depending on the direction in x-space). Extensions of previous results will then involve boundary values, in specified real regions Ω_α around P (of the form $|p' - P|^2 < \alpha$ or more generally $\Phi(p'; P) < \alpha$), of functions analytic in specified "local tubes." Boundary values in Ω_α are obtained here from the directions of a given cone Γ, independent of the real point in Ω_α.

On the other hand, decompositions in real regions into sums of boundary values of analytic functions, with boundary values obtained from directions depending on the real point considered, can also be established. (They correspond to "gluing together" decompositions obtained at each real point p.)

§3. Products and integrals of distributions, restrictions to submanifolds

The following results can be established either by direct methods using convolution arguments, or through decomposition theorems of Sect. 2.

(a) (Product of distributions)

If $\mathrm{ES}_P(f_2)$ contains no direction opposite to a direction of $\mathrm{ES}_P(f_1)$, the product $f_1 f_2$ is a well-defined distribution in the neighborhood of P and $\mathrm{ES}_P(f_1 f_2) \subset \{\mathrm{ES}_P(f_1) + \mathrm{ES}_P(f_2)\} \equiv \{x; x = x_1 + x_2,\ x_1 \in \mathrm{ES}_P(f_1),\ x_2 \in \mathrm{ES}_P(f_2)\}$.

In "$u = 0$" situations such that $\mathrm{ES}_P(f_1)$ and $\mathrm{ES}_P(f_2)$ do contain opposite directions, there is no general result on $\mathrm{ES}_P(f_1 f_2)$, even if the product $f_1 f_2$ can be defined in some way (product of functions) and if $\mathrm{ES}_P(f_1) + \mathrm{ES}_P(f_2)$ is not all $\mathbb{R}^n$. However, some results can be established under various regularity conditions on f_1 and f_2. $\mathrm{ES}_P(f_1 f_2)$ is not in general shown to be contained in $\mathrm{ES}_P(f_1) + \mathrm{ES}_P(f_2)$ itself, but in a larger set involving limiting procedures.

(b) (Restrictions to submanifolds)

Let us consider a distribution f defined in $\mathbb{R}^n$ and a smooth (analytic) submanifold M of $\mathbb{R}^n$. Given a point P of M, let $N(P)$ be the conormal set at P to M. The following result, closely related to that stated above on products, then holds: if $\mathrm{ES}_P(f) \cap N(P)$ is empty (apart from the origin), f can be restricted to M

(in the sense of distributions) in the neighborhood of P and

$$\mathrm{ES}_P(f_{|M}) \subset \mathrm{ES}_P(f)/N(P),$$

where the right-hand side is the set of points x in $\mathrm{ES}_P(f)$ modulo addition of points in $N(P)$. (This is indeed a cone with apex at the origin in the cotangent space at P to M.)

If $\mathrm{ES}_P(f) \cap N(P)$ is not empty ("$u = 0$" situation), analogous results can again be established, up to limiting procedures, under various supplementary regularity conditions.

(c) (Integrals)

Let $f(p, p')$ be a distribution defined on $\mathbb{R}^{n+n'}$ ($p \in \mathbb{R}^n$, $p' \in \mathbb{R}^{n'}$) whose support with respect to p' is contained in a given compact set K, when p lies in some neighborhood of a given point P. Let

$$g(p) = \int f(p, p')\mathrm{d}p'.$$

The following result then holds: x is outside $\mathrm{ES}_P(g)$ if $(x, 0)$ is outside $\mathrm{ES}_{(P,P')}(f)$ for all points P' in K.

§4. Holonomicity (introduction)

The nature of singularities of a distribution or hyperfunction f can be analyzed by studying differential or pseudo-differential systems satisfied by f locally.

By definition, f is holonomic (in the sense of Sato) if it satisfies a maximally overdetermined system of pseudo-differential equations.

In the simplest case of a function $f(p_1, \ldots)$ which has a singularity at $p_1 = 0$ and which can be analytically continued as a multisheeted analytic function around $p_1 = 0$ in complex space, holonomicity means essentially either that there is a finite number of sheets (the determination f_r of f obtained at $p_1 > 0$ after r turns around $p_1 = 0$ coincides with f for some integer r), or more generally, that for some r, f_r is a linear combination of $f_0, f_1, \ldots, f_{r-1}$. The space generated by all successive determinations is finite-dimensional ("Finite-determination property"). Some regularity conditions are also required. Equivalently, f is locally a *finite* sum of the form $\sum_\nu a_\nu(p_1, \ldots){p_1}^{\alpha_\nu}(\ln p_1)^{n_\nu}$, where a_ν is locally analytic or uniform around $p_1 = 0$, α_ν is a complex number and n_ν is a ≥ 0 integer.

§5. Phase-space decompositions

Given f defined in $\mathbb{R}^n$, results of Sect. 2 provide, as will now be explained, a class of phase-space decompositions of its Fourier transform $\tilde{f}$ of the form:

$$\tilde{f}(x) = \int a(x,p)\mathrm{d}p, \qquad (A.2)$$

where, for each given p:

(a) $a(x,p)$ depends essentially on values of f in a region around p whose width tends to zero as $|x|$ increases.

(b) $a(x,p)$ satisfies exponential fall-off properties in x-space that depend on p and are better in general than those of $\tilde{f}$ itself: more precisely, $a(x,p)$ will decay exponentially in x-space in all directions that do *not* belong to the micro-support $\mathrm{ES}_p(f)$ of f at p, with a rate of exponential decay linked to analyticity properties of f in local tubes around p.

For simplicity, we consider below a function f with sufficient decrease as $|p| \to \infty$. A first class of decompositions satisfying properties (i) and (ii) is obtained with the choice, for any $\gamma > 0$,

$$a(x,p) = (\gamma|x|)^{n/2} F(x,p;\gamma), \qquad (A.3)$$

where F is defined in Eq. (A.1).

However, this is not the best choice if one is interested in the best possible rates of exponential decay (outside $\mathrm{ES}_P(f)$). With the choice (A.3), this rate cannot in general exceed $(1/4\gamma)$, which is the result obtained if $f \equiv 1$. If we restrict for simplicity our attention to a function f analytic at $p(\mathrm{ES}_p(f)$ empty), this is also the rate obtained in all directions if f extends analytically in the local tube $\{p' + iq'; |p' - p| \leq 1/(2\gamma),$ $|q'| \leq (1/2\gamma) - |p' - p|\}$. Therefore, the above choice of $a(x,p)$ does not allow one to exploit possible analyticity properties of f in larger local tubes around various points p. In order to get better results, we then consider different choices of $a(x,p)$ in which γ will depend on p, of the form:

$$a(x,p) = \int f(p') \mathrm{e}^{-ip'x - \gamma(p)|x||p'-p|^2} g(x,p')\mathrm{d}p', \qquad (A.4)$$

where

$$g(x,p') = \left[\int \mathrm{d}p \; \mathrm{e}^{-\gamma(p)|x||p'-p|^2} \right]^{-1}. \qquad (A.5)$$

If, for example, f is analytic in $\mathbb{R}^n$, admits an analytic continuation in complex space in a domain of the form $\{k = p + iq, |q| < c|p|, c > 0\}$, is uniformly bounded in this domain, and has a sufficient decrease in real directions, it is convenient to choose $\gamma(p) = 1/|p|$, in which case $a(x, p)$ admits bounds of the form ($\forall p \in \mathbb{R}^n$)

$$|a(x, p)| < \text{cst } e^{-c'|p||x|}. \tag{A.6}$$

The result can be adapted to the Euclidean propagator $1/(p^2 + m^2)$ and provides, at $d = 4$, a decomposition of $C(x)$ of the form (A.2) with

$$|a(x, p)| < \frac{\text{cst}}{1 + |p|^2} e^{-\text{cst}|p||x|} \tag{A.7}$$

or the corresponding discrete version

$$C(x) = \sum_{i=1}^{\infty} C^{(i)}(x), \tag{A.8}$$

where $C^{(i)}(x) = \int_{M^{i-1} \leq |p| \leq M^i} a(x, p) dp$ satisfies the bounds

$$|C^{(i)}(x)| < M^{2i} e^{-\text{cst } M^i |x|}. \tag{A.9}$$

Bibliography

Books on axiomatic field theory and the theory of local observables

[SW] Streater, R.F. and Wightman, A., *PCT, Spin and Statistics and All That*, Benjamin, New York (1964).

[Jo] Jost, R., *The General Theory of Quantized Fields*, American Math. Society, Providence (1965).

[Ma] Martin, A., *Scattering Theory: Unitarity, Analyticity and Crossing*, Springer-Verlag, Heidelberg (1970).

[BLOT] Bogolubov, N.N., Logunov, A.A., Oksak, A.I., Todorov, I.T., *General Principles of Quantum Field Theory*, Nauka Publishers, Moscow (1987) and Kluwer Academic Publishers, Dordrecht (1990).

[Ha] Haag, R., *Local Quantum Theory*, in preparation.

Books on constructive field theory

[Si] Simon, B., *The $P(\varphi)_2$ Euclidean (Quantum) Field Theory*, Princeton University Press, Princeton (1974).

[GJ] Glimm, J. and Jaffe, A., *Quantum Physics: a Functional Integral Point of View*, Springer-Verlag, Heidelberg (1981, 1987).

[Ri] Rivasseau, V., *From Perturbative to Constructive Renormalization*, Princeton University Press, Princeton (1991).

S matrix and S-matrix theory

[Ch] Chew, G.F., *The Analytic S-matrix*, Benjamin, New York (1966).

[ELOP] Eden, R.J., Landshoff, P., Olive, D.I. and Polkinghorne, J.C., *The Analytic S-matrix*, Cambridge University Press, Cambridge (1966).

[I] Iagolnitzer, D., *The S-matrix*, North Holland, Amsterdam (1978).

General books on field theory

[ItZ] Itzykson, C. and Zuber, J.B., *Quantum Field Theory*, McGraw-Hill, New York (1980).

[ItD] Itzykson, C. and Drouffe, J.M., *Statistical Field Theory*, Cambridge University Press, Cambridge (1989).

[Zin] Zinn-Justin, J., *Quantum Field Theory and Critical Phenomena*, Clarendon Press, Oxford (1989).

Proceedings of conferences

[ChD] Chretien, M. and Deser, S. eds., *Particle Symmetries and Axiomatic Field Theory*, Gordon and Breach, New York (1966) (Brandeis, 1965).

[VW1] Velo, G. and Wightman, A.S. eds., *Constructive Quantum Field Theory*, Lecture Notes in Physics 25, Springer-Verlag, Heidelberg (1974) (Erice, 1973).

[BaI] Balian, R. and Iagolnitzer D. eds., *Structural Analysis of Collision Amplitudes*, North Holland, Amsterdam (1976) (Les Houches, 1975).

[VW2] Velo, G. and Wightman, A.S. eds., *Constructive Quantum Field Theory II,* Plenum Press, New York (1990) (Erice, 1988).

References

[AHR] Araki, H., Hepp, K. and Ruelle, D., On the asymptotic behaviour of Wightman functions in space-like directions, Helv. Phys. Acta **35**, 164 (1962).

[AIM] Arnaudon, D., Iagolnitzer, D. and Magnen, J., Weakly self-avoiding polymers in four dimensions: rigorous results, Phys. Lett. B *2+3*, 268(1991).

[BaF] Battle, G. and Federbush, P., A phase-cell cluster expansion for Euclidean field theories, Ann. Phys. **142**, 95 (1982).

[Bal] Balaban, T., Constructive gauge theory II, *Constructive Quantum Field Theory* II, Velo, G. and Wightman, A.S. eds., Plenum Press, New York (1990), p. 55.

[B1] Bros, J., Some analyticity properties implied by the two-particle structure of Green's functions in general quantum field theory, *Analytic Methods in Mathematical Physics*, Gilbert, R.P. and Newton, R.G. eds., Gordon and Breach, New York (1970), p. 85.

[B2] Bros, J., Propriétés algébriques et analytiques de la fonction de N-points en théorie quantique des champs, Thesis, Paris (1972).

[B3] Bros, J., r-particle irreducible kernels, asymptotic completeness and analyticity properties of several particle collision amplitudes, Physica **124 A**, 145 (1984).

[B4] Bros, J., Analytic structure of Green's functions in Quantum Field Theory, *Mathematical Problems in Theoretical Physics*, Osterwalder, K., ed., Lecture Notes in Physics **116**, Springer-Verlag, Berlin, Heidelberg, New York (1980), p. 166.

[B5] Bros, J., Derivation of asymptotic crossing domains for multiparticle processes in axiomatic quantum field theory: a general approach and a complete proof for $2 \to 3$ processes, Physics Reports **134**, n°5–6, 325 (1986).

[BD1] Bros, J. and Ducomet, B., Two-particle structure and renormalization, Ann. Inst. Henri Poincaré **45**, n°2, 173 (1986).

[BD2] Bros, J. and Ducomet, B., in preparation.

[BEG1] Bros, J., Epstein, H. and Glaser, V., A proof of the crossing property for two-particle amplitudes in general quantum field theory, Commun. Math. Phys. **1**, 240 (1965).

[BEG2] Bros, J., Epstein, H. and Glaser, V., Local analyticity properties of the n-particle scattering amplitude, Helv. Phys. Acta **43**, 149 (1972).

[BEG3] Bros, J., Epstein, H. and Glaser, V., On the connection between analyticity and Lorentz covariance of Wightman functions, Commun. Math. Phys. **6**, 77 (1967).

[BI1] Bros, J. and Iagolnitzer, D., Causality and local analyticity: some mathematical results, Ann. Inst. Henri Poincaré **18**, n°2, 147 (1973).

[BI2] Bros, J. and Iagolnitzer, D., Unitarity equations and structure of the S matrix at the m-particle threshold in a theory with pure $m \to m$ interaction, Commun. Math. Phys. **85**, 197 (1982).

[BI3] Bros, J. and Iagolnitzer, D., 2-particle bound states and asymptotic completeness in weakly coupled field theories, Commun. Math. Phys. **119**, 331 (1988).

[BL] Bros, J. and Lassalle, M., Analyticity properties and many-particle structure in general quantum field theory, Commun. Math. Phys. **54**, 33 (1977).

[BP] Bros, J. and Pesenti, D., Fredholm theory in complex manifolds with complex parameters: analyticity properties and Landau singularities of the resolvent, J. Math. Pures et Appl. **58**, 375 (1980).

[Br] Brydges, D., A rigorous approach to Debye screening in dilute classical Coulomb systems, Commun. Math. Phys. **58**, 313 (1978).

[Bu1] Buchholz, D., On particles, infraparticles and the problem of asymptotic completeness, *VIIIth International Congress on Mathematical Physics*, Mebkhout, M. and Seneor, R. eds., World Scientific, Singapore (1987), p. 381.

[Bu2] Buchholz, D., private communication.

[Bu3] Buchholz, D., private communication.

[BuPoSt] Buchholz, D., Poormann, M. and Stein, U., Particle weights: a new concept in quantum field theory, Hamburg preprint (1991).

[Ca] Cahill, K., Generalized optical theorems, *Structural analysis of collision amplitudes,* Balian, R. and Iagolnitzer, D. eds., North Holland, Amsterdam (1976), p. 137.

[dCdVMS] de Calan, C., da Veiga, P., Magnen, J. and Seneor, R., Constructing the 3-dimensional Gross-Neveu model with a large number of flavour components, Phys. Rev. Lett. **66**, 3233 (1991), and in preparation.

[dCR] de Calan, C. and Rivasseau, V., Local existence of the Borel transform in Euclidean φ_4^4, Commun. Math. Phys. **82**, 69 (1981).

[CD] Combescure, M. and Dunlop, F., Three-body asymptotic completeness for $P(\varphi)_2$ models, Commun. Math. Phys. **85**, 381 (1982).

[CS] Chandler, C. and Stapp, H.P., Macroscopic causality conditions and properties of scattering amplitudes, Journal of Math. Phys. **10**, 826 (1969).

[DE] Dimock, J. and Eckmann, J.P., Spectral properties and bound state scattering for weakly coupled $P(\varphi)_2$ models, Ann. Phys. **103**, 289 (1977).

[DIS] Duneau, M., Iagolnitzer, D. and Souillard, B., Properties of truncated correlation functions and analyticity properties for classical lattices and continuous systems, Comm. Math. Phys. **31**, 191 (1973), Strong cluster properties for classical systems with finite range interactions, Commun. Math. Phys. **35**, 307 (1974).

[EE] Eckmann, J.P. and Epstein, H., Time-ordered products and Schwinger functions, Comm. Math. Phys. **64**, 95 (1979).

[EGI] Epstein, H., Glaser, V. and Iagolnitzer, D., Some analyticity properties arising from asymptotic completeness in quantum field theory, Commun. Math. Phys. **80**, 99 (1981).

[EGS] Epstein, H., Glaser, V. and Stora, R., General properties of the n-point functions in local quantum field theory, *Structural analysis of collision amplitudes*, Balian, R. and Iagolnitzer, D. eds., North-Holland, Amsterdam (1976), p. 5.

[EMS] Eckmann, J.P., Magnen, J. and Seneor, R., Decay properties and Borel summability for the Schwinger functions in $P(\varphi)_2$ theories, Commun. Math. Phys. **39**, 251 (1975).

[Fe] Federbush, P., A phase cell approach to Yang-Mills theory, Commun. Math. Phys. **14**, 317 (1988) and references therein to a series of works of the author.

[FG] Felder, G. and Gallavotti, G., Perturbation theory and nonrenormalizable scalar fields, Comm. Math. Phys. **102**, 549 (1985).

[FMRS1] Feldman, J., Magnen, J., Rivasseau, V. and Seneor, R., A renormalizable field theory: the massive Gross-Neveu model in two dimensions, Commun. Math. Phys. **103**, 67 (1986).

[FMRS2] Feldman, J., Magnen, J., Rivasseau, V. and Seneor, R., Construction of infrared φ_4^4 by a phase space expansion, Commun. Math. Phys. **109**, 437 (1987).

[FMRS3] Feldman, J., Magnen, J., Rivasseau, V. and Seneor, R., Bounds on completely convergent Euclidean Feynman graphs, Comm. Math. Phys. **98**, 273 (1985) and Bounds on renormalized graphs, Commun. Math. Phys. **100**, 23 (1985).

[Fre] Fredenhagen, K., On the general theory of quantized fields, *Mathematical Physics X*, Schmüdgen K. ed., Springer-Verlag, Berlin, Heidelberg (1992), p.136.

[Fro] Frochaux, E., Probability axioms for quantum field theory, Helv. Phys. Acta **62**, 1038 (1989).

[Gal] Gallavotti, G., Renormalization theory and ultraviolet stability for scalar fields via renormalization group methods, Rev. Mod. Phys. **57**, 471 (1985).

[GJ] Glimm, J. and Jaffe, A., Positivity of the φ_3^4 hamiltonian, Fortschr. Phys. **21**, 327 (1973).

[GJS] Glimm, J., Jaffe, A. and Spencer, T., The particle structure of the weakly coupled $P(\varphi)_2$ model and other applications of high temperature expansions, Part II: the cluster expansion, *Constructive Quantum Field Theory*, Velo, G. and Wightman, A.S. eds., Lecture Notes in Physics **25**, Springer-Verlag, Heidelberg (1973).

[GK1] Gawedzki, K. and Kupiainen, A., Gross-Neveu model through convergent perturbation expansions, Commun. Math. Phys. **102**, 1 (1985).

[GK2] Gawedzki, K. and Kupiainen, A., Massless lattice φ_4^4 theory: rigorous control of a renormalizable asymptotically free model, Commun. Math. Phys. **99**, 197 (1985).

[GK3] Gawedzki, K. and Kupiainen, A., Renormalization of a nonrenormalizable quantum field theory, Nucl. Phys. **B262**, 33 (1985).

[GN] Gross, D. and Neveu, A., Dynamical symmetry breaking in asymptotically free field theories, Phys. Rev. **D10**, 3235 (1974).

[He] Hepp, K., On the connection between Wightman and LSZ quantum field theory, *Axiomatic Field Theory*, Chretien, M. and Deser, S. eds., Gordon and Breach, New York (1966), p. 133.

[Ho] Hormander, L., Acta Math. **127**, 79 (1971).

[I1] Iagolnitzer, D., Analytic structure of distributions and essential support theory, *Structural analysis of collision amplitudes*, Balian, R. and Iagolnitzer, D. eds., North Holland, Amsterdam (1976), p. 295 and references therein.

[I2] Iagolnitzer, D., The $u = 0$ structure theorem, Commun. Math. Phys. **63**, 49 (1978).

[I3] Iagolnitzer, D., Factorization of the multiparticle S matrix in two-dimensional space-time models, Phys. Rev. **D18**, 1275 (1978) and The multiparticle S matrix in two-dimensional space-time, Phys. Lett. **76B**, 207 (1978).

[I4] Iagolnitzer, D., Irreducible kernels and nonperturbative expansions in a theory with pure $m \rightarrow m$ interaction, Commun. Math. Phys. **88**, 235 (1983).

[I5] Iagolnitzer, D., Irreducibility, analyticity and unitarity or asymptotic completeness in massive quantum field theory: some general conjectures and results, Fizika **17** (3), 361 (1985).

[I6] Iagolnitzer, D., Renormalized Bethe-Salpeter kernel and 2-particle structure in field theories, Ann. Inst. Henri Poincaré **51** (2), 111 (1989).

[I7] Iagolnitzer, D., Microlocal analysis and phase-space decompositions, Lett. Math. Phys. **21**, 323 (1991).

[I8] Iagolnitzer, D., Causality in local quantum field theory: some general results, Saclay preprint SPhT/90-120 (1990).

[II] Iagolnitzer, D. and Iagolnitzer, L., Nonholonomicity of the S matrix and Green functions in quantum field theory: a direct algebraic proof in some simple situations, Commun. Math. Phys. **124**, 79 (1989).

[IM1] Iagolnitzer, D. and Magnen, J., Asymptotic completeness and multiparticle structure in field theories, Commun. Math. Phys. **110**, 51 (1987).

[IM2] Iagolnitzer, D. and Magnen, J., Asymptotic completeness and multiparticle structure in field theories II; Theories with renormalization: the Gross-Neveu model, Commun. Math. Phys. **111**, 89 (1987).

[IM3] Iagolnitzer, D. and Magnen, J., Bethe-Salpeter kernel and short-distance expansion in the massive Gross-Neveu model, Commun. Math. Phys. **119**, 567 (1988).

[IM4] Iagolnitzer, D. and Magnen, J., Large-momentum properties and Wilson short-distance expansion in nonperturbative field theory, Commun. Math. Phys. **119**, 609 (1988).

[IM5] Iagolnitzer, D. and Magnen, J., Polymers in a weak random potential in dimension four: rigorous renormalization group analysis, Ecole Polytechnique preprint (1992).

[IS] Iagolnitzer, D. and Stapp, H.P., Macroscopic causality and physical region analyticity in S-matrix theory, Commun. Math. Phys. **14**, 15 (1969).

[KK] Kashiwara, M. and Kawai, T., The theory of holonomic systems with regular singularities and its relevance to physical problems, *Complex analysis, microlocal calculus and relativistic quantum theory,* Iagolnitzer, D. ed., Lecture Notes in Physics **126**, Springer-Verlag, Heidelberg (1980), p. 5.

[KKO] Kashiwara, M., Kawai, T. and Oshima, T., A study of Feynman integrals by micro-differential equations, Commun. Math. Phys. **60**, 97 (1978).

[KKS] Kashiwara, M., Kawai, T. and Stapp, H.P., Microanalyticity of the S matrix and related functions, Comm. Math. Phys. **66**, 95 (1979).

[KS1] Kawai, T. and Stapp, H.P., Discontinuity formula and Sato's conjecture, Publ. RIMS, Kyoto Univ. **12**, Suppl. 155 (1977).

[KS2] Kawai, T. and Stapp, H.P., On the regular holonomic character of the S matrix and microlocal analysis of unitarity type integrals, Comm. Math. Phys. **83**, 213 (1982).

[Ko] Koch, J., Irreducible kernels and bound states in $\lambda P(\varphi)_2$ models, Ann. Inst. Henri Poincaré **31**, 173 (1979).

[KMR] Kopper, C., Magnen, J. and Rivasseau, V., in preparation.

[LSZ] Lehmann, H., Symanzik, K. and Zimmerman, W., On the formulation of quantized field theories, Nuovo Cimento (Ser.10) **1**, 205 (1955).

[LMPS] Logunov, A.A., Muzafarov, L.M., Pavlov, V.P. and Sukhanov, A.D., Analytic structure of the amplitude of the forward $3 \rightarrow 3$ process, Teoret. Mat. Fiz. **40**, 179 (1979).

[MRS] Magnen, J., Rivasseau, V. and Seneor, R., Construction of (Yang-Mills)$_4$ with an infrared cut-off, Ecole Polytechnique preprint (1991).

[MS] Magnen, J. and Seneor, R., The infrared behaviour of $\nabla\varphi_3^4$, Ann. Phys. **152**, 30 (1984).

[MW] Mitter, P., and Weisz, P.H., Asymptotic scale invariance in a massive Thirring model with $U(n)$ symmetry, Phys. Rev. **D8**, 4410 (1973).

[MP] Muzafarov, L.M. and Pavlov, V.P., Analyticity of the $3 \rightarrow 3$ forward amplitude in the neighborhood of the physical region, Teor. Mat. Fiz. **35**, 151 (1978).

[N] Nelson, E., Construction of quantum fields from Markov fields, Journ. Funct. Anal. **12**, 97 (1973).

[0] Omnes, R., Finite range of strong interactions and analyticity properties in momentum transfer, Phys. Rev. **146**, 1123 (1966).

[0S] Osterwalder, K. and Schrader, R., Axioms for Euclidean Green's functions, Comm. Math. Phys. **31**, 83 (1973).

[Ph1] Pham, F., Singularités des processus de diffusion multiple, Ann. Inst. Henri Poincaré **6**, n°2, 89 (1967).

[Ph2] Pham, F., *Introduction à l'Etude topologique des singularités de Landau,* Memorial des Sciences Mathématiques **164**, Gauthier-Villars, Paris (1967).

[Re] Renouard, P., Analytic interpolation and Borel summability of the $\{\frac{\lambda}{N}|\varphi_N|^4\}_2$ models, Commun. Math. Phys. **84**, 257(1982).

[Ru1] Ruelle, D., On the asymptotic condition in quantum field theory, Helv. Phys. Acta **35**, 147 (1962).

[Ru2] Ruelle, D., Connection between Wightman functions and Green functions in p-space, Nuov. Cim. **19**, 356 (1961).

[Ruij] Ruijsenaars, S.N.M., Scattering theory for the Federbush, massless Thirring and continuum Ising model, J. Funct. Anal. **48**, 135 (1982) and references therein.

[SKK] Sato, M., Kawai, T. and Kashiwara, M., *Microfunctions and pseudo-differential equations*, Lecture Notes in Math. **287**, Springer-Verlag, Heidelberg (1973), p. 265.

[Sp] Spencer, T., The decay of the Bethe-Salpeter kernel in $P(\varphi)_2$ quantum field theory, Commun. Math. Phys. **44**, 143 (1975).

[SpZ] Spencer, T. and Zirilli, F., Scattering states and bound states in $\lambda P(\varphi)_2$, Commun. Math. Phys. **49**, 1 (1976).

[Sta] Stapp, H.P., Discontinuity formulae for multiparticle amplitudes, *Structural analysis of collision amplitudes,* Balian, R. and Iagolnitzer, D. eds., North Holland, Amsterdam (1976) and references therein.

[Sy1] Symanzik, K., Many-particle structure of Green's functions, *Symposia on theoretical physics*, Ramakrishnan, A. ed., Plenum Press, New York (1967), p. 121.

[Sy2] Symanzik, K., Infrared singularities and small distance-behaviour analysis, Commun. Math. Phys. **34**, 7 (1973).

Index

absorptive parts, 66, 103
adjacent cells, 99
amputated functions, 71, 83, 84
analysis of Landau singularities, 44–51
analytic N-point functions, 59, 96–108
analytic S matrix, 36–44
analytic wave front set, 272
asymptotic causality, 81–89
asymptotic causal properties, 59
asymptotic completeness, 57, 237, 264, 265
asymptotic freedom, 151
asymptotic Haag-Ruelle theory, 57, 77
asymptotic localization, 25, 73, 76
asymptotic series, 150
asymptotic states, 77
axioms: Epstein-Glaser, 56; Euclidean, 55, 154, 184; Haag-Kastler, 56; Osterwalder-Schrader, 56; Wightman, 56, 67–68

bad sets, 43
bare mass. *See* mass, bare
bare coupling. *See* coupling, bare
basic discontinuity formulae, 42–44, 266
Bethe-Salpeter equation, 62, 223
Bethe-Salpeter kernel, 62, 128, 216; modified or direct, 246; renormalized (*see* renormalized Bethe-Salpeter kernel)
Bethe-Salpeter-type equations, 128
Bethe-Salpeter-type irreducible kernels, 62, 67
Borel procedure, 150
Borel summability, 153, 184–186
bosonic models, 203–207
bosons, 8, 55, 67
bound states, 55, 65, 214, 237

causal configurations, 87
causal diagrams, 16
causal directions, 17, 22
cell, 97
charges, 57, 69
chronological functions, 70
chronological operators, 56, 70
classical (multiple scattering) diagram, 15
cluster expansion(s), 150, 173; inductive, 175–177, 218; of order larger than one, 217–218; pairwise, 173, 218
cluster property: in energy-momentum space, 12; in space-time, 12, 59, 69, 155, 156
complex mass-shell, 36
conflicting plus $i\varepsilon$ rules, 24
connected functions, 71; S matrix, 12
conservation of probabilities, 9
contracted graphs, 21
contraction scheme, 160
correlation inequalities, 151
coupling: bare, 187; effective, 164, 196–197, 200; renormalized, 151
crossing, 5, 38, 60, 64–65

decomposition: of N-point functions, 64; of the S matrix, 35; theorems of, 273
degree of divergence: of a graph, 162
discontinuity formulae: of absorptive parts, 108–109, 127; of Feynman-type integrals, 256–264; of the S matrix, 266–269
dispersion relations, 39, 64
domination procedure, 206

edge-of-the-wedge theorem, 271; generalized (*see* generalized edge-of-the-wedge theorem)

effective coupling, 164
effective expansion, 164–170
energy-momentum conservation, 9
ergodicity, 155
essential support, 4, 271
Euclidean axioms. *See* axioms, Euclidean
Euclidean functions, 149
Euclidean invariance, 155
even theories, 58
expansion: effective (*see* effective expansion); Feynman-type, 249; perturbative, 160
expansions: of Euclidean functions, 179; in terms of holonomic contributions, 6, 47, 50, 51; in terms of the renormalized BS kernel, 235
exponential decay of connected N-point functions: in Euclidean space-time, 180–181, 197; in Minkowski space-time, 82, 84

face, 99
factorization equations, 6, 54
factorization property: of chronological operators, 81; in cluster expansions of order larger than one, 218; of the multiparticle S matrix, 6; of N-point functions, 82
Federbush model, 152
fermionic models, 152, 186
fermions, 8, 68
Feynman integrals: analytic structure of, 209
Feynman-type integrals, 62, 64, 249
finite-determination property, 48, 275
Fock space: of free-particle states, 6
forest, 163
Fredholm equations, 41, 62, 135
Fredholm solutions, 41
Fredholm theory in complex space, 64, 127
free-particle states, 6, 80
future (causal) cone, 82

$\mathcal{G}$-convolution, 126
gauge theories, 151
general N-point functions, 126
generalized edge-of-wedge theorem, 273–274

generalized optical theorems, 5, 42–44, 111, 112, 266
geometrical cell, 97
good sets, 42–43, 261
graph by graph unitary, 257
Gross-Neveu model, 151; massive, 151, 186–203

Haag-Kastler axioms. *See* axioms, Haag-Kastler
Haag-Ruelle theory. *See* asymptotic Haag-Ruelle theory
hermitean analyticity, 5, 37, 60
holomorphy envelope: of primitive domain, 102
holonomy or holonomicity, 5, 45, 47–49, 210, 275

inductive cluster expansion. *See* cluster expansion(s), inductive
infinite-volume limit, 180, 189
integrable models, 152, 186
irreducible graphs, 210, 251
irreducible kernels, 4, 49, 125–137, 212, 214, 255; 1-particle, 211; 2-particle, 131; 3-particle, 135
irreducible kernel satisfying a regularized Bethe-Salpeter equation, 125–137, 216, 225
irreducibility: r-particle (in a given channel), 227; total, 251;

K matrix, 50

Landau equations, 16
Landau surfaces, 4, 16
large-momentum properties, 198, 216, 231
linear program, 58
local analyticity, 61, 102–103
local decompositions, 61, 89–90
local discontinuity formulae, 4, 33, 268–269
locality, 55
local maximal analyticity, 5, 40
local momentum-space analyticity. *See* local analyticity
local tubes, 91, 272
loop equations, 16
Lorentz invariance, 156

Lorentz invariants, 37
LSZ reduction formulae, 59, 91

m-particle threshold, 47, 145
M_0 points, 15
macro-causal analysis: of 2-particle threshold, 145–148
macro-causal factorization, 4, 28–30, 138, 144
macrocausality or macroscopic causality, 4, 26, 143
macro-causal S matrix, 24–36
many-particle (structure) analysis, 64, 249–269
Martin pathologies, 26, 41
mass, 7; bare, 149; physical, 213; renormalized, 213
mass gap, 57
mass spectrum, 57
maximal analyticity, 36; local (*see* local maximal analyticity)
Mayer expansion, 179
Mayer links, 153
microsupport, 4, 271
Mitter-Weisz model, 151
mixed-α diagram, 16, 17
mixed-α Landau surfaces, 16
modified Landau surfaces, 15, 210
momentum assignment, 166
multiparticle dispersion relations, 42, 268
multiparticle S matrix, 3–54; in two-dimensional space-time, 52–54
multiple scattering graphs, 15
multisheeted structure, 39, 40, 112

Nelson argument, 203
Neumann series, 50, 124, 127, 131, 137, 238, 241
nonholonomic structure: of the S matrix, 48
nonlinear program, 58, 108–137
normal threshold, 262, 263

one-particle singularities, 104, 110, 111
one-particle states, 7, 73
optical theorems: generalized. *See* generalized optical theorems
orthogonal decompositions, 48
Osterwalder-Schrader analytic continuation, 149

Osterwalder-Schrader axioms. *See* axioms, Osterwalder-Schrader
Osterwalder-Schrader positivity, 156

$P(\varphi)_2$ models, 170–186
pairwise cluster expansion. *See* cluster expansion(s), pairwise
paracell, 96; functions of, 97; operators of, 97
partial wave analysis, 48
perturbative theory, 150, 160–170, 208–210
phase space analysis: Euclidean, 150, 192, 205–207, 216
phase space decompositions, 276
phase space expansions, 196–197
φ_2^4, 170
φ_3^4, 205
φ_4^4, 151; infrared, 151, 207
physical mass. *See* mass, physical
physical poles, 34
physical region, 10
physical sheet, 38, 60
physical side: of a surface, 21
plus-α Landau equations, 16
plus-α Landau surfaces, 16
plus $i\epsilon$ directions, 23–24
Poincaré group, 7, 55
Poincaré invariance, 7, 10
pole-factorization theorem, 34, 111
precell, 96
primitive domain, 101
probability measure, 154
product of distributions, 274
propagator: Feynman, 149
pseudoface, 99

r-particle irreducibility. *See* irreducibility, r-particle (in a given channel)
reconstruction theorem, 70
reducibility, 253: total, 253
reduction formulae. *See* LSZ reduction formulae
reflection positivity, 155
regularization: internal, 169
related Landau surfaces, 23, 34
renormalizable theories, 162
renormalization, 194; useful, 153, 168; useless, 170
renormalization group, 150

renormalizable theories, 162

renormalized Bethe-Salpeter kernel, 216, 217, 231

renormalized coupling. *See* coupling, reormalized

renormalized (Feynman) integrals, 163

renormalized mass. *See* mass, renormalized

restriction (of distributions) to submanifolds, 274

Ruelle identity, 99

S matrix, 6, 79

Schwinger functions, 157

second sheet analyticity, 39–41, 112–114

semi-axiomatic approaches, 247–248

short-distance expansion, 198, 202–203

Sine-Gordon model, 152

singular spectrum, 271

space-time diagram, 15

spectral condition, 58, 72

spin-statistics theorem, 8

Steinmann relations, 43, 99

structure equations, 249

sub specie aeternitatis (physical systems), 6

subtracted Bethe-Salpeter equation, 236–237

superficial degree: of divergence, 162

super-renormalizable theories, 162

superselection sectors, 58

Thirring model, 152, 186

total (r-particle) irreducibility. *See* irreducibility, total

total (r-particle) reducibility. *See* reducibility, total

trajectory: classical, 11, 15

transition amplitude, 9

transition probability, 9

triangle graph, 19, 44–46

trivial diagrams, 16

truss-bridge diagrams, 134

two-particle bound states, 237, 243

two-particle threshold, 47, 112

two-particle structure: of N-point functions, 128, 132, 237–249

$u = 0$ diagram, 15

U-kernel, 49–50, 114, 241

ultraviolet limit: in $P(\varphi)_2$, 203–205

undressed (6-point) function, 133

unitarity of the S matrix, 9

unitarity equations, 13, 14

unphysical sheets, 39

unstable particles, 26, 39, 66

unstable particle poles, 26, 39

useful renormalization. *See* renormalization, useful

vacuum vector, 55

velocity cone, 25, 74

vertices at infinity, 17

wave front: analytic. *See* analytic wave front

wave function, 7

Wick product, 159, 171

Wightman axioms. *See* axioms, Wightman

Wightman functions, 56, 69

Wigner's theorem, 7, 9

Wilson-Zimmerman short-distance expansion, 198, 202–203

Lightning Source UK Ltd.
Milton Keynes UK
UKHW030926030922
408250UK00008B/606